An Introduction to the Solar System

Compiled by a team of experts from The Open University, this textbook has been designed for introductory university courses in planetary science. It starts with a tour of the Solar System and an overview of its formation. The composition, internal structure, surface morphology and atmospheres of the terrestrial planets are then described. This leads naturally to a discussion of the giant planets and why they are compositionally different. Minor bodies are reviewed and the book concludes with a discussion of the origin of the Solar System and the evidence from meteorites. Written in an accessible style that avoids complex mathematics, and illustrated in colour throughout, this book is suitable for self-study and will appeal to amateur enthusiasts as well as undergraduate students. It contains numerous helpful learning features such as boxed summaries, student exercises with full solutions, and a glossary of terms. The book is also supported by a website hosting further teaching materials: **http://publishing.cambridge.org/resources/0521546206**

About the editors:

NEIL McBRIDE studied for a PhD at the University of Sheffield in the area of asteroids, comets and meteoroids, and subsequently worked on the Dust Impact Detection System which flew aboard the European Space Agency's cometary probe, Giotto. Dr McBride continued his researches at the University of Kent, working on the meteoroid environment in near-Earth space as well as minor bodies of the outer Solar System. Dr McBride joined the Planetary and Space Sciences Research Institute, at The Open University, in 2000 as Lecturer in Planetary and Space Science. His major teaching interests are in planetary science, physics and astronomy, and space science.

IAIN GILMOUR undertook PhD research in the Earth Sciences department at the University of Cambridge. He then spent several years in the US before returning to the UK to found the organic geochemistry laboratory in the Planetary and Space Sciences Research Institute at The Open University. His research interests are in astrobiology – primarily the origin of extraterrestrial organic matter, the Early Earth, large-scale planetary impacts and the biogeochemistry of stable isotopes. Dr Gilmour is now a Senior Lecturer and has authored several texts on the origin of life, in addition to winning a Europrix Multimedia Art Award for developing multimedia teaching materials.

D1510841

Background image: A false-colour image of Io, a satellite of Jupiter, taken by the Galileo spacecraft. This false-colour image uses near-infrared, green and violet filters to enhance the subtle colour variations of Io's surface. (NASA)

Thumbnail images: (from left to right) ultraviolet image of Venus taken from the Pioneer Venus orbiter; the main belt asteroid Gaspra (image taken by the Galileo spacecraft while en route to Jupiter); a topographic map of Mars produced by the Mars Orbiter Laser Altimeter on board the Mars Global Surveyor spacecraft; close-up of the planet Mercury showing impact craters on its surface. (NASA)

An Introduction to the Solar System

Edited by Neil McBride and Iain Gilmour

Authors:

Philip A. Bland

Neil McBride

Elaine A. Moore

Mike Widdowson

Ian Wright

PUBLISHED BY THE PRESS SYNDICATE OF THE UNIVERSITY OF CAMBRIDGE

The Pitt Building, Trumpington Street, Cambridge, United Kingdom

CAMBRIDGE UNIVERSITY PRESS

The Edinburgh Building, Cambridge, CB2 2RU, UK

40 West 20th Street, New York, NY 10011–4211, USA

477 Williamstown Road, Port Melbourne, VIC 3207, Australia

Ruiz de Alarcón 13, 28014 Madrid, Spain

Dock House, The Waterfront, Cape Town 8001, South Africa

http://www.cambridge.org

First published 2003

This co-published edition first published 2004

Edited, designed and typeset by The Open University.

Printed and bound in the United Kingdom by Bath Press, Blantyre Industrial Estate, Glasgow G72 0ND, UK

A catalogue record for this book is available from the British Library

ISBN 0 521 83735 9   hardback
ISBN 0 521 54620 6   paperback

This publication forms part of an Open University course S283 *Planetary Science and the Search for Life*. Details of this and other Open University courses can be obtained from the Course Information and Advice Centre, PO Box 724, The Open University, Milton Keynes MK7 6ZS, United Kingdom: tel. +44 (0)1908 653231, e-mail general-enquiries@open.ac.uk

Alternatively, you may visit the Open University website at http://www.open.ac.uk where you can learn more about the wide range of courses and packs offered at all levels by The Open University.

To purchase a selection of Open University course materials visit the webshop at www.ouw.co.uk, or contact Open University Worldwide, Michael Young Building, Walton Hall, Milton Keynes MK7 6AA, United Kingdom for a brochure. tel. +44 (0)1908 858785; fax +44 (0)1908 858787; e-mail ouwenq@open.ac.uk

1.1

CONTENTS

INTRODUCTION

In astronomical terms, the Solar System is our own backyard. Set against the vast numbers of stars in our Galaxy, the vast number of other galaxies in the observable universe and the incredible distances involved, our Solar System is an almost unimaginably tiny part of the Universe. However, this is where we live. It is where life on Earth developed, and it gives us our only vantage point from which to view the rest of the Universe.

Unlike other planetary systems, the objects in our Solar System are close enough to allow us to see 'close up' the splendour and diversity of the planets. We are close enough to study in detail the planets and other objects, and try and understand 'what makes the Solar System tick'. By doing this, we not only attempt to understand the system in which life evolved, but also gain an insight into how other planetary systems must evolve all over the Universe.

One of the more fundamental questions often asked is, 'Why is the Solar System the way it is?' In answering this question, we have to address more detailed questions such as, how were the planets made? What were the planets made of? Were all the planets made from the same material? Why do they look so different? Do the planets have the same internal structure? Does their surface appearance change with time? The answers to these questions lie in the physical and chemical *processes* that act on the bodies within the Solar System. Understanding these processes allows us to appreciate how the planets and the other Solar System bodies have formed and have been changed over time, and hence why they look the way they do today. In this book, you will be looking at these *processes* in detail.

CHAPTER 1
A TOUR OF THE SOLAR SYSTEM

A great way to start your study of the **Solar System** is to get an overview of what our planetary system looks like by taking a tour of the **planets**. In this introductory chapter, you will see the incredible diversity that the Solar System offers, made accessible by the use of spacecraft sent into space to gather scientific data (an important part of which is in the form of high-resolution images). The tour will set the scene and highlight the planetary features that will be explained by the processes considered in detail later on. So let us begin our tour of the Solar System.

1.1 A grand tour

There are nine planets in our Solar System. Each planet travels on an approximately circular **orbit** around the Sun, which lies at the heart of the Solar System. In order of increasing distance from the Sun, the nine planets are Mercury, Venus, Earth, Mars, Jupiter, Saturn, Uranus, Neptune and Pluto. The sizes of the planets vary greatly, but all are dwarfed by the Sun. Figure 1.1 shows the relative sizes of the planets and the Sun. In the figure, the planets are aligned in the correct order with increasing distance from the Sun, although the relative distance from the Sun is *not* shown (you will consider this later). You can see that there appears to be a broad division between the four small inner planets and the much larger outer planets (with the exception of Pluto, which is the outermost 'planet' and a relatively small body).

> The four inner bodies (Mercury, Venus, Earth and Mars) are called the **terrestrial planets**, whereas Jupiter, Saturn, Uranus and Neptune are usually referred to as the **giant planets**.

Many of the planets have **moons** (also called **satellites**) that orbit the planet in the same way as the Moon orbits the Earth. Some of the satellites that you will meet on our tour are similar to the terrestrial planets in terms of their composition or structure, and are sometimes called **terrestrial-like** bodies.

We start our tour by taking a closer look at each member of the Solar System. As the Sun lies at the centre of the Solar System, it seems a sensible place to start. Although the topic of this book focuses on **planetary bodies** (a term that refers not only to the planets, but also to their satellites and other small bodies such as asteroids) you should appreciate that the Sun is *not* a planet, but a **star**. As such, it is a huge ball of gas, consisting mainly of hydrogen and helium (although other elements are also present in smaller amounts). At the centre, nuclear reactions release energy. This is why the Sun is hot – about 5770 K at its surface and an amazing 15 000 000 K at its centre. (Note the SI unit of absolute temperature is the kelvin, K. 0 K = –273 °C, and 0 °C = 273 K.)

Figure 1.2 shows an impressive image of the Sun. Material, seen above the surface, can be lost to space. Ejection of material from the Sun can have consequences here on Earth (you will meet this in Chapter 5).

Some major missions are listed in Appendix A, Table A7.

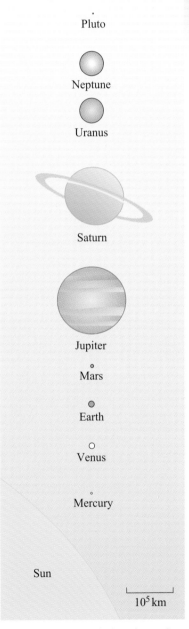

Pluto

Neptune

Uranus

Saturn

Jupiter

Mars

Earth

Venus

Mercury

Sun

10^5 km

Figure 1.1 The relative sizes of the planets and the Sun. The planets are shown in the correct order (with increasing distance from the Sun), although the relative distance from the Sun is *not* shown to scale.

Figure 1.2 The Sun (radius 695 500 km). The image, taken in ultraviolet light using the Soho spacecraft, shows that the Sun is rather complex and 'active'. (NASA)

1.1.1 Mercury

The first planetary body on our tour is Mercury (Figure 1.3). Mercury, being the closest planet to the Sun, can get very hot. In sunlight, parts of the surface can reach about 740 K (approximately 470 °C), whereas in darkness the temperature can drop to about 80 K (−190 °C). Clearly, the surface of Mercury is not a very hospitable place. Looking at Figure 1.3, perhaps the most striking features are the round 'scars' on the surface. These are **impact craters**. A close-up view of some impact craters on Mercury is shown in Figure 1.4 and you will consider impacts in detail in Chapter 4. These impact craters have been made by the 'leftovers' of the planetary formation process, namely **asteroids** and **comets** (both of which you will be looking at in detail in Chapter 7). Asteroids are predominantly rocky and metallic bodies, whereas comets have a large fraction of icy material in them. However, both have broadly similar effects when they slam into the surface of a planet at great speed – they leave impact craters. Any undisturbed surface of a planetary body will accumulate impact craters over time. Thus a very cratered surface implies that the surface is relatively old, whereas a lack of craters might indicate that the surface has been renewed in some way, wiping out the craters from the surface. The most prevalent mechanism for resurfacing is **volcanism**, whereby **lava** (the melted rock we are familiar with on Earth) flows and covers pre-existing terrain. Furthermore, the surface can develop fractures and cracks. These concepts will be dealt with in more detail in Chapters 2 and 3.

Returning to Mercury, the cratering over the surface appears reasonably uniform, although there are some relatively small areas that look smoother, indicating some volcanic resurfacing has taken place. There are also linear features, indicating that the surface has 'cracked' at some time in the past. You can see from the clarity of the images in Figures 1.3 and 1.4 that Mercury does not have an obscuring atmosphere. In fact, Mercury does have some *extremely* tenuous atmosphere, but it is 10^{15} (a thousand million million) times less dense than the atmosphere on Earth, which is actually a better vacuum than any vacuum we can create in a laboratory.

Before you leave Mercury, there is another property of interest to be considered, which is mean **density** (see Box 1.1). Mercury's mean density is about $5.4 \times 10^3 \, \text{kg m}^{-3}$, which is almost as high as that of Earth. You will consider densities of the planetary bodies at the end of this chapter, and you will see that this is quite a surprising result considering that Mercury is the smallest of the terrestrial planets. It indicates that Mercury must include a relatively large proportion of dense material.

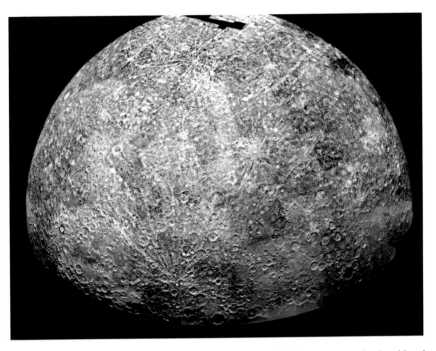

Figure 1.4 A close-up of the planet Mercury. Impact craters are clearly visible. The region shown is approximately 550 km across. (NASA)

Figure 1.3 The planet Mercury (radius 2440 km). This image was obtained by the Mariner 10 spacecraft in 1974 (so far the only spacecraft to go to Mercury). Impact craters cover the surface. (NASA)

BOX 1.1 DENSITY

Density (sometimes called bulk density) is a measure of the mass per unit volume of a substance:

$$\text{density} = \frac{\text{mass/kg}}{\text{volume/m}^3} \tag{1.1}$$

The SI units of density are thus kilograms per cubic metre (kg m^{-3}). Density values of common materials can cover quite a wide range. Water has a density of $1.0 \times 10^3 \text{ kg m}^{-3}$, whereas a rock such as granite is around $2.7 \times 10^3 \text{ kg m}^{-3}$, and iron is $7.9 \times 10^3 \text{ kg m}^{-3}$. In other words, a cubic metre of granite would weigh 2700 kg, or 2.7 tonnes! Since one cubic metre is somewhat larger than, for instance, the average pebble or rock you might pick up on a beach, these large numerical values for density are often difficult to grasp. Instead, it is often convenient to think of densities in smaller units, so you may come across, or prefer to think of, density values expressed as grams per cubic centimetre (i.e. g cm^{-3}). Thus a density of $2.7 \times 10^3 \text{ kg m}^{-3}$ could be expressed as 2.7 g cm^{-3}. However, when making calculations involving density, always ensure you use the SI units for density.

It is important to appreciate that a planetary body might be made of layers of material that have quite different densities, for example it may have high-density material (such as iron) at its core and somewhat lower-density material (such as rock) nearer the surface. The calculation of mass/volume gives rise to a value of *mean density* for the body.

1.1.2 Venus

The next planet on our tour is Venus. The chances are that you have seen Venus with the naked eye, even if you didn't realize it at the time. Venus is often seen as an extremely bright 'star' an hour or two before sunrise or after sunset, depending on the relative positions of Venus and the Earth in their orbits. A small telescope can resolve Venus as a disc. The planet may also look like a crescent or a gibbous object, depending on the Earth–Venus–Sun geometry at the time. Even powerful telescopes tend to show Venus as a featureless planet due to the presence of a thick atmosphere. In terms of its size and mean density, and the fact that it has a significant atmosphere, Venus could be considered as the 'twin' of Earth. In fact, there are very important differences, particularly regarding the composition of the atmosphere and the resulting surface environment. Figure 1.5 shows an image of Venus, which picks out some cloud structure that is not normally apparent. The clouds are made from tiny droplets of sulfuric acid, hinting that Venus might not be the most welcoming environment for us to visit!

A view of the surface terrain can be obtained using cloud-penetrating radar. One such image is shown in Figure 1.6. The surface of Venus is very complex, with far fewer impact craters than on Mercury, but with many volcanoes and lava plains suggesting significant surface renewal. The only images obtained from the surface of Venus were taken from a series of Soviet Union spacecraft, called Venera. Taking images of Venus was an impressive technical feat considering the hostility of the surface environment. The surface atmospheric pressure was almost a hundred times that on Earth, and the temperature was around 670 K (400 °C). A high-pressure oven is not a good place for sensitive scientific instruments! However before the equipment expired, the Venera spacecraft returned their precious images.

Figure 1.5 The planet Venus (radius 6052 km). This image, taken by the Galileo spacecraft, is falsely coloured to highlight the subtle structure of the clouds which are not usually seen (normally Venus looks more of a uniform white in appearance). (NASA)

Figure 1.6 Details of the surface of the planet Venus, which is usually totally obscured by clouds, taken by the Magellan spacecraft using cloud-penetrating radar. (NASA)

Figure 1.7 shows one of the few colour images obtained by the Venera 13 spacecraft. The surface shows evidence of old lava flows, with a cracked and rugged appearance. The action of the atmosphere has also given rise to surface erosion. The atmosphere of Venus is mostly (97% by volume) carbon dioxide (unlike Earth which is mostly nitrogen and oxygen). The carbon dioxide gives rise to a strong 'greenhouse effect' that traps heat below the lower layers of the atmosphere – hence the very high surface temperature. Venus, while an Earth-twin in some respects, would definitely not be a hospitable place to visit.

Figure 1.7 The surface of Venus obtained by the Venera 13 spacecraft in 1982. Part of the spacecraft is seen at the bottom of the image. (Courtesy of the Russian Academy of Sciences/ RNII KP/IPPI/TsDKS)

1.1.3 Earth

The next planet on our itinerary is Earth. The familiar blue planet is shown splendidly in Figure 1.8. Although Earth may seem rather familiar, and even rather boring in the context of exploring exciting new worlds in the Solar System, it is worth pausing for thought when looking at Figure 1.8. Consider just how important this image is for showing us our home planet and its place in the Solar System. Before the space age, we could only imagine seeing our planet from afar. But now we have an appreciation of the Earth as a finite, isolated and even rather fragile planet in space. This view of Earth is probably very important in appreciating global ecology issues and the need to look after our planet.

The Earth also allows us to study at close quarters many of the mechanisms that influence and characterize the other bodies in the Solar System. Our understanding of the internal structure of large terrestrial-like bodies, volcanism and atmospheres, is greatly enhanced by looking at what happens on (or *in*) the Earth, and using this

Figure 1.8 The planet Earth (radius 6371 km). This image was taken from Apollo 16 in 1972. The image shows the oceans (blue), land (brown) and cloud (white) as well as the Antarctic ice-cap (uniform white at the bottom of the figure). (NASA)

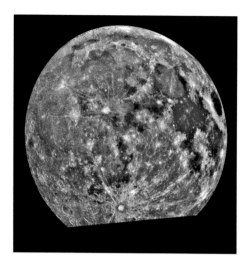 

Figure 1.9 The Moon (radius 1738 km). This image was taken by the Galileo spacecraft. The various colours of the image highlight the dissimilar mineralogy of the regions. (NASA)

Figure 1.10 An image of the Moon, taken from Apollo 16, showing some of its heavily cratered far side (right-hand part of the image). (NASA)

knowledge to consider what must happen elsewhere. For this reason much of the material considered in the following chapters looks closely at the Earth to enhance our understanding of the other planets.

The atmosphere of the Earth (which you will look at in more detail in Chapter 5) is crucial for the survival of life on the planet. Our atmosphere causes a significant rise in surface temperature because of a 'greenhouse effect', which is mainly due to carbon dioxide and water vapour (modest compared to Venus, but still accounting for a 33 K higher temperature than an atmosphereless Earth would have). This means that the *mean* temperature at the surface is 288 K (15 °C), allowing liquid water to exist over much of the planet. The atmosphere also carries heat away from the Equator, so that the Equator is not as hot as it might be and the polar regions are not as cold as they might be. This allows life to survive at a greater range of latitudes than would otherwise be the case if we didn't have atmospheric circulation. The Earth's atmosphere comprises (by volume) 78% nitrogen and 21% oxygen, with other gases (including carbon dioxide) being just a small part. It is perhaps a sobering thought to bear in mind that Venus is an example of what could happen if greenhouse gases such as carbon dioxide became a really significant proportion of the Earth's atmosphere.

The other familiar planetary body on our tour is the Earth's only natural satellite, the Moon. Often ignored due to over familiarity, the Moon offers a spectacular (if rather monochromatic) terrain. Viewed with the naked eye or, better still, through binoculars or a small telescope (best viewed when *not* at full Moon to obtain the best visual contrast), the relatively bright 'highland' regions peppered with impact craters, and the darker and less cratered *mare* (pronounced mar-ray) regions, are clearly seen.

The Moon orbits the Earth about thirteen times each year, and presents the same face to us all the time. This means its **rotation period** (the time it takes to turn once on its axis) exactly matches its **orbital period** (the time it takes to travel once around the Earth). This is called **synchronous rotation**, and is common among satellites throughout the Solar System. Figure 1.9 shows an image of the Moon. False colours have been used in this figure to differentiate predominant surface minerals. Figure 1.10 shows a photograph that reveals some of the far side of the Moon. Fewer mare regions are seen and more impact craters are obvious in this image. The mare regions are younger formations formed by the flooding of lava that wiped away ancient impact craters. You will consider the formation and composition of the Moon in Chapter 2, and the historical impact record of the Moon in Chapter 4.

1.1.4 Mars

Continuing our tour outwards from the Sun, we next encounter Mars (Figure 1.11). It is not hard to understand why it is often referred to as 'the red planet' (although in fact, most people would probably describe it as orange). Mars can often be seen with the naked eye as a 'star' that has a very obvious orange hue to it. Figure 1.11 shows some striking features. The image shows a huge canyon system (called *Valles Marineris*), which represents a fracture in the planet's surface that extends about 4000 km across the planet. This canyon dwarfs the Earth's Grand Canyon, having regions that are 11 km deep and 200 km wide. Also very obvious are the dark, circular features near the left-hand side of the image. These are enormous, old volcanoes. The largest volcano on Mars, *Olympus Mons* (not shown in Figure 1.11), which is also the largest volcano in the Solar System, is 24 km high and has a volume a hundred times greater than Mauna Loa in Hawaii – the largest equivalent feature on Earth.

Figure 1.11 The planet Mars (radius 3390 km). This is a composite image produced from images obtained by the Viking Orbiter spacecraft. (US Geological Survey)

Figure 1.12 The surface of the planet Mars, as imaged by the Mars Pathfinder lander mission in 1997. (NASA)

Figure 1.12 shows an image from the surface of Mars. Large boulders embedded in dust and 'soil' can be seen in the rather barren landscape. Lava flows have altered the terrain in other regions of the planet, and even evidence of ancient running water has now been identified. The atmosphere of Mars is mainly carbon dioxide (95% by volume) and in this respect it is similar to the atmosphere of Venus. However, on Mars, the atmospheric pressure at the surface is much reduced, being only about 0.006 times that on Earth. This rather tenuous atmosphere means that the greenhouse effect you might expect from the high carbon dioxide content is very modest, adding only about 6 K to the mean temperature of the planet (which is 223 K). The surface is desert-like – it is very dry and can get reasonably warm during the day and extremely cold at night. Although the atmosphere is thin, winds on the planet can be formidable, giving rise to large dust storms that can last for weeks or months.

Mars has two relatively tiny satellites, Phobos (Figure 1.13) and Deimos (Figure 1.14), which are thought to be asteroids that have been captured by the gravitational influence of Mars. The bodies are irregularly shaped: Phobos is approximately 26 km × 18 km in size, Deimos is approximately 16 km × 10 km in size. Phobos is shown in Figure 1.13, and you can see a large (relative to the size of the body) impact crater on the left-hand side, as well as other smaller impact craters. Phobos orbits only 6000 km above the surface of Mars, and will probably collide with Mars within the next 50 million years. Deimos is shown in Figure 1.14. There are few craters seen, and the surface may be covered in fine dust, or **regolith**.

Figure 1.13 Phobos (26 km × 18 km), a satellite of Mars. This body is thought to be a captured asteroid. (NASA)

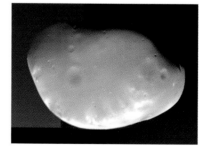

Figure 1.14 Deimos (16 km × 10 km), a satellite of Mars. As with Phobos, this body is also thought to be a captured asteroid. (NASA)

Phobos and Deimos are tiny compared to the Moon, and to the other satellites you will encounter in the outer Solar System part of our tour. However they lead nicely into the next objects on our tour, asteroids.

1.1.5 Asteroids

Between Mars and Jupiter is the **asteroid belt**, which is a 'swarm' of rocky and metallic bodies. The asteroid belt extends all the way round the Sun and each asteroid orbits the Sun. It is from this reservoir of bodies that Phobos and Deimos may have originated. Figure 1.15 shows an image of the asteroid Gaspra. You can see that impact craters pepper the surface, just like on Mercury and the Moon. Even small bodies like Gaspra cannot avoid getting hit by other objects. You will look at asteroids, and their orbits, in much more detail in Chapter 7.

Our grand tour now takes you from the terrestrial planets to the outer Solar System and into the domain of the giant planets. You will explore the giant planets and their many satellites. In comparison to the terrestrial planets, some of these satellites are quite significant in size. Figure 1.16 shows the relative sizes of the major satellites of the giant planets (only bodies with a radius larger than about 200 km have been included) compared to the terrestrial planets. You can see that satellites such as Io, Ganymede, Callisto and Titan are really like small planets, being comparable in size to Mercury and not too far from the size of Mars. The figure also indicates the timescale of the most recent surface-altering 'activity', such as volcanism, on each planet and satellite shown.

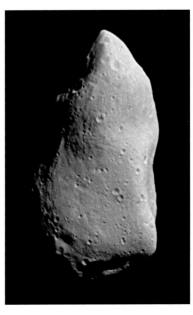

Figure 1.15 Gaspra (19 km × 11 km), a main-belt asteroid, i.e. an asteroid in the asteroid belt. This image was taken by the Galileo spacecraft while en route to Jupiter. (NASA)

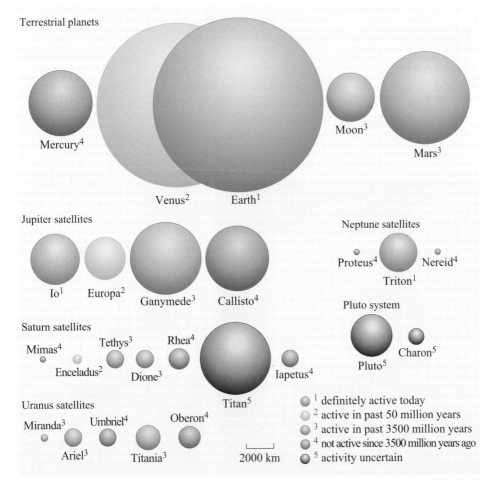

Figure 1.16 The terrestrial planets and the satellites of the giant planets (with radii greater than about 200 km). Pluto and its satellite, Charon, are also shown. The colour and number coding shows how recently the surfaces have been affected by surface-altering 'activity' such as volcanism, fracturing or buckling.

1.1.6 Jupiter

Continuing our tour, Jupiter is the next stop (Figure 1.17). This planet is massively different from what we have seen so far. Jupiter is the largest planet (see Figure 1.1). Its radius is over ten times that of Earth and its mass about three hundred times that of Earth. Its density therefore is about a quarter of that of Earth; the reason for this is that Jupiter consists mostly of gas. The other giant planets also share this property, hence they are sometimes referred to as **gas giants**.

The most striking features of Jupiter are the colourful bands and the swirling clouds. What you are seeing in Figure 1.17, is the top of a dynamic gaseous atmosphere that extends deep into the planet. The whole planet rotates in just 9.9 hours, although the atmosphere at the poles and the equator rotates at slightly different rates, giving rise to 'winds' of $150 \, \mathrm{m \, s^{-1}}$ at the equator. The largest feature is the Great Red Spot (clearly shown in Figure 1.17, with a close-up shown in Figure 1.18), which is around 20 000 km across. It is a huge storm that has been observed for the last few hundred years, continuously rotating in an anti-clockwise direction with wind speeds of around $100 \, \mathrm{m \, s^{-1}}$ at the edges. The atmosphere itself is 90% hydrogen and 10% helium, with traces of methane, ammonia and water vapour. The cloud-top temperature is around 120 K ($-150\,°\mathrm{C}$), although beneath the surface the temperature and pressure must rise rapidly due to gravitational compression. At a depth of 10 000 km, the pressure is likely to be a million times that on Earth, with a temperature of 6000 K (approximately the temperature of the surface of the Sun). At Jupiter's core, the pressure is likely to be an amazing 40 million times that on Earth, and the temperature 16 000 K. You will discover much more about Jupiter (and the other giant planets) in Chapter 6.

The many satellites of Jupiter (there are at least 39) offer an incredibly diverse selection of planetary bodies. Figure 1.19 shows an image of Io, a satellite of Jupiter. Its incredible surface (sometimes described as pizza-like in appearance) shows evidence of a vast amount of volcanic activity. An example of a volcanic region is shown in Figure 1.20. Indeed, Io is the most volcanically active body in the Solar System. To be this active, Io must have an input of energy to heat the body. But how? The answer lies in its orbit. Io is the innermost major satellite of Jupiter, and as such

Figure 1.17 The planet Jupiter (radius 69 910 km). The image was taken by the Cassini spacecraft in 2000, while en route to Saturn. The Great Red Spot is by far the largest feature of the planet. (NASA)

Figure 1.18 Jupiter's Great Red Spot. The spot is about 20 000 km across. (NASA)

undergoes major **tidal heating**. You will return to this concept in Chapters 2, 3 and 7. Io is mainly 'rocky' in composition, making it a terrestrial-like body. Temperatures in the region of 1000 K and more are required to melt these rocky materials. This is in stark contrast to the cold cloud tops of Jupiter!

The next satellite we consider could hardly be more of a contrast to Io. Europa (Figure 1.21) has an icy surface that is covered in cracks, with hardly any impact craters, indicating a very young surface. Europa is a predominantly rocky body. However, it has a layer of ice, about 100 km deep, on the surface. Europa would also be expected to undergo significant tidal heating (albeit to a much lesser extent than Io as it is farther from Jupiter). This presents a fascinating possibility that beneath Europa's icy 'crust' there is a global ocean of melted ice (i.e. water). This also leads to speculation that, where there is liquid water, there might be primitive life!

The possibility of an icy crust on top of a sub-surface ocean does appear consistent with the surface features on Europa, which look remarkably like broken ice packs and fractured ice plains (see Figure 1.22). The surface of Europa also leads us to a new concept – that of **cryovolcanism**. This is the name given to the effect where cold slurries of ice and liquid erupt and flow across the surface like 'cool volcanoes', just as hot molten rock erupts and flows as lava. It appears that cryovolcanism is commonplace amongst the icy satellites of the giant planets, and the surface of Europa is one such case. The relatively recent activity has wiped clean the impact craters that would otherwise have been visible on Europa's surface.

Jupiter has two even larger satellites than Io and Europa. These are Ganymede and Callisto (Figures 1.23 and 1.24). Ganymede is the largest satellite in the Solar System and is actually bigger than the planet Mercury. However, it contains much less mass than Mercury because, like Callisto, it is a predominantly icy body. Both these large satellites are heavily cratered, showing that their surfaces are much more ancient than those of Io and Europa.

Figure 1.19 Io (radius 1821 km), a satellite of Jupiter. This image was taken by the Galileo spacecraft. (NASA)

Figure 1.20 A close-up of the surface of Io, a satellite of Jupiter, taken by the Galileo spacecraft. This image shows a volcanic region about 250 km across. The bright red–orange feature at the left-hand side of the image indicates a recent flow of hot lava. (NASA)

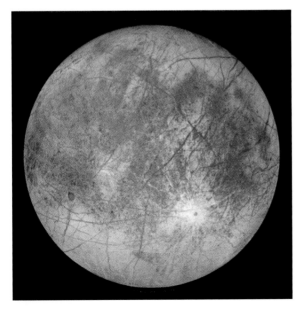

Figure 1.21 Europa (radius 1565 km), a satellite of Jupiter. (NASA)

Figure 1.22 A close-up of Europa, showing the fractured icy surface. The region shown is approximately 35 km × 50 km. (NASA)

Figure 1.24 Callisto (radius 2403 km), the outermost large satellite of Jupiter. The whole surface is heavily cratered. (NASA)

Figure 1.23 Ganymede (radius 2634 km), the largest satellite of Jupiter. The belts of paler terrain are younger than the darker terrain and have been resurfaced by a combination of cryovolcanism and tectonic processes. (NASA)

1.1.7 Saturn

We leave the Jupiter system and move on to Saturn and its satellites. Figure 1.25 shows the planet, instantly recognizable because of the prominent system of rings around it. Saturn itself is not quite as massive as Jupiter, being 'only' 95 Earth masses and about 15% smaller than Jupiter in radius. The clouds seen in Figure 1.25, with somewhat enhanced colours, form bands across the disc that are parallel to its equator. They are rather like the bands and clouds of Jupiter, although obvious storms and swirling structures are not apparent. The planet rotates in just 10.7 hours. This gives rise to the atmosphere bulging at the equator. You may have noticed in Figure 1.25 that the planetary disc does not appear to be perfectly circular, but is somewhat

Figure 1.25 The planet Saturn (radius 58 230 km). This Voyager image has had the colours enhanced slightly to make the banding in the clouds more obvious. The apparent break in the rings just to the right of the planetary disc is simply where the rings are in shadow. (Dr Bradford A. Smith, National Space Science Data Center)

flattened. This is a real effect and not just a distortion in the image. Winds on the equator reach speeds of around $500\,\text{m s}^{-1}$, and storms and even 'spots' can evolve (although not to the extent seen on Jupiter).

The rings of Saturn are remarkable, and beautiful. They are not solid, but are made of icy particles and boulders, the majority of which are between about a centimetre and a few metres in size. These particles most probably originate from the catastrophic break-up of a satellite due to impact. The ring particles all orbit the planet in the same plane (Saturn's equatorial plane), creating an amazingly thin disc – the inner rings are only about 100 m thick. The ring particles, being icy, are highly reflective. This makes them easy to see in reflected sunlight. If you could gather up all the particles in the entire ring system, and put them together into one object, this object would be about 3000 km in diameter. Figure 1.26 shows a close-up image of the rings. Notice that there are distinct 'gaps' in the rings. These arise because of the gravitational influence of Saturn's satellites 'shepherding' the ring particles. You will consider the mechanism for this in Chapter 7. You may be surprised to learn that the other giant planets also have ring systems, albeit not on the same scale as Saturn, making them much harder to see.

Saturn has at least 30 satellites. On this tour, we will consider the seven major satellites (those with radii greater than about 200 km). Figure 1.27 shows the innermost major satellite, Mimas. Although the image is not particularly sharp, you can see that the icy surface appears heavily cratered and that one relatively massive crater is a quarter of the diameter of the entire body. This crater, 130 km in diameter, will have been produced by the impact of a body about 10 km across. If it had been slightly larger it would have broken Mimas apart.

Titan, Saturn's largest satellite, is also one of the larger satellites in the Solar System, being a little under half the size of Earth. What makes it very interesting is that it has a thick, predominantly nitrogen, atmosphere – a description that also fits Earth. Although the exact composition is not yet known, it is likely that nitrogen accounts for 82–99% by volume (Earth's atmosphere is 78% nitrogen by volume), with the rest being mainly methane and other hydrocarbons (which gives Titan its orange colour). Because the atmosphere is opaque, we have very little information

Figure 1.26 The rings of Saturn. This Voyager image shows clearly the gaps between the rings. (Dr Bradford A. Smith, National Space Science Data Center)

Titan is the sixth major satellite of Saturn. It lies between the orbits of Rhea and Iapetus.

Figure 1.27 Mimas (radius 199 km), a satellite of Saturn. The large crater *Herschel*, 130 km in diameter, is evidence of an impact that came close to breaking Mimas apart. (NASA)

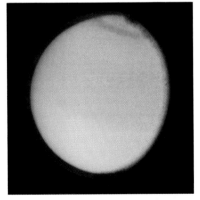

Figure 1.28 Titan (radius 2575 km), a satellite of Saturn. Titan is potentially one of the most interesting satellites in the Solar System although, as seen in this Voyager image, its thick atmosphere makes it look featureless and somewhat unexciting. (NASA)

about the surface. It is likely to be icy, with a surface temperature of around 90 K (−180 °C), and an atmospheric pressure about 1.5 times that of Earth. It is possible that ethane and methane in the atmosphere could condense into droplets, producing a hydrocarbon rain. This would give rise to lakes or seas of liquid methane and ethane. One can imagine that the surface might bear some resemblance to a deep frozen Earth.

The European Space Agency (ESA) Huygens probe, a small spacecraft about one metre across carried aboard the NASA Cassini spacecraft, is designed to answer many questions about Titan. Upon arrival at Saturn in 2004, the Huygens probe is released from the Cassini spacecraft in order to enter the atmosphere of Titan and make a 2-hour descent to the surface by parachute. Various instruments on board Huygens are designed to measure the properties of the atmosphere and the icy, rocky or liquid nature of the surface, thus allowing us to learn a great deal about Titan.

The next satellite on our tour is Enceladus. Its icy surface is smoother than the surface of Mimas. In fact there are some areas that are only weakly cratered and have long ridges and fractures. This is evidence of cryovolcanism with cryolavas filling some of the regions at some time in the past (Figure 1.29). Enceladus is one of the most highly reflective bodies in the Solar System, probably indicating that the surface is coated in a thin, reflective frost (possibly as a result of some cryovolcanic eruptions).

Moving on to the larger satellites, Tethys and Dione (Figures 1.30 and 1.31), you can see rugged icy terrains with prominent impact craters. Fractures and ridges on Tethys are consistent with cryovolcanic activity, although Dione shows more evidence of flooding by cryogenic lavas. Dione's pale, wispy streaks appear to be fault lines. Dione has probably undergone at least two major episodes of cryovolcanic activity in the past, which has wiped many smaller craters from the surface. Tethys has a very low density, indicating very little rock content in the body.

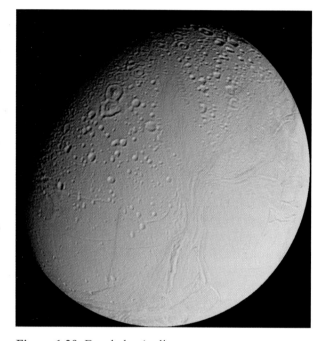

Figure 1.29 Enceladus (radius 249 km), a satellite of Saturn. The cratered region (upper left) is in contrast to the smoother region (lower right), which shows evidence for cryovolcanism. (US Geological Survey)

Figure 1.30 Tethys (radius 530 km), a satellite of Saturn. (NASA)

Figure 1.31 Dione (radius 560 km), a satellite of Saturn. This Voyager image shows several reasonably large craters (the largest is near the top, being 160 km across), with evidence of cryogenic lavas flooding some regions. (Courtesy of Calvin J. Hamilton)

Figure 1.32 Rhea (radius 764 km), a satellite of Saturn. (Courtesy of Calvin J. Hamilton)

Figure 1.33 Iapetus (radius 718 km), a satellite of Saturn. This Voyager image shows the dark leading edge of the body, and the reflective trailing edge. (NASA)

The satellites Rhea and Iapetus are a similar size, but are quite different in appearance (Figures 1.32 and 1.33). Rhea shows quite a uniformly cratered surface indicating that the satellite has been relatively inactive, with many small craters being maintained. However, Iapetus has one hemisphere that is very bright and highly reflective (consistent with the icy surfaces you have seen in other satellites), and one hemisphere that is very dark, as if covered by an obscuring 'sooty' deposit.

1.1.8 Uranus

Our tour now leaves the Saturnian system and continues outwards towards the next giant planet, Uranus. Uranus (and Neptune) are smaller than Jupiter and Saturn. Uranus has a mass of almost fifteen times that of Earth and is about four times larger in size than Earth. The planet was discovered by Sir William Herschel in 1781. Images show a rather featureless blue–green planet, as shown in Figure 1.34. One oddity of Uranus is that its spin axis is tilted over, 98 degrees from the 'vertical'. This is most likely due to a huge impact event early in the planet's history that literally knocked Uranus over on its side. It means that one pole of the planet can point towards the Sun for long periods of time. The resulting polar heating drives the atmospheric flow on

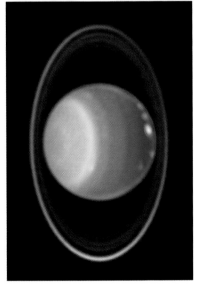

Figure 1.34 The planet Uranus (radius 25 360 km). This image is from Voyager 2, taken in 1986. The planet has a rather featureless blue–green atmosphere. (NASA)

Figure 1.35 The planet Uranus, showing its rings. This image was taken using the Hubble Space Telescope. (NASA)

Figure 1.36 Miranda (radius 236 km), a satellite of Uranus. This Voyager image shows the cratered surface covered with a layer of dust or 'snow' and distinct regions (left- and right-hand sides of the image) related to some icy volcanic activity. (Dr Bradford A. Smith, National Space Science Data Center)

the planet, with winds flowing from the (south) pole to the equatorial regions. The prominent banding and storms seen on Jupiter and Saturn are not seen on Uranus, although some subtle structure is exposed in some false-colour images.

You may recall all the giant planets have ring systems, although they are only really obvious in the Saturnian system. The rings of Uranus (discovered in 1977) are the next most obvious. Figure 1.35 shows an image obtained using the Hubble Space Telescope. The particles in the rings of Uranus are extremely dark (unlike Saturn's rings). There is evidence of small dust particles and metre-sized boulders. As in the Saturnian system, the rings are separated by gaps, and will be gravitationally influenced by satellites.

Uranus, like Saturn, has a multitude of satellites (at least 21 and probably more), although only five are considered major bodies (radii greater than 200 km). The first of these to consider is Miranda (Figure 1.36). This body has an extremely rugged and exotic terrain. There are areas that are mostly craterless, and are distinctly different from the rest of the body. It is likely that these regions have been formed by some cryovolcanism processes. The other regions on the satellite have more impact craters and therefore a somewhat older surface. The craters (except for a few sharply defined relatively recent craters) appear to be blanketed with dust or 'snow', presumably as a result of some explosive cryovolcanic eruptions.

Ariel and Umbriel (Figures 1.37 and 1.38) are similar in size, and each is over double the size of Miranda. Ariel shows impressive trough-like fractures that break up the impact-scarred surface. Many of the troughs will have been subsequently filled by icy lavas, although it is unlikely that Ariel has suffered any major cryovolcanism in the recent past. The surface of Umbriel appears quite different, being much darker than Ariel (and other icy surfaces) and the surface appears to be heavily cratered. The most notable feature is the bright structure at the top of the image. This is an impact crater that is about 100 km in diameter. It is lighter in colour because the impact has penetrated, and excavated material from, the subsurface layers, which appear to be of a somewhat different composition from the surface material.

Figure 1.37 Ariel (radius 579 km), a satellite of Uranus. This Voyager image shows large trough-like fractures in the surface of the body. (Dr Bradford A. Smith, National Space Science Data Center)

Figure 1.38 Umbriel (radius 585 km), a satellite of Uranus. Although this Voyager image is of quite low resolution, the heavily cratered surface is clearly visible. The bright patch (top) is an impact crater that is 100 km in diameter. (NASA)

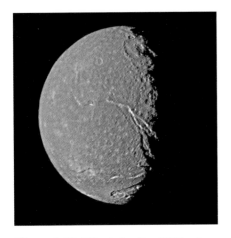

Figure 1.40 Oberon (radius 761 km), a satellite of Uranus. This Voyager image shows an 11 km high mountain range, silhouetted against space, in the lower-left region. (NASA)

Figure 1.39 Titania (radius 789 km), a satellite of Uranus. This satellite has an appearance quite similar to Ariel. (NASA)

The satellites Titania and Oberon (Figures 1.39 and 1.40) are larger than Ariel and Umbriel. Titania is similar in appearance to Ariel, although it is more heavily cratered. There are fault regions on the surface that are 1500 km long and 2–5 km high. There are no signs of lava flooding in the faults although Titania has certainly had some cryovolcanic activity in the past. Oberon is more like Umbriel (although not quite as dark). Bright patches are seen on the surface that are most likely to be regions of ejecta from relatively recent impact craters (similar to the one obvious bright crater on Umbriel).

1.1.9 Neptune

Figure 1.41 The planet Neptune (radius 24 620 km). This Voyager 2 image, showing a 'Great Dark Spot', was obtained in 1986. (NASA)

Figure 1.42 Proteus (radius 209 km), a satellite of Neptune. This image shows a heavily cratered surface with one huge (relative to the size of the body) impact crater. (NASA)

Our tour now arrives at the last of the giant planets, Neptune (Figure 1.41). Neptune has a mass of about seventeen times that of the Earth and is only slightly smaller than Uranus (being about four times larger than Earth). The existence of Neptune was suspected for some time, after detailed observations of the motion of Uranus suggested that there must be another large planet in the outer Solar System gravitationally influencing the orbit of Uranus. Neptune was discovered in 1846 by astronomers Johann Galle and Heinrich D'Arrest. Its atmosphere has a glorious 'electric blue' colour. More features are apparent in the atmosphere than for Uranus, with some banding and pale clouds being evident. Neptune also appears to have its own version of Jupiter's Great Red Spot. Neptune's 'Great Dark Spot' is also an oval-shaped storm system, although it is much less long-lived than Jupiter's spot (Hubble Space Telescope images reveal that the Great Dark Spot has already dispersed).

Neptune has at least eight satellites, although only three are of any great size. Proteus, shown in Figure 1.42, is a non-spherical body, and although the image is of quite low resolution, the surface appears to be heavily cratered. In addition, and reminiscent of Saturn's satellite Mimas, Proteus has one very large impact crater. As with Mimas, the impact must have come close to breaking up the entire body.

Triton (Figure 1.43) is Neptune's only large satellite. The surface is an icy mixture of nitrogen, carbon monoxide, methane and carbon dioxide. It has a temperature of just 40 K (−233 °C). One of the unusual things about Triton is that it orbits Neptune in the

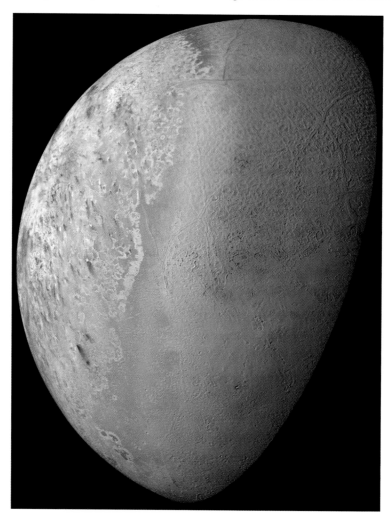

Figure 1.43 Triton (radius 1353 km), a satellite of Neptune. Triton is Neptune's only large satellite. (US Geological Survey)

opposite direction to most other satellites. This is evidence that Triton was captured rather than forming near Neptune from 'leftover' material. This capture process will probably have involved violent impacts with other existing satellites, and will have left Triton in an orbit that gave rise to significant tidal heating. This all means that the surface of Triton has undoubtedly had a complex geological history, and that the surface is probably heavily modified from its original appearance.

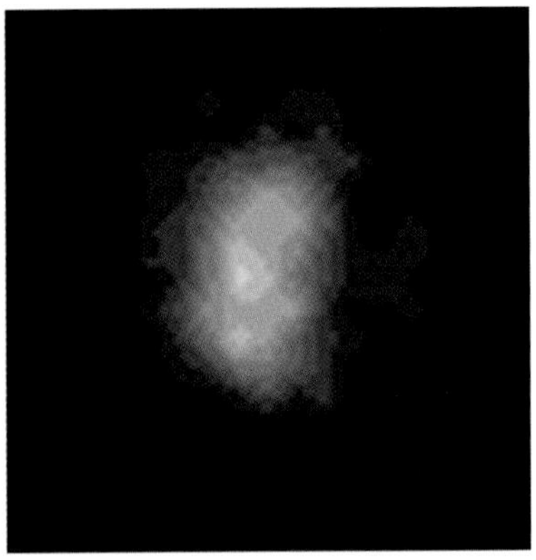

The terrain on Triton (Figure 1.43) appears divided. The area on the left is Triton's southern polar cap. Some dark streaks are apparent in this region. These are due to cryogenic geyser-like eruptions that send plumes of dust about 10 km above the surface, and then leave dark stains across the surface. These geyser plumes were actually seen by Voyager 2, showing that activity is ongoing. This sort of activity gives rise to a very tenuous nitrogen atmosphere. The right-hand part of the image (Figure 1.43) shows a different texture, which has been likened to the skin of a cantaloupe melon. Cryogenic lavas have probably flooded much of this region. Triton has probably got much in common with Pluto (the last planet on our tour) and will have originated from the same region of the outer Solar System.

The third Neptunian satellite we will consider is Nereid (Figure 1.44). It is somewhat smaller than other satellites that our tour has included but is noteworthy as its orbit around Neptune traces out a path that is a very elongated ellipse. This suggests that Nereid might be a collisional fragment, thrown onto an unusual orbit. Whether it is related to the capture of Triton is unclear, but it does hint at the violent history that the satellites of Neptune had to endure.

Figure 1.44 Nereid (radius 170 km), a satellite of Neptune. This Voyager 2 image is of very low resolution, but hints at a significantly non-spherical body. This irregular shape, combined with the fact that its orbit around Neptune is very elliptical, suggests that Nereid is probably a collisional fragment. (NASA)

1.1.10 Pluto

Our grand tour now moves on to the last planet, Pluto. It is the only planet that has not yet had a spacecraft mission fly near it (although hopefully this will change over the next ten years or so). For this reason we know less about Pluto than the other planets. However, we do know that it would be an extremely interesting place to visit. Pluto was discovered in 1930 after a huge photographic search by astronomer Clyde Tombaugh. This little planet offered a big surprise in 1978 when a satellite, named Charon, was discovered. However Charon is about half the size of Pluto, and so the system is not so much a planet with a satellite, as a binary planetary system (or 'double planet'). Figure 1.45 shows a Hubble Space Telescope image of the Pluto–Charon system. The bodies orbit each other, separated by about 19 400 km. Although this image does not show any detail on the bodies, some important properties have been determined by various other means. Both objects have synchronous rotation, such that the same hemispheres face each other all the time. The Charon-facing side of Pluto is a lot brighter (more reflective) than the other side. Pluto also has a very tenuous (mainly nitrogen) atmosphere, similar to that seen on Triton. Indeed, it is likely that Pluto is quite similar to Triton (although

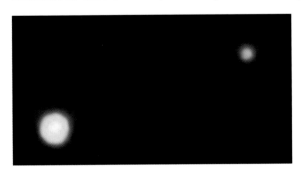

Figure 1.45 Pluto (radius 1137 km) and its satellite Charon (radius 586 km). This image was obtained using the Hubble Space Telescope and shows the two bodies 19 400 km apart. This is the closest our Solar System comes to a binary planetary system. (NASA)

The Kuiper Belt is sometimes also referred to as the 'Edgeworth–Kuiper Belt'.

less geologically processed). Both Pluto and Charon are believed to have formed beyond Neptune in the region called the **Kuiper Belt**. The Kuiper Belt refers to a belt of planetary bodies beyond Neptune, of which Pluto appears to be the largest member. You will be considering the Kuiper Belt in detail in Chapter 7.

1.1.11 Comets

Our tour of the Solar System is almost complete. However one type of body has not had much of a mention so far – comets. Comets are small bodies that pervade the entire Solar System and, along with asteroids, are responsible for many of the impact craters found on the other planetary bodies. The usual view of a comet from Earth is that of a long, wispy structure (Figure 1.46). The long tail is seen because of tiny dust particles reflecting sunlight and can extend millions of kilometres into space. However, the actual source of the dust comes from a rather small object, which looks a bit like an asteroid. Figure 1.47 shows Halley's comet as imaged by the Giotto spacecraft that flew within 600 km of the body in 1986. The actual comet, which was about 15 km by 7 km in size, is very dark and irregularly shaped. Gas and dust particles stream from its surface. These are released as the sunlight heats the icy surface. You will take a closer look at comets in Chapter 7.

Our grand tour is now complete. You have seen the diverse nature of the planetary bodies in our Solar System, and have seen that clues to their geological history and internal structure can be gathered from their surfaces. You will be considering the origin of the Solar System in detail in Chapters 8 and 9, particularly looking at the evidence from meteorites, which tells us about how, and when, Solar System bodies were formed. You will also look at how they came to have particular chemical compositions. But first we will give a brief *overview* of the formation process, to put our observations of the current Solar System into context.

Figure 1.46 A beautiful example of a comet, called Ikeya–Zhang, as seen in April 2002. (Copyright © 2002 by Michael Jager (Austria))

Figure 1.47 Halley's comet, as seen from the Giotto spacecraft that flew within 600 km of the comet in 1986. The dark irregularly shaped body (15 km × 7 km) is emitting large amounts of dust, and gas produced from melting ices. (European Space Agency)

1.2 The formation of the Solar System

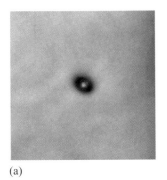
(a)

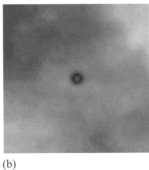
(b)

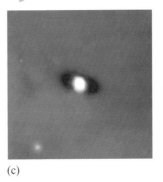
(c)

Figure 1.48 Images, taken with the Hubble Space Telescope, showing young stars with protoplanetary discs around them. These discs will probably result in solar systems. (NASA)

Theories of Solar System formation are by no means complete, with details of some particular processes being poorly understood. But the overall formation scenario is well accepted, and was first put forward in its original form in 1796 by the French scientist Pierre-Simon Laplace. Laplace suggested that the Solar System was formed by the collapse of a large, initially spherical, rotating cloud of gas and dust. Modern work has refined this idea, but the overall theme remains.

The formation process started with a huge tenuous cloud of gas and dust, which underwent contraction due to mutual gravitational attraction. The contracting cloud inevitably had some rotational motion and, as it contracted further under gravity, the rate of spin would have increased, rather like a spinning ice skater pulling in their arms. As the resulting cloud spun, it flattened into a disc, with the young Sun at the centre. This structure is referred to as a **protoplanetary disc** (or **solar nebula**). Such discs can be observed around some other stars, as shown in Figure 1.48.

The term *dust* in this context refers to solid particles or 'grains' which have formed by the agglomeration of individual atoms, molecules, or smaller dust grains. They can be thought of as fragments of rock and ice, often with rather randomized structures.

In the protoplanetary disc, the dust particles collided and stuck together. As more and more stuck together, larger particles were formed. This is called **accretion**. Then objects several metres across, and before long kilometres across, would have formed. These objects, called **planetesimals**, were the 'building blocks' of the planets. Collisions between the numerous planetesimals then built a fewer number of larger bodies, called **planetary embryos**. Figure 1.49 shows an artist's impression of this process. Eventually, one planetary embryo will have dominated and accreted all the other significantly sized bodies in its region of the disc. When all the material in that region had been exhausted, the planet was 'complete'. This whole process would have taken in the region of 10^8 years, which is *relatively* quick when compared to the present age of the Solar System (4.6×10^9 years).

It is important to broadly consider the temperature distribution of the protoplanetary disc while all this was going on. The material was hotter near the young Sun, and the temperature fell towards the edges of the disc. In the inner, hotter regions (that had temperatures of $>500\,\text{K}$), small ice particles could not exist and so accretion of rocky and metallic material dominated. In the outer regions however (which had temperatures of $<300\,\text{K}$), ices could remain frozen and go to form planets – the ices could of course be melted in the formation of the planet, but would be retained as gas due to the gravitational field of the planet.

Figure 1.49 An artist's impression of accretion within the protoplanetary disc. The end result of this process would be finished planets. (Dana Berry, Space Telescope Science Institute)

Thus, we have a brief overview of the formation process and of why rocky and metallic planets (the terrestrial planets) are generally observed in the inner Solar System and gaseous planets in the outer Solar System. It explains, if somewhat briefly, why the planets formed in a disc (in the same plane) and why all the planets go around the Sun in the same direction of motion.

1.3 The layout of the Solar System

For millennia, the stars visible in the night sky have been the subject of study. Naturally, the Sun and Moon dominated thinking about the Earth's place in the cosmos, but five planets (Mercury, Venus, Mars, Jupiter and Saturn) were well known to early civilizations because of their brightness and rapid motions relative to the 'fixed' stars. Indeed, quite sophisticated knowledge of planetary motions developed in several early cultures, defining the idea of a Solar System (as opposed to the fixed stars). However, it was only in 1512 that Polish astronomer Nicholas Copernicus, after studying records of naked-eye observations of the motions of the planets against the starry background, realized that the Earth went around the Sun. Furthermore, the German astronomer Johannes Kepler deduced in 1604 that the planets move in elliptical, rather than circular, orbits. It was not until the Italian scientist Galileo Galilei discovered four satellites of Jupiter in 1610, with one of the first telescopes ever made, that the existence of other bodies in the Solar System became known, and it was 1801 before the first asteroid was observed.

Since the invention of the telescope, the number of known bodies in the Solar System has increased enormously. Technological progress has led to occasional surges in that number. The construction of larger and larger telescopes during the 19th and early 20th centuries led to the discovery of many planetary satellites and huge numbers of asteroids. In the 1970s and 1980s, spacecraft explorations of the planets revealed yet more satellites and entire planetary ring systems. Developments in the technology of light-sensitive semiconductors such as charge-coupled devices (CCDs), which form the basis of digital cameras, meant that photographic film for astronomical work was mostly abandoned by the 1990s. Using CCDs, astronomers could record virtually all the light passing through the telescope, allowing incredibly faint objects to be detected. Using this technology, a new belt of asteroid-like bodies was discovered at the edge of our Solar System in 1992. Called the Kuiper Belt, this huge reservoir of 'mini planets' represents the furthest reaches of our planetary disc.

We will now consider the actual orbits of the planets and the layout of the Solar System.

The Sun, being by far the most massive body, lies at the heart of the Solar System, with the planets orbiting around it. When considering the distances of the planets from the Sun, we inevitably find we are using huge numbers. For example, the Earth lies about $150\,000\,000\,000$ m (1.5×10^{11} m, or 150 million km) from the (centre of the) Sun. It is much more convenient to define this distance as one **astronomical unit** (AU). Thus Earth is at 1.0 AU, Jupiter at 5.2 AU and Neptune at 30.1 AU from the Sun. Note however, that one must take care to use SI units when appropriate.

■ How long does it take sunlight, travelling at 3×10^8 m s^{-1}, to make its journey from the Sun to the Earth?

❏ The Earth is at 1 AU, which is 1.5×10^{11} m from the Sun, thus the light travel time is $(1.5 \times 10^{11}$ m$)/(3 \times 10^8$ m s$^{-1})$, which is 500 s, or about 8.3 minutes.

The nine planets in the Solar System (Mercury, Venus, Earth, Mars, Jupiter, Saturn, Uranus, Neptune and Pluto) orbit the Sun very close to the same plane, known as the **ecliptic plane**. As we saw in Section 1.2, this is a consequence of the formation process. Planetary bodies orbit the Sun on elliptical orbits although seven out of the nine planets have orbits that are only very slightly elliptical, and indeed the orbits can often be approximated to a circle. Pluto and Mercury have orbits that are significantly elliptical. In fact Pluto's orbit takes it within the orbit of Neptune at its closest point to the Sun, and far beyond Neptune at its farthest point from the Sun (thus Pluto is the outermost planet only *some* of the time). Pluto's orbit is also significantly more inclined to the ecliptic plane than the other planets.

The asteroid belt (sometimes referred to as the asteroid *main belt*) lies between Mars and Jupiter, and the Kuiper Belt lies beyond Neptune, although it should be appreciated that **minor bodies** (asteroids and comets) can be found outside these limits. Figure 1.50 illustrates graphically the relative positions of the major bodies (the figure also includes a reminder of the relative sizes of the planets).

Figure 1.50 (a) Diagrammatic summary of the Solar System, showing that the planets orbit the Sun close to the ecliptic plane. Note that this is an oblique view – the orbits are actually nearly circular. (b) The relative sizes of the planets and the Sun. (c) The relative distances of the planets from the Sun.

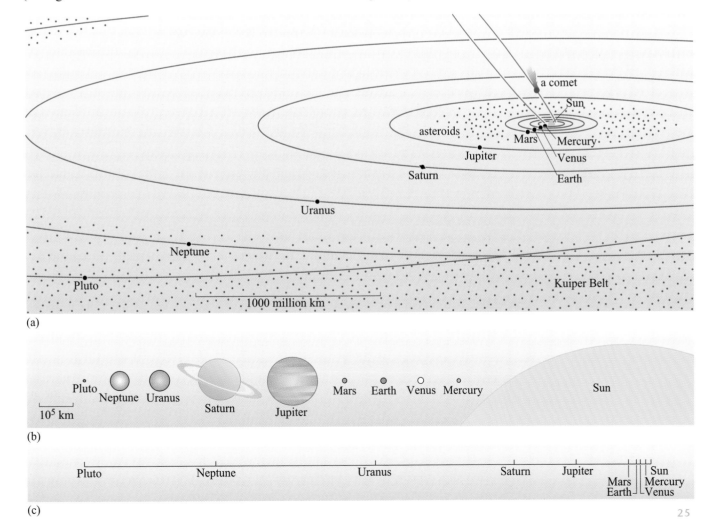

(a)

(b)

(c)

Although the relative distances of the planets from the Sun are shown in Figure 1.50c, it is perhaps hard to appreciate the relative distances involved from a small diagram. Here is a great way of gaining an intuitive appreciation of how much further from the Sun the outer planets are when compared to those in the inner Solar System. Get a ball of string and knot it at distances scaled to represent the distance of each planet – say 10 cm to represent every 10^8 km. (The orbital distances of the planets are given in Appendix A, Table A1.)

◼ If you need 5.9 m of string to reach Pluto, how much would you need to reach the nearest star, 4.2 light years (4×10^{16} m) away?

❑ As 10 cm (0.1 m) of string represents 10^8 km (10^{11} m), then 1 m of string represents 10^{12} m. Thus 4.2 light years would take ($4 \times 10^{16}/10^{12}$) m, i.e. 4×10^4 m, or 40 km.

Although you have seen that the Solar System extends out to Pluto and the Kuiper Belt, strictly speaking this is not the full picture. We are aware of comets being found in our inner Solar System, but in fact the whole Solar System is surrounded by a huge spherical cloud of comets, at distances stretching up to tens of thousands of AU. This cloud, first postulated by Dutch astronomer Jan Oort, is likely to contain over 10^{11} comets that were formed in the inner Solar System, but then 'thrown out' by the influence of gravity from the giant planets. The inner **Oort cloud** occasionally supplies us with errant comets entering our Solar System.

The Oort cloud is sometimes also referred to as the Oort–Öpik cloud.

1.4 Physical properties of Solar System bodies

You have now looked at the layout of the Solar System, the orbits and distances associated with Solar System bodies, and briefly, the sizes of the planets. Let us look more closely at one important physical property, density, and how it varies between the planetary bodies. Tables A1, A2 and A3, in Appendix A, list the sizes and densities of the planets, their major satellites and a selection of asteroids. To investigate how planetary radius varies with density, we could plot a graph of these quantities. This sounds reasonably straightforward, but looking at the range of radii you have to consider, perhaps it is more complicated.

◼ Why would it be difficult to plot the radii of planetary objects on a linear graph?

❑ The range in radius values is so enormous (from a few kilometres to about 70 000 km) that the data would not all usefully fit on an ordinary linear graph – the smaller-sized objects would all be crammed up at one end.

The problem is overcome by using a graph with one logarithmic axis for radii and one 'ordinary' axis, for densities. Such graphs are often termed 'log–linear'. On the logarithmic axis, each major interval is a power of ten, for example 10^1, 10^2, 10^3 and so on, and these intervals are equally spaced. Note, however, that the divisions within each large square are not equally spaced (Figure 1.51). Such graphs are a convenient way of plotting data that range across many powers of ten.

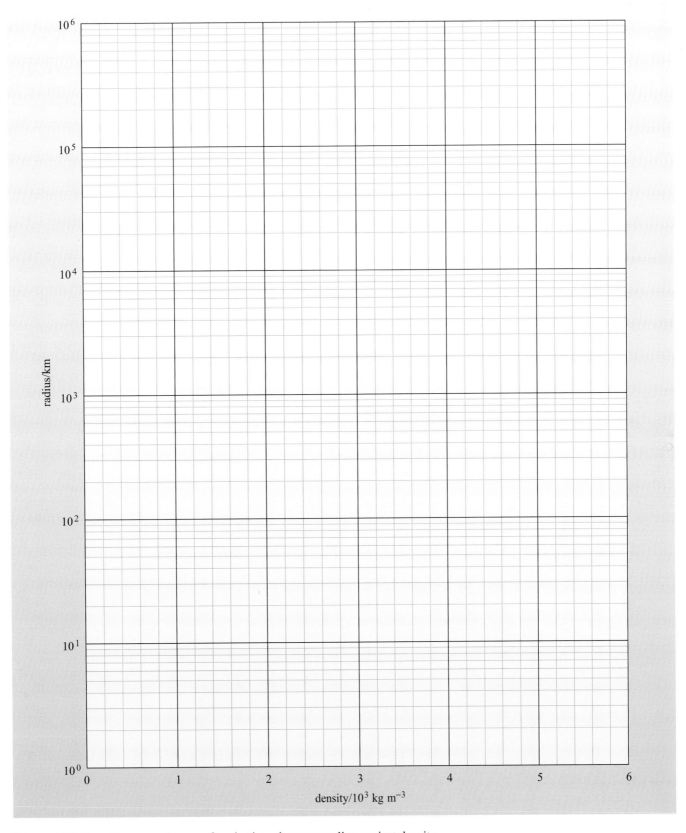

Figure 1.51 Log–linear graph paper for plotting planetary radius against density.

QUESTION 1.1

Show how the radii of the planetary bodies vary with density.

To answer this question, use the graph shown in Figure 1.51, and plot the radius against the density of each of the planetary bodies given in Tables A1, A2 and A3 (planets, satellites and asteroids). Write the name of each object (abbreviated if necessary) alongside its position on the graph to help you identify it.

QUESTION 1.2

On the graph you have prepared for Question 1.1, try to identify groupings of the objects. Identify these groupings on your graph. For example, you could identify the icy satellites, terrestrial planets (or terrestrial-like bodies) etc.

Four planets form an obvious family, characterized by their huge dimensions (radii greater than 10 000 km) and low densities. These are the giant planets (or gas giants), which are dominated by Jupiter, whose mass is more than that of all the other planets put together.

A second group straggles off towards the high-density area at the right-hand side of the graph. This group includes rather dense, but fairly large bodies, with radii of more than 1000 km, notably the terrestrial planets, Mercury, Venus, Earth and Mars, and three dense satellites – our own Moon, plus Jupiter's satellites Io and Europa. Their densities suggest that these bodies are predominantly rocky, similar to the Earth (terrestrial-like bodies).

■ The minerals that make up rock have densities between about $2.5 \times 10^3 \, \text{kg m}^{-3}$ and $3.5 \times 10^3 \, \text{kg m}^{-3}$. Is it likely that the terrestrial planets could consist *exclusively* of rocky material?

❑ No, they must contain a denser component as their densities (particularly the densities of Earth, Venus and Mercury) are considerably higher.

This denser component is thought to be mostly metallic iron. In the next chapter you will consider in detail why this is the case.

Clustered towards the left-hand side of the graph are a large number of satellites, which form a third group, with radii of a few hundred to a few thousand kilometres, and densities between $1 \times 10^3 \, \text{kg m}^{-3}$ and about $2 \times 10^3 \, \text{kg m}^{-3}$. These satellites consist mostly of ice, with varying proportions of rocky materials. Low-density satellites such as Mimas, Tethys and Iapetus must consist almost entirely of ice, whereas denser examples such as Ganymede, Callisto and Titan must contain substantial proportions of rock. Pluto and Charon also lie in this icy satellite group. Europa comes closest to straddling the icy satellites and terrestrial-like bodies group. This is due to its icy surface layer, which is about 100 km thick and overlies its rocky interior.

The asteroids that you plotted form another group. The asteroids listed in Table A3 are relatively small bodies, but we might expect that larger asteroids would lie further up the graph, towards the icy satellite and terrestrial groups. Note that the two tiny satellites of Mars (Phobos and Deimos) also plot right at the bottom of the graph within the small asteroid group, consistent with the 'captured asteroids' idea that you met earlier.

You have now completed our tour of the Solar System, considered in broad terms its formation, and looked at one aspect of 'comparative planetology' by considering the sizes and densities of the planetary bodies. In the next chapter you will look at the internal structure of the terrestrial planets and consider in detail how the internal structure is not only related to the formation processes, but also dependent on ongoing geological processes – processes that largely explain the appearance of the planetary bodies you have studied in this introductory chapter.

1.5 Summary of Chapter 1

- The Solar System formed due to the gravitational contraction of a huge cloud of gas and dust, forming a protoplanetary disc (or solar nebula). Planetesimals grew from the material in the protoplanetary disc, and planetesimals accreted to form planetary embryos, and thus, eventually, the planets.

- The Solar System is broadly separated into terrestrial planets (predominantly rocky and metallic in composition) in the inner region, and giant planets (predominantly gaseous) in the outer regions. This is a consequence of the formation process where lower temperatures existed in the solar nebula farther away from the (young) Sun.

- The planets orbit the Sun on elliptical orbits, although the orbit of each planet, with the exception of Pluto and Mercury, can be approximated to a circle.

- The planets, in order of increasing distance from the Sun, are Mercury, Venus, Earth, Mars, Jupiter, Saturn, Uranus, Neptune and Pluto (although Pluto is closer to the Sun than Neptune for part of its orbit).

- The distance between the Sun and the Earth (more correctly the mean distance between their centres) is defined as one astronomical unit (AU).

- There is a belt of asteroids (rocky and/or metallic minor bodies) lying mainly between the orbits of Mars and Jupiter (although asteroids can be found elsewhere in the Solar System albeit in smaller numbers).

- There is another belt of minor bodies lying beyond Neptune called the Kuiper Belt.

- Comets (icy minor bodies) can be found throughout the Solar System. Comets also make up the Oort cloud (a massive spherical shell of comets at great distances from the inner Solar System).

- Images of planetary bodies reveal a great diversity in the surfaces. Impact craters (made by asteroids and comets) are common on most bodies. A lack of impact craters implies that there has been some resurfacing (the surface is relatively young).

- Evidence of volcanism and lava flows can be seen on the terrestrial planets, and evidence of cryovolcanism can be seen on many of the icy satellites.

- Io, a satellite of Jupiter, is volcanically active, due to the body undergoing significant tidal heating.

CHAPTER 2
THE INTERNAL STRUCTURE
OF THE TERRESTRIAL PLANETS

2.1 Introduction

Section 1.2 briefly outlined how a planet-sized body can form through the process of accretion, but this represents the completion of only the earliest stage of a planet's evolution. In this chapter we examine the processes that occur during the next developmental stages of the terrestrial planets, and how we can obtain and interpret information regarding their internal composition.

Earth scientists often refer to the Earth's core, **mantle** and **crust**, indicating different layers exist within our planet. These layers display distinct properties based on the various **mineral** and rock compositions within the Earth (Figure 2.1). It could also be argued that the atmosphere and hydrosphere (i.e. the oceans and ice-caps) are also layers within our planet (the evolution and properties of planetary atmospheres will be dealt with in later chapters).

We live on planet Earth, so it is only natural that we have a wealth of knowledge regarding Earth's composition and structure. Much of this information is derived from the study of rocks. However, we do not know precisely what the Earth is composed of, since direct geological evidence from rock samples tells us only about its outermost part.

The different layers within the Earth have been identified through a variety of methods and observations. It is believed that these layers owe their origins to the early stages of our planet's history that followed accretion. If we can begin to understand how to detect and interpret the evolution of layering within the Earth, it will help us in examining the evidence for layering within the other terrestrial planets, for which data are much more limited.

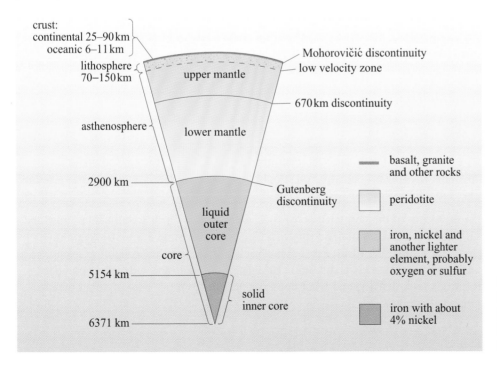

Figure 2.1 The compositional layers within the Earth. The thicknesses of the crust and lithosphere have been exaggerated for clarity.

So how does a planet-sized ball of matter, created from the raw materials of the early Solar System, attain a layered structure? Is the layering observed within Earth peculiar to our planet, or is it part of the normal pattern of evolution of the terrestrial planets? Can the presence of a layered structure within Mercury, Venus, Mars and the other terrestrial-like bodies, be detected? If these other bodies are indeed layered, then what properties and structures does the layering confer upon them? All these are questions that will be investigated in this chapter, but first the evidence for Earth's layering will be explored.

2.2 Investigating Earth's internal structure

Earth's radius is about 6400 km, yet even the deepest mines can only provide us with samples from a maximum depth of about 15 km, in other words only the outermost 0.25% of our planet! Nevertheless, geologists can find examples of materials formed at much greater depths, up to 100 km for instance. This is because, geologically speaking, Earth is a highly active planet. Lateral movement in our planet's outermost layers creates folding and buckling in crustal rocks, especially on the margins of the huge areas that make up the continents. Such areas, together with those forming the ocean floors, comprise immense 'plates', and the processes and effects of the relative motion of these plates is known as **plate tectonics** and is described in detail in Section 2.4.5. Plate tectonic processes, together with deep-seated volcanic activity, can bring rock material, which originally formed at greater depths, nearer to the surface. For instance, **plutonic** rocks, which are those formed at depth, can be broken off and carried to the surface in molten lava as exotic nodules or **xenoliths** (from the Greek *xenos* meaning strange or foreign, and *lithos* meaning rock) and then preserved inside lava flows when this molten rock solidifies at the Earth's surface. Alternatively, the buckling, folding and faulting of the Earth's crust, together with surface erosion, can eventually lead to the exhumation of materials which originated at a similarly deep level. However, if we relied upon this material alone, we would still know very little about Earth's composition and structure below 100 km depth, therefore much of our understanding of Earth's deeper structure is necessarily based upon indirect evidence.

2.2.1 Density and the messages in the rocks

Density (Box 1.1) is the first piece of evidence that indicates the rocks at deeper levels of the Earth may be different from the rock materials that can be examined at the surface. Earth's density can be calculated very accurately because both its size and mass can be determined.

QUESTION 2.1

(a) The mass of the Earth has been determined as 5.9737×10^{24} kg. Calculate the Earth's bulk (i.e. mean) density. You should assume the Earth is a sphere with a radius of 6371 km. Quote your answer to 3 significant figures.

(b) Examination of the composition of rocks comprising Earth's continental crust and ocean floor crust gives a bulk density of 2.7×10^3 kg m^{-3} and 3.0×10^3 kg m^{-3}, respectively, whilst that of the most dense rocks (i.e. xenoliths brought up from depths of up to 100 km) is about 3.5×10^3 kg m^{-3}. How do these densities compare with Earth's bulk density, and what does your result from (a) suggest about density variation with depth?

Question 2.1 demonstrates that the interior of the Earth must be more dense than that of the materials found at its surface. Two key factors contribute to this argument. The first of these factors is **self-compression**, resulting from the increase of pressure with depth (Box 2.1).

QUESTION 2.2

Using the density values of ocean floor crust and a mantle xenolith given in Question 2.1, what would be the density of material at 2900 km depth (where Earth's core begins) if we could assume that density increases linearly with depth?

So how does the density of rock material respond to the increasing pressure within the Earth, and is the composition of this rock material always similar at different depths? As you might expect, there are other important factors controlling the density and composition of materials within the Earth, and self-compression alone is not the entire story.

The atomic structures of minerals are sensitive to pressure. When pressure is increased a mineral becomes slightly more dense because its constituent atoms are forced closer together. However, eventually a point is reached where the degree of compression becomes too much for a mineral to retain its original crystal structure, so forcing a major readjustment of the atoms to form a more stable arrangement. This more stable form can be achieved in two ways. First, the atoms can arrange themselves into a more compact crystal structure – this is what happens with carbon when diamonds are created industrially by compressing graphite (another crystalline form of carbon) to very high pressures. As might be expected, the density of diamond ($3.5 \times 10^3 \, \mathrm{kg \, m^{-3}}$) is much greater than that of graphite ($2.0 \times 10^3 \, \mathrm{kg \, m^{-3}}$). The second type of readjustment involves a chemical reaction between the constituent minerals within a rock to produce a denser assemblage of minerals.

BOX 2.1 UNITS OF PRESSURE

The SI unit of pressure is the pascal, abbreviated to Pa.

$$1 \, \mathrm{Pa} = 1 \, \mathrm{N \, m^{-2}} = 1 \, \mathrm{kg \, m^{-1} \, s^{-2}}.$$

Very high pressures are experienced at depth within planetary bodies (Chapters 2 and 3), or during impact events (Chapter 4). In instances such as these where high pressures are usual, values are more commonly expressed as megapascals (MPa) or gigapascals (GPa).

For instance:

$$5 \times 10^9 \, \mathrm{pascals \, (Pa)} = 5000 \, \mathrm{megapascals \, (MPa)} = 5 \, \mathrm{gigapascals \, (GPa)}.$$

Alternatively, pressure at depth within the Earth (and similarly within other planetary bodies) is often given in kilobars (kbar). Atmospheric pressure is 1 bar at sea-level, so a value of 1 kbar is a thousand times atmospheric pressure. As 1 bar = 10^5 Pa, then the conversion of kilobars to gigapascals is given by:

$$10 \, \mathrm{kilobar \, (kbar)} = 1 \, \mathrm{gigapascal \, (GPa)}$$

To illustrate how self-compression resulting from the increase of pressure with depth can fundamentally affect the nature of a rock's constituent minerals, we can consider the physical and chemical changes which occur in the rock type **peridotite**. This is a rock composition commonly occurring as xenoliths, and which is regarded as being typical of the Earth's mantle. Table 2.1 shows some of the properties of minerals commonly found in peridotite. Three different aluminium-bearing minerals, plagioclase feldspar, spinel and garnet, can exist in peridotite according to the depth at which the xenolith originated. Each of these minerals has a different composition which means that, in order for them to develop, a chemical reaction must have taken place using elements derived from the other minerals present in the peridotite. Importantly, as pressure on the rock increases with depth, the chemical composition and internal crystal structures of the constituent minerals adjust accordingly by generating more compact configurations. One effect of this adjustment is that even greater temperatures become necessary to break apart the mineral structure in order to allow melting. This is why much of Earth's interior is solid, even though temperature is known to increase with depth.

The idea of mineral structure changes and associated *compositional variation* with increasing pressure, is the second key factor that helps explain why Earth's density varies with depth. It is perhaps surprising, but we will demonstrate that the composition of Earth must vary significantly with depth, by looking at the abundances of major and trace elements (Box 2.2) of Earth's surface rocks, and comparing these abundances with those found in meteorites. (See Box 2.3.)

Looking at the abundances of major and trace elements of Earth's surface rocks and of meteorites may at first appear to be an odd comparison, so let us begin by first considering the differences between peridotite and crustal rocks (Table 2.2).

BOX 2.2 MAJOR AND TRACE ELEMENTS IN ROCK MATERIALS

Conventionally, geochemical data are divided into major and trace elements. The major elements are those which predominate in any rock analysis, so the term is used to describe those elements whose concentrations can be measured in percentages (%) of the element present in the whole rock. Trace elements are defined as those whose concentrations are present at less than the 0.1% level, and are therefore conveniently expressed in parts per million (ppm) or, in some instances, parts per billion (ppb). For example 0.01% is equivalent to 100 ppm. Some elements can behave as a major element in one group of rocks, and as trace elements in others, these may include K, Cr, Ti, and Ni (Table 2.2).

Following a convention inherited from early 20th century techniques of analytical chemistry, the concentrations of major elements are usually expressed as 'weight per cent (wt %) oxide'. However, this *does not* necessarily mean they are actually present as oxides in the rock. In fact they are most commonly combined together in the form of **silicate minerals** (see Table 2.1 for chemical formulae of some typical rock-forming minerals). It is important to note that since these major elements are given as percentages, the concentration of these 'oxides' in a given rock should sum to 100%. Therefore, it follows that if, for instance, one rock has a very high MgO content, and a second rock a much lower MgO content, then one or more other elements in the second must be correspondingly higher in concentration.

Table 2.1 Chemical formulae, density and stability range of minerals commonly found in peridotite.

Mineral	Chemical formula	Density (10^3 kg m^{-3})	Stability range (GPa)	Depth (km)
olivine	$(Mg,Fe)_2SiO_4$	3.2–4.4	to >12	to >400
pyroxene	$(Ca,Mg,Fe)_2Si_2O_6$	3.0–4.0	to >9	to >300
plagioclase feldspar	$(NaSi,CaAl)AlSi_2O_8$	2.77	0.75–0.9	25–30
spinel	$(Mg,Fe)(Al,Cr,Fe^{3+})_2O_4$	3.55	0.9–2.4	30–80
garnet	$(Ca,Mg,Fe)_3Al_2Si_3O_{12}$	3.58	>2.25	>75

Table 2.2 Chemical compositions of average oceanic and continental crust, mantle (peridotite) and meteoritic (CI carbonaceous chondrite) materials.

Wt% oxide	Xenolith (mantle peridotite)	Oceanic crust (basalt)	Continental crust (average)	Meteorite (CI carbonaceous chondrite)
SiO_2	45.35	48.18	59.67	34.68
TiO_2*	0.16	1.4	0.78	0.07
Al_2O_3	4.26	15.4	15.94	3.28
Fe_2O_3	8.24	10.75	7.48	30.42
MnO	0.14	0.17	0.11	0.19
MgO	38.17	9.58	4.42	25.07
CaO	3.39	11.06	6.43	2.67
Na_2O	0.29	3.36	3.21	0.5
K_2O*	0.03	0.07	1.91	0.05
NiO*	0.31	0.02	0.01	2.1
Cr_2O_3*	0.43	0.05	0.02	0.6
Totals:	100.77	100.04	99.98	99.63
Element (ppm)				
Ba	7	12	707	2.41
Cr*	2970	370	120	3975
K*	180	600	15773	545
Nb	0.713	2.5	13	0.25
Ni*	2460	138	51	16500
Rb	0.64	3	61	2.32
Sr	21.1	136	503	7.26
Th	0.084	0.2	5.7	0.03
Ti*	960	8400	4676	445
U	0.021	0.1	1.3	0.01
Y	4.55	35	14	1.57
Zn	29.5	70	74	462
Zr	11.2	88	124	3.87

* Elements that behave as a major element in one group of rocks, and as trace elements in others.

Note: due to analytical variation, the oxide totals do not always add up to precisely 100%.

Comparing terrestrial rock types

A straightforward numerical comparison of concentrations in Table 2.2 reveals that many major compounds, such as SiO_2, TiO_2, Al_2O_3, CaO, Na_2O and K_2O, and many trace elements such as Ba, Nb, Rb, Sr, Th, U, Y, Zn and Zr, are more abundant (i.e. show relative enrichment) in both the oceanic and continental crust compared with xenoliths (i.e. peridotite). In particular, TiO_2, Al_2O_3, CaO and Na_2O are significantly concentrated in these crustal materials and, conversely, both crustal types display much lower concentrations (i.e. relative depletion) of MgO compared with peridotite, together with significantly lower concentrations of the trace elements Cr and Ni.

A similar comparison of the two types of crustal material reveals a further relationship. For instance, the continental crust is more enriched in SiO_2, Al_2O_3 and K_2O, together with the trace elements such as Ba, Rb, Sr, Th, U and Zr. However, compared with oceanic crust, it is relatively depleted in TiO_2, Fe_2O_3, MgO and CaO, and the trace elements Cr and Ni.

Clearly, chemical segregation has occurred between the rocks comprising the deeper and shallower levels of the Earth (i.e. peridotite and crustal materials). There are also important compositional differences between the less dense continental crust and its denser oceanic counterpart. These broad patterns of relative enrichment and depletion commonly occur between these types of terrestrial rocks, and reveal important clues regarding the behaviour of constituent elements during the evolution of the planet, and the melting of one rock to form another.

BOX 2.3 METEORITES AND THE ORIGINAL COMPOSITION OF THE SOLAR NEBULA

Most people are familiar with the term 'shooting star' or **meteor**, but few are aware of its importance. Shooting stars are actually caused by small pieces of solid matter, called **meteoroids**, colliding with the atmosphere. The friction created as the meteoroid enters the Earth's atmosphere causes its surface to heat up, and the brilliant flash of light records its passage. However, if the object survives this fiery plunge and hits the ground, it is then called a **meteorite**. Therefore, meteorites are bits of extraterrestrial material that have survived descent through the Earth's atmosphere. Meteorites of various types and origins occur. Most of these originate from asteroids, but a small number have been shown to be of lunar or Martian origin. When scientists study a meteorite they look for clues to its origin by studying its mineral, chemical and isotopic composition. These data reveal information about the conditions in which the meteorite formed, such as from where in the Solar System it came, how long it has been in space, and its age since formation.

Mineralogically, meteorites consist of varying amounts of nickel–iron alloys, silicates, sulfides, and several other minor minerals. Detailed analyses of collected meteorites reveal they display a wide range of chemical compositions and mineralogical textures. Due to these differences, they are classified into three main groups: irons, stony-irons, and stones.

Stony meteorites are the most abundant of the three meteorite groups and come closest to resembling rocks found at the Earth's surface in their appearance and composition. The major portion of these meteorites consists of the silicate minerals olivine, pyroxene, and plagioclase feldspar (Table 2.1). Metallic nickel–iron occurs in varying percentages and is accompanied by an iron sulfide mineral. In addition, they can contain 2–5% carbon in the form of carbon compounds, and up to 20% by mass of water in chemical combination with these carbon compounds and with the silicate minerals. The presence or absence of **chondrules** (glassy, sub-spherical 'droplets' 0.1–2.0 mm in diameter) divides the stony meteorites into two main subgroups: *chondrites* (having chondrules) and *achondrites* (without chondrules). Scientists believe that the material in which these small, rounded, nearly spherical chondrules occur (i.e. the meteorite matrix) represents the chemical composition that condensed from the solar nebula, and therefore provides a record of some of the earliest material in the Solar System. Chemical analysis of the **CI carbonaceous chondrites** (Table 2.2), thought to be the most **primitive** of all the meteorite classes, reveals that the elements they contain are present in the same proportions as those determined for the Sun (Figure 2.2) and other stars in the solar neighbourhood by spectrographic analysis (less the gaseous elements such as hydrogen, of course). This similarity with the Sun's composition (see below) demonstrates that they have not had any secondary (i.e. later) alteration of their chemistry throughout the entire history of the Solar System, and so gives an important insight as to the composition of the primitive materials from which the terrestrial planets and terrestrial-like bodies originally accreted.

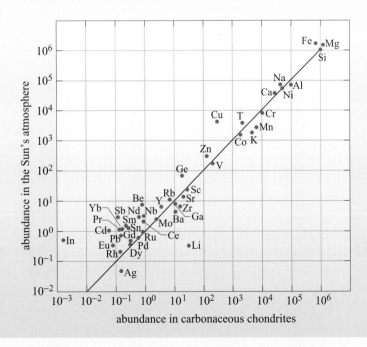

Figure 2.2 A comparison of element abundances in the Sun's atmosphere and in CI carbonaceous chondrite meteorites. By convention, the abundances are plotted as the number of atoms of each element for every 10^6 atoms of silicon present, and are shown on logarithmic scales. The straight line indicates where elements would plot if there were identical abundances present in both the Sun and the meteorites.

Comparing terrestrial rocks and meteorites

You may remember from Section 1.2 how planetesimals and eventually planetary embryos coalesced from material condensing in the solar nebula. Importantly, some of this primitive material left over after planetary accretion continues to arrive on Earth today in the form of meteorites (Box 2.3). Given that the Earth and CI carbonaceous chondrite meteorites are thought to have originally formed from solar-nebula material of similar composition (Figure 2.2), then it might be expected that they should display broadly similar bulk compositions. A simple comparison similar to that described above for mantle and crustal rocks can be performed to illustrate differences between these terrestrial rocks and meteorite composition. Geochemists usually present these types of enrichments and depletions in concentration on a 'spider diagram' or **spidergram.** This graphical method is particularly helpful because it can show all the differences and similarities at a glance. In order to do this, the relative concentrations of each element in the two rock types must first be determined. This can be expressed as the relative concentration ($conc_{rel}$) as follows:

$$conc_{rel} \text{ of rock A compared to } conc_{rel} \text{ rock B} = \frac{conc_{rel} \text{ of rock A}}{conc_{rel} \text{ rock B}} = n$$

when $n > 1$, there is a relative *enrichment* in A compared with B.

when $n < 1$ there is a relative *depletion* in A compared with B.

This procedure of comparing data by calculating a numerical value that indicates enrichment or depletion is known as 'normalizing' the data. If we were to perform this series of calculations comparing a terrestrial rock type (rock A) with carbonaceous chondrite meteorite composition (rock B), the resulting data is said to be 'chondrite normalized'.

QUESTION 2.3

(a) Using the data provided in Table 2.2, complete Table 2.3 by calculating the relative concentration ($conc_{rel}$) values for (i) mantle (peridotite) rocks, (ii) oceanic crust rocks, and (iii) continental crust rocks, compared with carbonaceous chondrite composition. Indicate whether each of the calculated values represents relative enrichment or relative depletion for the different elements in the terrestrial rocks. Express your results to 3 decimal places where appropriate.

(b Plot your results from Table 2.3 on the graph in Figure 2.3. Consider Cr, Ni, K and Ti as ppm values only. Data for chondrite normalized ocean crust, together with the first two points for chondrite normalized peridotite (mantle) and continental crust, have already been plotted. The order of elements has been chosen to illustrate the main differences and similarities in these three chondrite normalized (i.e. spidergram) patterns.

(c) What can you deduce must have happened to the element distribution within the Earth since its condensation and accretion from the solar nebula?

Table 2.3 CI carbonaceous chondrite normalized values of oceanic and continental crust, and mantle materials (to be completed).

	Mantle/chondrite	Oceanic crust /chondrite	Continental crust/chondrite
SiO_2	1.31 enriched	1.39 enriched	1.72 enriched
TiO_2*			
Al_2O_3			
Fe_2O_3		0.35 depleted	
MnO			
MgO			
CaO			
Na_2O			6.42 enriched
K_2O*			
Ba	2.90 enriched		
Cr*			0.030 depleted
K*			
Nb			
Ni*			
Rb			
Sr			
Th	2.80 enriched		
Ti*			
U			130.00 enriched
Y			
Zn			
Zr		22.74 enriched	

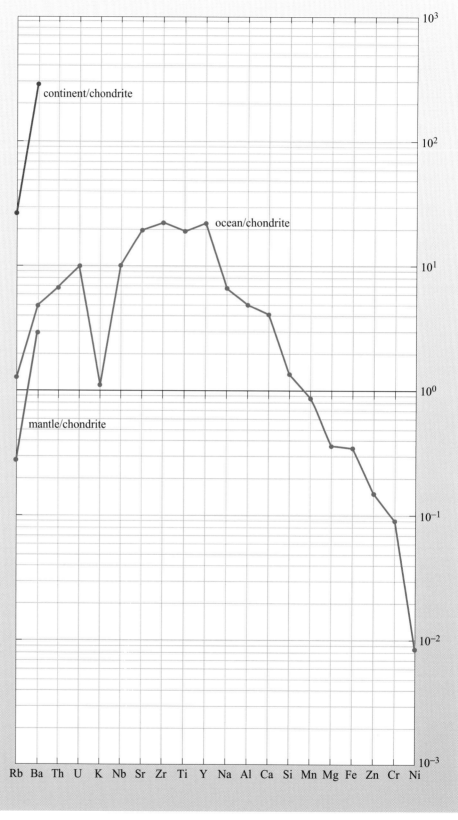

Figure 2.3 Chondrite normalized spidergram (to be completed). Note $10^0 = 1$.

To summarize, when compared with primitive meteorites, elements such as Fe and Ni are much *less* abundant in the Earth's mantle and crust. Others, such as Si, Ti, Ba, Nb, Th, U and Zr are much *more* abundant. Furthermore, compared with CI chondritic compositions the majority of elements in the mantle are depleted whereas the majority of the elements in crustal rocks show a relative enrichment.

If both meteorites and the Earth (and, by analogy, the other terrestrial planets and comparable bodies) originally started with similar bulk compositions of broadly chondritic abundances, then clearly some elements appear to have since become more concentrated towards the Earth's surface. This must mean that the other elements have become concentrated deeper down. Most importantly, all of these relative enrichments and depletions indicate that there is now a significant compositional variation within the Earth. Moreover, since Fe and Ni are depleted in *both* the peridotite and crustal materials (i.e. oceanic and continental rocks) compared with chondritic composition, this variation in their abundances cannot be simply explained in terms of an exchange of these elements between Earth's mantle and crust. Evidently, a large proportion of the original Fe and Ni must have migrated elsewhere, and this particular issue will be explored further in Section 2.4.1. However, the question next examined is whether the compositional variation observed between mantle and crustal rocks changes gradually with increasing depth, or whether it occurs as discrete compositional zones or layers within our planet.

2.2.2 Seismic evidence for layering

Most of the evidence regarding Earth's internal structure and composition comes not from rocks, but indirectly from **seismic waves** (from the Greek *seismos* meaning earthquake). These waves are vibrations or shocks that can pass through the Earth, and can be recorded by geophysical monitoring equipment. They are created by either natural earthquakes or artificial explosions. Since these vibrations often pass through the deep interior of the Earth, monitoring them provides estimates of important physical properties such as the velocity of seismic waves deep within the planet. Seismic wave velocity and the manner in which the waves pass through the deeper levels are related to the mineralogical composition and hence the chemical composition. If, for example, Earth's composition and density were uniform, then the velocity of propagation of seismic waves would be the same throughout the Earth's structure. However, experiments that monitor and record wave velocities from both artificial and natural seismic events demonstrate that this is not the case (Figure 2.4). There are two main types of wave that travel through the Earth. **P-waves** are pressure or compression waves and can pass through both solid and liquid. **S-waves** are created by a shearing movement and so can only travel through solids (Figure 2.5).

P-waves, also known as primary waves, typically are the fastest waves throughout the Earth. S-waves, also known as shear waves or secondary waves, travel more slowly at 60% to 70% of the speed of P-waves. The relationships between depth and velocity of P-waves and S-waves within the Earth are given in Figure 2.4, and on this basis three major seismic divisions can be recognized and provide a basis for defining layering within the Earth. This tripartite division is most obvious for the recorded P-wave velocities because the Earth's outermost 'skin' is characterized by low P-wave velocities and this corresponds to the crust, which is 25–90 km thick in continental areas, and 6–11 km thick in the ocean basins. At the base of the crust, the seismic velocity of P-waves increases dramatically from about 5 km s^{-1} to 8 km s^{-1}, thus marking the top of the mantle. This change in velocity is created by a compositional boundary between crustal and mantle materials. This boundary is known as the Mohorovičić discontinuity, or 'Moho' (Box 2.4). This important seismic boundary provides excellent evidence for a fundamental compositional layering structure within our planet.

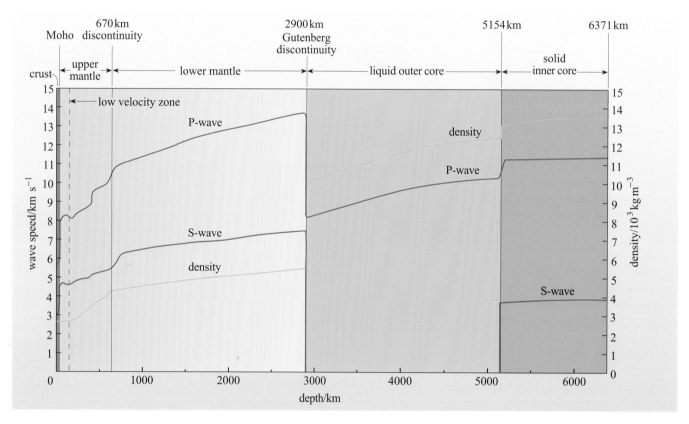

Figure 2.4 Velocity profiles of P-waves and S-waves within the Earth, and inferred densities. The term 'velocity profile' refers to the changes in velocity of seismic waves with increasing depth.

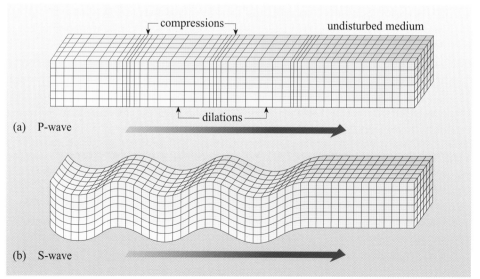

Figure 2.5 Propagation of (a) P-waves and (b) S-waves through a medium.

BOX 2.4 ANDRIJA **MOHOROVIČIĆ** AND THE 'MOHO'

Figure 2.6 Andrija Mohorovičić.

Andrija Mohorovičić (1857–1936) was a prominent Croatian scientist (Figure 2.6). Although he devoted most of his energy to meteorology, he had most success in the field of seismology. After becoming the head of the Meteorological Observatory in Zagreb in 1892, he gradually extended the activities of the observatory to include the field of geophysics, including seismology, geomagnetism and gravitation.

After the turn of the century, Mohorovičić's scientific interest focused on the problems of seismology. He advanced insight into the way in which seismic waves, created by shallow-depth earthquakes, spread through the Earth's interior. Using these seismic studies, he was the first to establish a velocity discontinuity 'surface' that separated the crust of the Earth from the mantle beneath. Soon afterwards, scientists elsewhere confirmed the existence of this discontinuity under all the continents and oceans and, once recognized, it was named the 'Mohorovičić discontinuity' in his honour. Mohorovičić's thoughts and ideas were far-sighted, and many came to be developed years later, such as the importance of using deep-focus earthquakes in constraining models of the Earth's interior, locating earthquake epicentres through the use of seismograph arrays, and the importance of building design in the mitigation of earthquake damage.

The term 'Moho' has now become the abbreviated form of Mohorovičić discontinuity. For many years the Moho was considered to be a well-defined, continuous plane that divided the crust from the mantle. However, more recently, detailed deep seismic studies have revealed it to be a much more complex feature that is flat in some areas, but laterally discontinuous and jumping between depths in others. Whatever its expression, this discontinuity represents a fundamental change in chemical composition at depth, and defines an important boundary in the form of a sharp seismic velocity discontinuity that separates the compositionally distinct layers of Earth's crust and underlying mantle.

P-wave velocities typically increase continuously through the mantle to a depth of nearly 3000 km, producing a rising velocity profile that is punctuated by a series of 'steps' at shallower depths, indicating more rapid increases in velocity. This rising profile is largely due to the increasing density of mantle material created by self-compression. One such increase, at about 670 km (known as the 670 discontinuity), marks the boundary between the upper and lower mantle (Figures 2.1 and 2.4), and also corresponds to the depth to which earthquakes associated with tectonic processes may be recognized. However, at the base of the mantle, around 2900 km, there is a sharp fall in P-wave velocity, creating another discontinuity (known as the Gutenberg discontinuity) in the seismic-wave velocity profile, and marking the interface between the mantle and the Earth's core. Interestingly, whilst the slower S-waves also increase in velocity through the mantle they are not transmitted through the outer part of the core.

■ Can you suggest a reason why S-waves might not be transmitted through the outer core?

❑ Since S-waves cannot travel through liquids, the outer core must be liquid. In fact, seismic evidence indicates that at least 95% of its volume is liquid. Nevertheless, under very special circumstances, S-waves have been recorded passing through the *innermost* part of the core, revealing that this centremost portion of the Earth must be solid.

Additional evidence that part of the core is liquid comes indirectly from the fact that the Earth has a magnetic field. In fact, Earth's magnetic field is the strongest of all the terrestrial planets and is similar in its properties to a **magnetic dipole**, which means it appears to act like a bar magnet (Figure 2.7). Traces of ancient magnetism, known as remnant palaeomagnetism, have been frozen into some of Earth's earliest volcanic rocks and demonstrate that this type of dipole field has been in existence for much of Earth's history.

Because there is no known solid that has magnetic properties above 1200 °C and, since the lower mantle and core are known to be considerably hotter than this, the only way that the Earth's magnetic field may be generated and maintained is if the core can conduct electricity and is in motion. Thus, the dipole field is thought to be an interaction between the Earth's rotational motion (i.e. once every 24 hours) and core convection, which together generate currents within the liquid part. It is this constant 'stirring up' of the liquid core that generates Earth's strong magnetic field.

A consequence of this strong field is that high-energy, ionized particles emitted from the Sun (i.e. the *solar wind*) that pass near to Earth are affected and attracted into Earth's magnetic field. In some instances these 'trapped' particles interact with the high atmosphere to produce aurorae (i.e. the northern and southern lights). The space around a planet in which these ionized particles become affected by the planet's magnetic field is known as the **magnetosphere**. You will look at this in more detail in Chapter 5.

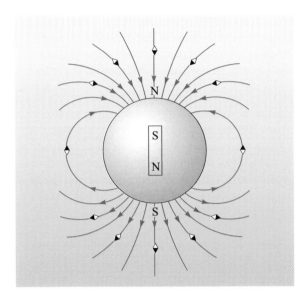

Figure 2.7 Schematic cross-section through the Earth showing the form of the dipole magnetic field created by convection currents within the liquid outer core (lines show the attitude of freely suspended compass needles at different latitudes).

It is intriguing to note that Venus, although similar in size to Earth and presumably also having a liquid core, does not possess a dipole field. In fact, the magnetic field of Venus is very weak. An explanation may lie in the fact that Venus has a much slower rotation rate (i.e. once every 243 days), and may also lack a solid inner core. It is thought that the presence of a very hot, solid central core may be necessary in order to heat the surrounding liquid core to a point where it begins to convect.

2.2.3 The composition of Earth's layers

Core

There are no methods available for sampling or obtaining core material. However, using seismic and magnetic data, together with information concerning the relative abundances of elements during the condensation process of the early Solar System, it is now believed that the *inner* core is probably an alloy, that is a mixture of iron with about 4% nickel in similar proportions to those found in iron meteorites. Iron is likely to be present since it is cosmically abundant, conducts electricity, and is relatively depleted in the overlying mantle and crust compared with meteorites (Table 2.2). The liquid *outer* core is thought to be broadly similar in composition (i.e. predominantly nickel and iron), but its lower density suggests that other lighter elements such as sulfur, potassium or oxygen must also be present.

Mantle

As mentioned in Section 2.2.1, both tectonic and volcanic processes can bring samples of the upper mantle to the Earth's surface. These rocks typically have the composition of peridotite. Peridotite is dominated by the ferromagnesian (Fe–Mg-rich) minerals olivine and pyroxene, together with either garnet, spinel or plagioclase feldspar. The bulk composition of this mantle material is typically lower in SiO_2, K_2O and Na_2O and higher in MgO than crustal rocks (Section 2.2.1). In fact, over 90% of peridotite comprises the ferromagnesian minerals olivine and pyroxene, together with small quantities of garnet, spinel or plagioclase feldspar. By contrast, pyroxene and olivine are less common in the crustal materials, whilst plagioclase feldspar is much more abundant. The fundamental differences in mineralogy, composition and bulk density of the mantle compared with crustal material (Tables 2.1 and 2.2) is the reason why the mantle has significantly higher seismic velocities than the overlying crust. Moreover, the junction between these two types of composition and mineralogy can be defined seismically as the Moho.

Although the mantle is thought to have a characteristic peridotite composition, an examination of Figure 2.4 shows that the seismic velocity profile of the mantle does not increase uniformly with depth as might be expected if increasing pressure were the only controlling factor. Instead, there are a series of steep rises in velocity in the topmost 1000 km of the mantle. These changes are interpreted as depths at which the pressure causes the crystalline structure of the constituent minerals to change to a more compact and denser form. For instance, in the shallower levels of the mantle (i.e. up to 100 km depth), the changes in velocity occur through chemical reactions involving aluminium-bearing minerals that generate increasingly more compact (i.e. more dense) mineral forms. As a consequence, plagioclase feldspar (density: 2765 kg m^{-3}) is replaced by spinel (density: 3550 kg m^{-3}), and then spinel by garnet (density: 3580 kg m^{-3}). In general, plagioclase feldspar is stable only in the uppermost 25 km of the mantle, spinel is stable at 28–70 km depth, and garnet is stable at 70–80 km depth (Table 2.1).

Crust

The crust is the geologically highly active outer part of Earth's layered structure. It is the least dense layer having a higher proportion of the silica- and aluminium-rich materials (Table 2.2). It is composed of two distinct types: the material that comprises the continents, and the material that forms the floor of the ocean basins. Both of these crustal compositions are dominated by silicate minerals, but the ocean crust (typically 6–11 km in thickness) in which ferromagnesian minerals are more common, is more homogeneous, being largely composed of **basalt**. By contrast, the continental crust (25–90 km in thickness) is much more variable in its composition because it is composed of many different rock types (for instance, **metamorphic**, **sedimentary** and **igneous** rocks) that have been added to the continental areas throughout geological time. However, continental crust may be considered to be broadly similar to that of **granite** in its composition. Ferromagnesian minerals are generally less common in the continental crust than in the ocean crust, so continental areas are the least dense part of Earth's layers.

Other terrestrial bodies

Most of the information regarding the nature of the deep layering of the Earth is derived from seismic studies (see Figure 2.1), with additional chemical and mineralogical information of the topmost 100 km of mantle and crust coming from rock samples. In the case of the other terrestrial planets, we have little or no seismic or direct compositional information, though important information about their internal structure and composition can be obtained from density and gravity field data collected by orbiting spacecraft and spacecraft fly-bys. Nevertheless, by assuming that similar processes operated during the evolution of these rocky planets, and by comparing their bulk densities with the composition of surface materials determined by both orbiting and lander probes, we can deduce they also have layered structures (Figure 2.8).

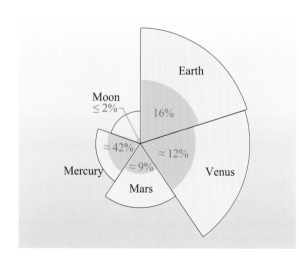

Figure 2.8 Comparison of the sizes and core radii of the terrestrial planets (including the Moon), showing the percentage of each planet's volume occupied by its core. The models of Mercury, Mars, and Venus are inferred from density and gravity field data.

BOX 2.5 CHEMICAL COMPOUNDS, IONIC AND MOLECULAR SUBSTANCES

Atoms of most elements form chemical compounds, and a compound can be expressed as a specific formula involving simple multiples of its constituent elements. Whether a compound or substance exists as a gas, liquid or solid (i.e. its **phase**) depends on some intrinsic properties of that substance, and these properties confer a vast range of behaviour. For instance, at normal atmospheric pressure, the element helium melts at a temperature of about 1 K, whereas the mineral corundum (an oxide of aluminium) does not melt until over 1750 K. Further illustration of this range is given by the materials in Table 2.4, most of which are relevant to this and the following chapters. Those substances with high melting temperatures and boiling temperatures are called *refractory*, whilst those with low values are called *volatile*. However, in order to understand why boiling temperatures and melting temperatures span such a large range, we must consider the ways in which the atoms are held together in these various compounds and substances.

Table 2.4 Simplified condensation sequence for important substances forming from the solar nebula. The most refractory (i.e. least volatile) are at the top of the table, with progressively less refractory materials toward the bottom.

Substance	Formula	Temperature of condensation[a] (K)
ionic substances		
corundum	Al_2O_3	1758
perovskite	$CaTiO_3$	1647
spinel	$MgAl_2O_4$	1513
nickel–iron metal	Ni, Fe	1471
pyroxene (diopside)	$CaMgSi_2O_6$	1450
olivine (forsterite)	Mg_2SiO_4	1444
alkali feldspars	$(Na,K)AlSi_3O_8$	<1000
troilite	FeS	700
hydrated minerals[b]	(variable)	550–330
molecular substances		
water	H_2O (as an ice)	180
ammonia	$NH_3.H_2O$ (ice)	120
methane	$CH_4.6H_2O$ (ice)	70
nitrogen	$N_2.6H_2O$ (ice)	70

[a] The temperatures of condensation given are those that would occur if the pressure in the nebula had been about 10^{-3} bar (10^2 Pa). At lower pressures, the condensation temperatures would have been reduced slightly.

[b] Hydrated minerals are chiefly silicates with OH or H_2O in their formulae.

Metallic bonding occurs when atoms *donate* one or more outer electrons to an electron '*cloud*' (Figure 2.9a). This negatively charged cloud flows between the positively charged metal ions and acts as a bonding agent that can operate in any direction, so holding them tightly together. As a result, metallic substances are dense and closely packed.

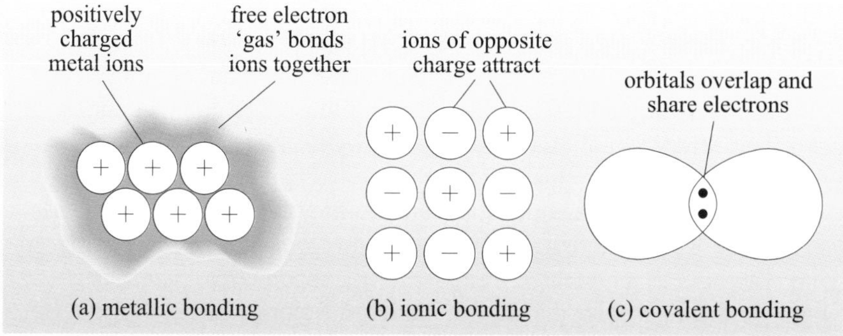

positively charged metal ions

free electron 'gas' bonds ions together

ions of opposite charge attract

orbitals overlap and share electrons

(a) metallic bonding

(b) ionic bonding

(c) covalent bonding

Figure 2.9 Bonding between atoms: (a) metallic bonding, (b) ionic bonding and (c) covalent bonding.

Ionic bonding occurs by the *transfer* of one or more electrons from one atom to another (Figure 2.9b). This interaction produces two ions of opposite electrical charge which are then strongly attracted to each other, such as a positively charged sodium ion (Na^+) and a negatively charged chlorine ion (Cl^-). Ionic bonds tend to be stronger than metallic bonds and, in common with metallic bonds, the binding force can operate and exist in any direction. A substance that consists of ions held together in this manner is known as an ionic substance and such substances tend to have fairly dense, closely packed structures.

Sodium chloride (the mineral *halite*, best known as common salt) is an archetypal ionic solid. Each sodium atom has lost one electron, rendering it positively charged and denoted by the symbol Na^+. Each chlorine atom has gained one electron, becoming negatively charged and denoted by the symbol Cl^-. In the sodium chloride crystal structure each Na^+ is surrounded by six Cl^- ions, and each Cl^- is surrounded by six Na^+ ions (Figure 2.10). Since the electrical forces holding this structure together are strong, it is difficult to break down sodium chloride either physically or by heating.

Naturally occurring compounds that are found within rock are referred to as minerals, and the different minerals comprising a given rock type is known as the *mineral assemblage*. Most minerals are ionic solids. One of the most refractory of these is perovskite, or calcium titanate ($CaTiO_3$) to give its chemical name (Table 2.4). In this substance, the interwoven networks of ions of calcium, titanium and oxygen (oxide) are held particularly strongly so that the melting temperature of $CaTiO_3$ is one of the highest known of all naturally occurring substances.

However, many ionic substances that are minerals do not obey a precise formula. For example, the chemical formula of pyroxene is given as $(Ca,Mg,Fe)_2Si_2O_6$ (Table 2.1) to indicate that it can have any proportion of iron

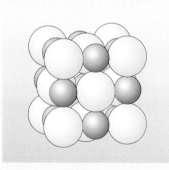

Figure 2.10 Model of sodium chloride (halite) structure.

(Fe) ions substituted in place of magnesium (Mg), and both iron and magnesium can substitute for calcium (Ca). All these ions have a charge of +2, so their relative abundances do not affect the balance of positive and negative ions. The mineral pyroxene can therefore be represented by the general formula $X_2Si_2O_6$, where X represents an atom of either Ca, Fe or Mg, (alternatively we can write the chemical formula explicitly as $(Ca,Mg,Fe)_2Si_2O_6$ so long as there is a total of 2 atoms of X for every 2 atoms of silicon (Si) and 6 of oxygen (O). A mineral having Si and O in its formula, which is typical of most common minerals in rock, is called a silicate.

Covalent bonding is formed when the atoms of two substances *share* two electrons through the overlap and merging of electron orbitals from each atom (Figure 2.9c). Covalent bonding requires precise overlap of the electron orbitals, and so bonds are usually restricted to specific directions. Consequently, covalently bonded atoms must be precisely positioned relative to one another, and the resulting covalent substance is not as closely packed, and so tends to have a lower density relative to ionic or metallic substances.

Molecular bonding occurs in many instances. Here, a number of previously bonded atoms (i.e. molecules) are themselves held together by covalent bonds. Such arrangements are known as molecular substances. However, since the bonds holding together the constituent molecules tend to be relatively weak, molecular substances are easily broken apart, and so tend to be characterized by high volatility. Water, for instance, is a relatively volatile molecular substance (Table 2.4) because, although individual water molecules consist of two hydrogens covalently bonded to a single oxygen, the water molecules themselves are held together by even weaker bonds, allowing ice to melt and water to boil at relatively low temperatures.

In which types of substances did elements exist during the cooling of the solar nebula?

As you may now realize, the answer to this question depends mainly on the temperature within the nebula. Temperatures must have been greatest near the centre, but must also have changed over time once the Sun and its associated protoplanetary disc began to evolve. Unfortunately, astronomical observations of analogous circumstellar dust clouds have proved better at determining the sizes of the constituent grains rather than their compositions, and so it is unclear whether such grains are dominated by ices, carbon-rich material or silicate minerals. However, the compounds that are thought to have been important in the evolution of the solar nebula, and which gave rise to our Sun and its system of planets, include water (H_2O), methane (CH_4), troilite or iron sulfide (FeS), corundum (Al_2O_3), and pyroxene $(Ca,Mg,Fe)_2Si_2O_6$. The story of the solar nebula, and of how solid material might have been gathered together, will be considered further in Section 2.3.1.

2.3 Origins of planets and of planetary layering

Before we consider layering in other planets, we should first consider how the planets became to be assembled in the first place. In Section 1.2 you were introduced to the idea that the accretion of planetesimals, followed by the subsequent collision of larger planetary embryos, is the most likely process by which the planets were built.

Due to the nature of the accretion process, the early planetesimal-sized bodies would have had an homogenous internal structure. This is because the coalescence of impacting grains and clumps of condensed nebular material would have resulted in a churning and mixing during planetesimal growth. Once the planetesimals began to attain the proportions approaching those of planetary embryos, it is likely that the heat generated by collisions would have been sufficient to allow both melting and, as denser materials began to sink inwards, segregation of the original constituents. So, by the time bodies had reached the proportion of planetary embryos, a crude layering would have already begun to evolve. However, the development of a more evolved layering, such as that within our planet, would require this separation process to have reached its completion once giant impacts had ceased. The sequence of events leading to the assembly of the terrestrial planets is summarized in Box 2.6.

BOX 2.6 THE PROCESS OF PLANET FORMATION

The evolution from condensation of grains in the solar nebula, through planetesimals and planetary embryos, and into planets, is summarized below and in Figure 2.11.

Condensation in the solar nebula

Our present understanding of how the Solar System came into being is that about 4.6 Ga ago the Sun formed by gravity-driven contraction of an interstellar gas cloud containing mostly hydrogen, but also traces of 'dust'. The dust consisted of ices, carbon, metallic and silicate substances (Figure 2.11a). Once the Sun formed and began to shine (i.e. began to release energy from nuclear reactions), nearly all the original dust remaining in the solar nebula was vaporized by its heat. The nebula then began to cool and new 'dust grains' started to condense within the surrounding nebula disc (Figure 2.11b).

The accretion process begins

Collisions between the newly condensed grains and particles of rocky and icy material were relatively common, and when they collided they tended to stick together (Figure 2.11c). Over a period of about 10 000 years, these random collisions of dust and particles built

up into clumps up to 10 mm across. These clumps of matter would have been of silicate composition in the inner orbits, and their composition is preserved within the most primitive chondritic meteorites. This accretion process continued further, building to clumps of 0.1–10 m in size (Figure 2.11c).

Formation of planetesimals

After about 100 000 years, further random collisions during the accretion process had produced a profusion of bodies of 0.1–10 km across, termed planetesimals ('tiny planets'). Once formed, the increased gravitational attraction exerted by the larger ones would become the major factor for continuing growth (Figure 2.11d). It would have meant that larger planetesimals began to grow preferentially at the expense of their smaller neighbours. This is the beginning of the phase known as 'gravitational focusing'. At this time a significant temperature increase of the developing planetesimals would have arisen through the release of heat from impacts. The mineralogy and texture of the less-primitive material formed at this stage, and subsequently preserved in some meteorites, indicates a complex history that may relate to fragmentation and heating of planetesimals during these impacts.

Accretion and the development of planetary embryos

Above 10 km, the gravitational focusing created by the larger planetesimals would have resulted in increasingly frequent collisions (Figure 2.11e). In this fashion, nearby planetesimals would be 'swept up' by larger bodies. Where growth of one planetesimal outpaced that of its neighbouring rivals, it may have developed into a planetary embryo, which is the term given to accreted bodies reaching a few thousand kilometres in diameter (Figure 2.11f). It is estimated that within a few thousand years the growth of these planetary embryos would have captured and incorporated most of the smaller planetesimals and, as a result, the initial profusion of planetesimals would have been replaced by perhaps only a few hundred large planetary embryos.

Planetary embryos and giant impacts

Having exhausted the nearby supply of planetesimals, the era of frequent collisions was over. The next stage of growth would have been slower and characterized by chance collisions between planetary embryos (Figure 2.11g). Giant impacts between embryos could have resulted in fragmentation of both, with the debris then recombining into a single mass. Moreover, the heat released during such massive impacts was, in some cases, capable of melting the newly combined mass, creating a body with a molten mantle termed a 'magma ocean' (Figure 2.11h). A thin crust would have developed as the amalgamated body began to cool. This crust would have been continually destroyed and reformed as impact debris fell back to the surface. Convection currents and degassing from within the body would also have disturbed and ruptured any solidified surface areas.

Differentiation of planetary embryos, and planetary assembly

Collision of planetary embryos would have been totally devastating. Importantly, the associated impact melting of the planetary embryos would have allowed denser materials to segregate inwards and lighter materials to work their way outwards, leading to the crude separation into a nickel–iron rich core and an outer rocky, silicate-rich mantle (Figure 2.11h).

Giant impacts would have continued to occur between these larger, partially differentiated bodies, resulting in the amalgamation of core material and the assembly of even larger bodies. It is estimated that the terrestrial planets would have taken about 10 million years to reach half their mass, and about 100 million years to fully complete their growth and build an Earth-sized planet through a process of chance collisions between planetary embryos such as these. The later evolution and further internal differentiation of these planets was from then on driven mainly from processes within the body.

Completion of terrestrial planet formation

Once the last giant impact in the inner Solar System had occurred, there were just four surviving terrestrial planets (Mercury, Venus, Earth and Mars). The total goes up to five if you count the Moon. The Moon is a planet in a geological sense because of its size and character, though not in an astronomical sense because it orbits the Earth rather than the Sun. The Moon apparently owes its origin to a giant impact in which much of the debris from a fragmented impactor, plus some ejecta from the outer layers of the larger 'target' body, ended up in orbit about the larger body. The larger body subsequently cooled and developed into the Earth, and the orbiting debris accreted to form the Moon (Figure 2.12).

Figure 2.11 (This and facing page) The process of planet formation.

(c) 10 mm   0.1–10 m

(d) 0.1–10 km

(e) 100–1000 km

(f) 1000–5000 km

(g)

(h) 3000–12000 km

Figure 2.12 (see caption opposite) Note: the distance between the Earth and newly formed Moon is not shown to scale.

Figure 2.12 The formation of the Moon. Collision and aftermath of a giant impact collision between two planetary embryos (a)–(c). Both embryos had already differentiated bodies. The larger one is thought to have been about four times the mass of the smaller one. Following the collision (b and c), the cores of the planetary embryos combined, and the mantle of both bodies became mixed, whilst some became fragmented, vaporized, and scattered into an arc (c). Some of this debris showered back to Earth, but the remainder came together as a clump under its own gravity and achieved a stable orbit around the neo-formed Earth, and from this material the molten Moon coalesced (d). The Moon then cooled and differentiated into its own crust, thick mantle, and (possibly) a small core. The Moon itself experienced bombardment and the formation of large craters, and the associated flooding of some craters by basalt lavas to form the mare (e). Over time, the Moon's orbit has slowly decayed, taking it progressively further away from Earth (f).

2.3.1 Condensation, accretion and collision

The material that gave rise to the accretion process first had to condense into grains from the components of the solar nebula (i.e. protoplanetary disc). This occurred once the nebula had cooled below the point at which the elements and compounds could only exist in a gaseous (volatile) state. The initial grains would have consisted of the most **refractory** substances (i.e. those most resistant to heat, and which have the highest point of condensation or vaporization), with progressively more **volatile** components (i.e. those with the lowest condensation or vaporization points), condensing as temperatures within the protoplanetary disc began to fall. This is known as the **condensation sequence**. Moreover, at any given instant during the cooling of the solar nebula, there would also have been a progression of condensing compositions away from its centre. This is because the more refractory compounds would be able to condense nearest to the young Sun, with the more volatile materials condensing further away (Table 2.4 and Box 2.5). This progression away from the Sun of the most refractory to the most volatile substances is thought to be the principal reason why the terrestrial or rocky planets occupy the inner Solar System, whilst the gaseous and icy examples are located further out.

In the early stages of accretion, once the planetesimals had begun to form, they would have gravitationally attracted smaller bodies occupying similar or nearby orbits (Figure 2.11c). The combined gravity of these coalesced bodies would have attracted yet more small bodies through a phenomenon known as gravitational focusing. This would then have led to a situation where the growth of one larger planetesimal outpaced and eventually captured its rivals, producing a planetary embryo of a few thousand kilometres in diameter, and a mass of up to 10^{22}–10^{23} kg (Figure 2.11f). Numerous planetary embryos must have developed through this accretion process.

It is important to note when considering planetary formation from planetesimals and planetary embryos that modern models reject the idea that one particular planetary embryo was somehow destined to become a fully formed planet by preferential growth through the addition of planetesimal-sized bodies. Rather, once planetesimal accretion had reached the stage where numerous planetary embryos had been formed, relatively few planetesimals would have remained because only a small number would have escaped the gravitational attraction of these larger planetary embryos. By this stage the embryos could not grow any further through the process of planetesimal capture. Current scientific consensus considers collision and amalgamation of planetary embryos as the likely process that led to the assembly of the planets (Figure 2.11g). If this model is correct, fully assembled, planet-sized bodies would not have emerged until the final stages of this collision and amalgamation process because, in most instances, numerous embryos would have been required to generate masses equivalent to those of the larger terrestrial planets (Figure 2.11h).

QUESTION 2.4

Estimate how many planetary embryos of mass 5×10^{22} kg would have been required to assemble (a) Mercury, and (b) Earth. The mass of each planet is available in Appendix A, Table A1.

Of course it should not be assumed that planetary bodies developed by the consecutive addition of planetary embryos of one particular size. It is much more probable that embryo collision first produced a number of larger embryos of varying sizes. These, themselves, then collided to create even more massive bodies. Eventually a few large amalgamated bodies, resulting from either a few or many such collisions, finally assembled into the planets. Although we can perhaps never know the sequence of such collisions, and the relative sizes of the bodies involved, what is certain is that the resulting giant impacts would have been cataclysmic events. Moreover, by considering the effects of such events, an insight into some of the key issues regarding the evolution of the terrestrial planets may be gained, such as the development of iron-rich planetary cores, why Mercury has such an anomalously large core, or the formation of the Moon (Sections 2.3.2 and 2.3.3).

In summary, the assembly of the terrestrial planets is perhaps best considered as having come about through a few tens of cataclysmic giant impacts involving planetary embryos, rather than by a constant incoming 'rain' of millions of small planetesimals that gradually built up over time into a planet-sized body. The implications of the preferred 'embryo impact' model are profound, both in terms of the way in which we should think about the origin of Earth and the other terrestrial planets, and in terms of their subsequent evolution.

2.3.2 Assembly, melting and differentiation

To generate the type of compositional layering displayed by the Earth requires a differential mobility of the primary constituents. Such differences must occur as a result of the differences in the physical and chemical properties arising from the varying chemistries and mineralogies of these constituents. The process of separating out different constituents as a consequence of their physical or chemical behaviour is known as **differentiation**.

An obvious way of mobilizing constituents is by allowing them to begin to melt. When a rock is heated, different minerals within the rock will melt at different temperatures. This phenomenon is known as **partial melting** and is a key process in the formation of liquid rock or **magma**, as you will see in Chapter 3. Once constituents have been mobilized in this manner, they will begin to migrate under the influence of pressure or gravity.

■ Imagine a body the size of a planetary embryo that had accreted from nickel–iron and silicate minerals. Nickel–iron has a density of about 7.9×10^3 kg m^{-3}, and a melting point some hundreds of degrees higher than silicates. What would happen if temperatures within this planetary embryo were increased to a point at which the melting of silicates began?

❑ Since nickel–iron has a higher melting point it would remain solid after the silicates had begun to melt and, because it is much denser than any silicate minerals, it would begin to sink towards the centre of the body.

If temperatures then continued to rise beyond the melting point of nickel–iron itself, an emulsion of liquid silicate and liquid nickel–iron would form (an emulsion is a mixture in which one substance is suspended in another). However, separation would still occur because the globules of nickel–iron would not mix into the silicate (because the two are immiscible) and so would continue to sink under gravity towards the centre of the body. In effect, the nickel–iron would 'rain out' from the emulsion towards the centre of the body. This is known as the 'rain-out' model and is a process of differentiation that is thought to lead to the first stage of core formation and the development of associated layering.

- Consider what would happen if a fast-moving object such as a planetesimal collides with a larger one, such as a planetary embryo, during the accretion process. (Hint: Think of the energy involved.)

- Whilst some of the kinetic energy will be retained by fragments flung away from the impact site, a considerable amount of the original kinetic energy of the colliding bodies will be converted to heat.

The possibility of such melting processes operating during the final stages of planetary growth is likely because of the heat generated by the collision of planetary embryos. The energy released is dependent upon their relative velocities and masses. Consequently, the energy released from collisions between planetary embryos would, in many instances, be sufficient to cause a considerable degree of melting. In those instances where particularly large embryos are involved, it may have been enough to melt entirely their amalgamated masses. In those cases, the resulting combined body would be able to differentiate due to its molten state, and so begin to develop a core–mantle layering (Figure 2.11h).

According to the 'rain-out' model, it is estimated that once core formation had begun, it was an extremely rapid process taking only a few tens of thousands of years to reach completion. Such rapidity, however, has important implications regarding the timing of core formation. It is probable that long before the planets were fully assembled, some differentiation and separation of a nickel–iron component had already begun within the planetary embryos (Figure 2.11f). This process became accelerated through melting resulting from earlier embryo–embryo collisions. However, the formation of the core of an assembled planet, such as Earth, could not have been fully completed until all of the iron delivered via impacting bodies had arrived. Therefore, the formation of planetary cores must have occurred simultaneously with the assembly of the planet. It would have begun at the point at which the larger planetary embryos became molten and began to differentiate, and continued until the stage where amalgamation of these embryos had assembled a planet in which the separation of all the delivered nickel–iron could reach its completion.

Given that the other terrestrial planets are thought to be of a broadly similar bulk composition to Earth (i.e. silicates, iron and nickel), albeit in different proportions, the processes that formed Earth's core are also likely to have operated within these planets. However, such separation processes are not restricted to silicates and metals, but could also operate on bodies composed of rock and ice, for example most of the satellites of the outer planets. In these cases, it would be the denser rocky material that would gravitate centrewards and form a rocky core when the ice components became molten or soft.

It is important to realize that the difference in composition between the Earth's crust and mantle, although significant, is much less than that between the mantle and the nickel–iron core. In both mantle and crust the dominant minerals are silicates. Chemical differences between the two arise from later partial melting of the mantle which occurred, and continues to occur, under special circumstances (these will be discussed in Chapter 3). When the mantle is partially melted, certain elements are mobilized more readily than others into the resulting liquid fraction. This process of differentiation involving exchange of elements between these solid and molten states is known as **element partitioning**. It is an important mechanism by which certain elements first become concentrated into the melt and then added into the overlying crust through migration via magma or other fluids. If continued over time, this concentration and migration will gradually augment the difference in composition between crust and mantle. This process has been fundamental in creating the chemical and mineralogical differences currently observed between Earth's mantle and crust.

QUESTION 2.5

Referring back to the completed spidergram (Figure 2.28), can you now offer an explanation as to why iron and nickel should be depleted in both crustal and mantle material when compared with chondrite composition?

2.3.3 Evidence from the Moon

The origin of Earth's Moon has long been the subject of debate. Seismic data collected by detectors deployed during the Apollo landings, which began in 1969 reveal that if the Moon has any core at all, it must be very small. If it is present, it is probably composed of nickel–iron, or nickel–iron and sulfur – like Earth's core. This lack of a core, or presence of only a small core, represents a significant difference between the Moon and the other terrestrial planets (Figure 2.8). Moreover, this lack of a large core yields important information regarding the timing of planetary accretion and differentiation of the Earth's layered structure. Analyses of lunar rock samples reveal that they share some very similar geochemical and **isotopic signatures** to analogous materials found on Earth, so lending support to the idea that both were created from material condensing in the same area of the solar nebula. However, their compositions differ markedly in other ways (Table 2.5).

The ratio of element abundances in the right-hand column of Table 2.5b demonstrates that the Moon is relatively depleted in volatile elements and enriched in refractory elements. Similarly, it is highly depleted in siderophile elements, since these would have already been scavenged by an earlier differentiation within the planetary embryos prior to the Moon-forming impact.

Relative to the Earth, the Moon is depleted in volatile elements such as K, Na and Rb but enriched in refractory elements such as Sr and U. The current theory to explain these similarities and differences is that the Moon was generated from the accretion of debris that was ejected from a late-stage giant impact between two planetary embryos (Figure 2.12). This seems to have been a very unusual result of a giant impact because in most cases the embryos should have merged into a single body. Moreover, it is likely that this Moon-forming impact, was the final giant impact to have affected Earth, since any subsequent giant impact would have almost certainly destroyed the Moon, or allowed it to escape. Whatever the case, this model has important implications for the origin and timing of layering in both bodies since the theory requires that both embryos had already become at least partially differentiated due to earlier impact heating (Figure 2.11f). In fact, the low abundances of siderophile

Table 2.5a Comparison of oxide abundances in primitive meteorites, Earth and Moon compositions (average crust + mantle material).

Wt% oxide	CI chondrite (primitive meteorite)	Earth (crust + mantle)	Moon (crust + mantle)
SiO_2	34.68	49.9	43.4
TiO_2*	0.07	0.16	0.3
Al_2O_3	3.28	3.64	6
Fe_2O_3	30.42	8.8	14.3
MgO	25.07	35.1	32
CaO	2.67	2.89	4.5
Na_2O	0.5	0.34	0.09
K_2O*	0.05	0.02	0.01

* Elements that behave as a major element in one group of rocks, and as trace elements in others.

Table 2.5b Comparison of elemental abundances in primitive meteorites, Earth and Moon compositions (average crust + mantle material).

	CI chondrite (primitive meteorite)	Earth (crust + mantle)	Moon (crust + mantle)	Ratio of trace element abundance Moon/Earth
Volatile[a] elements				
K (ppm)	545	180	83	0.46
Rb (ppm)	2.32	0.55	0.28	0.51
Cs (ppb)	279	18	12	0.67
Moderately volatile				
Mn (ppm)	1500	1000	1200	1.20
Refractory elements				
Cr (ppm)	3975	3000	4200	1.40
Th (ppb)	30	80	112	1.40
Eu (ppb)	87	131	210	1.60
La (ppb)	367	551	900	1.63
Sr (ppm)	7.26	17.8	30	1.69
U (ppb)	12	18	33	1.83
Siderophile[b] elements				
Ni (ppm)	16500	2000	400	0.200
Mo (ppb)	1380	59	1.4	0.024
Ir (ppb)	710	3	0.01	0.003
Ge (ppb)	48000	1200	3.5	0.003

[a] 'Volatile' refers to elements such as potassium (K) rather than gases.
[b] Siderophile elements are those which, when molten, have an affinity to, or combine with (i.e. scavenge) iron (for instance Ni), and so are removed from the silicates during differentiation processes.
Note: some elements are ppm and some are ppb.

elements such as Ni (Table 2.5b) within the Moon's mantle and crust can be attributed to the fact that these smaller concentrations were derived from the already depleted outer layers (i.e. mantle material) of the two impacting embryos.

▓ How is the sequence of events described above consistent with the observation that the Moon's core (if any) is very small relative to the size of the whole Moon?

❑ The Moon formed from mantle material derived from the colliding, and partially differentiated planetary embryos. This mantle material was already depleted in Ni, Fe, and S due to the development of cores within the embryos. There would have been relatively little Ni, Fe and S left to differentiate inwards once the Moon had formed.

For the Moon's crust and mantle, this depletion was probably further augmented by the fact that it also became further differentiated immediately after its formation. Moreover, the Moon's depletion in volatile elements, relative to the Earth, can be explained if the Moon accreted from the partially vaporized debris coalescing after the impact. In these circumstances the more volatile elements would have had the opportunity to escape into space prior to accretion.

Interestingly, a giant impact following partial differentiation of planetary embryos may also explain the anomalously large size of Mercury's core. In this case the impact between two such embryos, which had already developed a core–mantle layering, could have resulted in the amalgamation of the core materials, and partial vaporization of their silicate-rich mantles. A proportion of the vaporized mantle materials may subsequently have been dispersed and lost into space. This may explain why Mercury became relatively enriched in core-forming (i.e. Ni and Fe) materials, depleted in the lighter, mantle-forming silicates and, unlike the Earth, developed no moon.

▓ Given the arguments regarding planetary accretion, volatile elements, and their behaviour following giant impacts, can you suggest a reason why the concentrations of Rb, K, and Na differ between chondrite and Earth's mantle? (Hint: refer back to the spidergrams in Figures 2.3 and 2.28).

❑ Compared with chondritic composition, Earth's mantle is depleted in Rb, K and Na. This cannot be explained in terms of segregation of these elements into the core, as was the case for Fe and Ni depletion, since the core does not contain these elements. Instead, these depletions can be best explained using a similar argument to that offered to explain the Moon's depletion in volatiles. During the final stages of the assembly of Earth, the amalgamated planetary embryos were beginning to differentiate into layered bodies with dense cores, and with elements such as K, Na and Rb segregating toward the surface. Continuing impacts would have partially vaporized the surface layers, and some of the more volatile elements, including Rb, K and Na, (and probably Zn) may have been lost to space. Later in Earth history, the segregation of material to form oceanic and continental crust would have further augmented removal of these elements from the mantle. However, the spidergram patterns reveal that dips at Rb and K, and to a lesser extent Na and Zn, also occur in the continental and oceanic crust patterns. These dips reflect the fact that the mantle had already become depleted in volatiles prior to the segregation that formed the crust.

2.4 Turning up the heat – how to 'cook' a planet

The preceding sections have indicated that a planetary embryo, or planet, needs to be heated before differentiation can occur and layering can begin to develop. We have already discussed the role of impact heating during the accretion and assembly of the terrestrial planets, but there are several sources of heat that can occur during the evolution of a planetary body. The most important are, on the one hand, the so-called 'primordial heat sources' which develop in the early stages of planetary evolution (i.e. those associated with accretion, collision and, as will be described, core formation), and on the other hand, radiogenic and tidal heating processes that can operate long after the planet has been formed.

2.4.1 Origins of primordial heat, differentiation and core formation

We can observe the effects of heating from more recent impacts on Earth from its results in impact craters and the formation of droplets of glass called **tektites**, which were formed by instantaneous melting of target rock (i.e. Earth's surface) and an impactor (this process will be examined in detail in Chapter 4). However, even the largest of these impacts for which we have evidence on Earth today, such as the Chicxulub impact site in the Gulf of Mexico (Figure 2.13), would have been tiny compared with those experienced during planetesimal–embryo and embryo–embryo collisions. These giant impacts would have been the main process of heating and melting these primordial bodies.

During their formation and early existence, all the terrestrial planets must have experienced this **accretional heating**. However, the intensity of this bombardment would have waned over time as the remaining debris was collected by the growing planetary embryos. This final stage of bombardment was probably the result of capturing debris derived from earlier giant impacts during the accretion phase, together with any remaining planetesimals and smaller debris. Once the planets had become assembled, the later impacts would, on average, have been progressively smaller and less frequent. Material delivered during this later bombardment stage could, therefore, really only have delivered heat to the planet's newly formed surface crust.

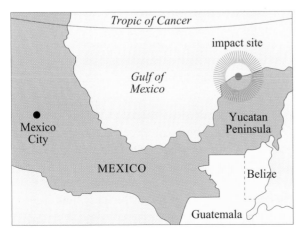

Figure 2.13 Locality map of Chicxulub, an impact feature 200 km in diameter, which occurred when a meteorite about 10 km in diameter hit the Yucatan peninsula of Mexico approximately 65 Ma ago. The effects of this huge impact may have been responsible for the extinction of the dinosaurs and up to 50% of the faunal and floral species living at that time.

Throughout the later bombardment, the surface of the magma ocean would have continually chilled and solidified through radiative heat loss to space. But this thin skin would have been continually ruptured by vigorous convection in the underlying molten fraction, the associated escape of gases, and by other incoming impactors or previous impact debris falling back to the surface. Escaping gases would have supplied an early atmosphere, or else helped replenish or replace that which had been blasted away by the shock of further impacts. Accordingly, early crusts on magma oceans must have been created and destroyed many times. Whether such remnants of early planetary crusts survive today depends on how active the planet surface has been since major bombardment ended. Active planets such as the Earth are continually re-surfaced, and any ancient material, including the evidence of impact sites, is likely to have since been substantially modified or more probably entirely destroyed. However, tracts of ancient crust on the Moon and analogous terrains on several other planetary bodies do appear to record the waning phases of bombardment.

One of the consequences of 'cooking' a body in this manner, and allowing the core and mantle to separate, is that the 'falling inwards' of the nickel–iron-rich fraction to form the core would release gravitational potential energy (in the same way that much of the energy of a dropped stone is released in the form of heat and sound when it hits the ground). The gravitational energy lost by the inward movement of nickel–iron within the differentiating molten planetary body would be converted first to kinetic energy and then into thermal energy. It is estimated that the core-forming process would have contributed significantly to a planet's primordial heating (although it would still have been an order of magnitude less than that generated by a collision with a large planetary embryo). This core-forming event is thought to have been largely completed very early in Earth's history but, rather than being a single catastrophic event, it is likely to have been spasmodic, its rate increasing each time the planet's mass was increased by addition of partially differentiated material from another giant impact.

Since both bombardment and core formation relate to events that occurred early in the history of planetary evolution, they are primordial processes and the heat generated by them was primordial heat. However, if these primordial heat sources had remained the only way of 'cooking' the planets, their intensity would have waned through time due to continual radiative heat loss to space. As you will learn later, the efficiency of this heat loss is also proportional to the size of the body (Section 2.4.3). The Moon, for instance, has lost most of its primordial heat. Even so, primordial heat stored from these early planet-forming processes is still thought to contribute a significant fraction of the internal heat energy of larger planets, such as Earth, but it is by no means the only source of energy within these planetary bodies. Since heat drives fundamental processes such as volcanism (Chapter 3), the fact that some terrestrial-like bodies have remained volcanically active for several billions of years whilst other have not, requires further explanation. In order to understand how such volcanic activity might have continued late into a planet's history, rather than diminishing over time, we need to consider other processes of heat generation.

2.4.2 Tidal heating

One heat source known to be generated within planetary bodies is tidal heating. This is created by the distortion of shape resulting from mutual gravitational attraction. These effects are readily observed upon the Earth's oceans where the attraction created by the Sun and Moon produce 'bulges' in the ocean water-masses which are then dragged around the planet as the Earth rotates. This produces the ebb and flow of tides observed at coastal locations. In much the same way, the solid Earth is also distorted by these forces so producing 'tides', with a maximum vertical displacement of its rocky surface of up to 1 m. This deformation causes heating within the planet, though precisely where this heating is concentrated depends upon a planet's internal properties. In the Earth's case, it is thought to occur largely within the crust and mantle.

QUESTION 2.6

The current rate of heating generated within the Earth by tidal distortion is estimated at 3.0×10^{19} J yr^{-1}. Given the mass of the Earth is approximately 6.0×10^{24} kg, determine the rate of tidal heating. Express your answer in W kg^{-1} (1 W = 1 J s^{-1}).

One consequence of tidal interaction in the Earth–Moon system is to have slowed down the Earth's rotation over time. This effect may be confirmed by counting growth rings on fossil organisms from the distant past (approximately 400 Ma ago), and comparing them with those recorded in modern living analogues. 400 Ma ago there were about 21.5 hours in a day, and 400 days in a year. An additional effect of this slowing of Earth's rotation is that the Moon's orbit is receding from the Earth at about 3 cm per year. Since tidal heating is proportional to the Earth–Moon distance, its influence must have diminished over time. Even so, tidal heating, past or present, probably never accounted for a major proportion of Earth's heat budget. However, its importance elsewhere should not be underestimated because it is known to be of great importance in bodies such as the satellites of the giant planets Jupiter and Saturn. The most spectacular example is Io (Figure 1.19), a satellite of Jupiter, where active volcanism is generated by tidal effects created by the planet.

2.4.3 Radiogenic heating

During the latter half of the 19th century, the eminent physicist William Thomson (Lord Kelvin, 1824–1907) attempted to determine the age of the Earth. He attempted this by considering that it had cooled slowly after its formation from a molten body. In effect, he was assuming the main sources of energy were from primordial heat and tidal heating. Taking many factors into account, including the mass of the Earth, the current rate of heat loss at its surface (i.e. surface heat flux), and the melting points of various constituents, he concluded that our planet could not be much older than about 20–40 Ma. This conclusion was contrary to the ideas of eminent geologists such as James Hutton (1726–1797) and Charles Lyell (1797–1875) who had already argued that immense spans of time were required to complete the changes produced by the action of **tectonic**, erosional and depositional processes. It was also greatly at odds with the emerging theories of the evolution of life because scholars such as Darwin also argued for much longer periods based upon the evidence of species. As a consequence, a 'heated' controversy continued for many years. Even though Kelvin's calculations did not gain wide acceptance with geologists, he was a powerful scientific influence of the time, and it was not until much later in the 1950s that accurate **radiometric dating** experiments eventually proved him wrong. These experiments were conducted on primordial lead in meteorites (Box 2.7) and demonstrated that the formation of the Earth occurred at about 4600 Ma ago (4.6 Ga ago). So why did Kelvin get the answer so wrong? The answer lies in the decay of unstable isotopes of certain radioactive elements, the discovery of which was not made until some decades after Kelvin's initial calculations. An Irish physicist John Joly (1857–1933) was one of the first to suggest radioactive decay, leading to radiogenic heating, was an important heat source within the Earth. It is now known that this radioactive decay creates an important, independent source of heat within the Earth which supplements that remaining from the primordial sources. This radiogenic heating is something which Kelvin could not possibly have known about when making his calculations.

BOX 2.7 RADIOACTIVE DECAY AND HALF-LIVES

The 'half-life' of a radioactive element is defined as the time taken for half of the original radioactive atoms of that element to decay spontaneously to form another element or isotope. If the rate of decay is rapid, then the half-life will be relatively short, as is the case for ^{14}C (half-life 5700 years) which is often used for radiocarbon dating of archaeological artefacts and materials. It takes only a few half-lives for the concentrations of an original radiogenic element to fall to negligible quantities. This is why ^{14}C dating is no use for dating ancient geological materials – after only 8 half-lives (45 600 years), less than 0.4% of any original ^{14}C would remain. If radiogenic elements are not replenished by the decay of other radiogenic isotopes, then they will eventually be lost altogether.

This is thought to have been the case with the decay of ^{26}Al, which has a half-life of 0.73 Ma. Studies of CI carbonaceous chondrites suggest that a significant proportion of the aluminium present at the time of condensation of the solar nebula was the unstable isotope ^{26}Al. This was originally created during a supernova explosion pre-dating the birth of our Solar System, and cannot be replenished by the spontaneous decay of any other radiogenic elements. The decay of ^{26}Al may have contributed significantly to the heating of planetary embryos but, because of its relatively short half-life compared to the age of the Solar System, any remaining ^{26}Al has long since vanished within the terrestrial planets. Whilst such short-lived isotopes will have been important heat sources during the early stages of terrestrial planet evolution, study of Earth materials indicates that it is the isotopes of the elements uranium (U), thorium (Th) and potassium (K) which are responsible for most of the radiogenic heating that has occurred throughout the history of our planet. These isotopes all have particularly long half-lives (see Table 2.6), and were present in sufficient quantities after condensation and accretion to ensure that they have remained abundant within present-day Earth.

Table 2.6 Half-lives of common isotopes in the Earth's crust and mantle. The continuing decay of these isotopes is an important internal heat source within the Earth today.

Isotope	Half-life (10^9 yr)	Present rate of heat generation in the Earth (10^{-12} W kg^{-1})[a]
^{235}U	0.71	0.04
^{238}U	4.50	0.96
^{232}Th	13.9	1.04
^{40}K	1.30	2.8

[a] Values are in kg of average Earth materials, *not* of the isotope concerned.

The heating effect of radioactive decay can be illustrated by the example of ^{235}U. The parent isotope ^{235}U decays through a series of α- and β-particle emissions to the daughter isotope ^{207}Pb. α-particle collisions and β-particle collisions with adjacent nuclei during this process create heating through the loss of kinetic energy. The data in Table 2.6 indicate that, after 0.71×10^9 years, half of the ^{235}U originally present will have decayed to ^{207}Pb and the remaining ^{235}U will continue to halve every 0.71×10^9 years. By determining the proportion of parent and daughter isotopes within a mineral originally containing only parent isotope, it is possible to calculate how long has elapsed since decay began. This is the basic principle of radiometric dating. For instance if a radioactive element X has experienced two half-lives since the formation of the Earth (about 4600 Ma ago) the amount remaining (X_{rem}) can be expressed as:

$$X_{rem} = \frac{1}{2} \times \frac{1}{2} \text{ or } \left(\frac{1}{2}\right)^2 \qquad (2.1)$$

Hence $X_{rem} = 0.25$ of the original amount after 2 half-lives.

Most of the common minerals contain small amounts of unstable isotopes, the most important of which are summarized in Table 2.6. Their decay to form more stable isotopes releases tiny increments of heat, as described in Box 2.7. This decay has produced a continuous source of heat within our planet since Earth's formation and, by analogy, also within the other terrestrial planets and the Moon. Of course, the total amount of heat produced will depend upon the concentration and type of radiogenic isotopes present in the constituent layers of the planet, and the mass of suitable material present in its different layers. Or, to put it another way, a smaller planetary body with fewer radiogenic elements will produce significantly less internal radiogenic heat than a larger, more radiogenic planetary mass. In addition, the efficiency of heat loss through the planet's surface will depend upon available surface area, which means that larger planets will cool more slowly than smaller ones (Box 2.8). Finally, whilst the rate of radiogenic decay is constant for each individual isotope system (Box 2.7), the total amount of radioactive decay, and hence heat generation, will decline over time as the reserves of the original radioactive materials are gradually used up. This gradual depletion of radioactive materials is expressed in 'half-lives', and each isotopic decay system has its own unique half-life.

QUESTION 2.7

(a) What proportion of the original ^{40}K and ^{232}Th currently remains since the formation of the Earth? (Hint: using the data in Table 2.6, determine how many half-lives of ^{40}K and ^{232}Th have expired since the Earth was formed.)

(b) Based upon the data in Table 2.6, what was the amount of radiogenic heating in the Earth 4.6 Ga ago, and how does this value compare with that of today?

BOX 2.8 SURFACE AREA TO VOLUME RATIO

It is a geometrical fact that larger bodies have a lower surface area per unit volume than smaller bodies. This can be expressed as the surface area to volume ratio. Consider two spheres, one with a radius (r) of 1 m, another with a radius of 10 m.

$$\text{Surface area of a sphere is } 4\pi r^2 \tag{2.2}$$

$$\text{Volume of a sphere is } \frac{4}{3}\pi r^3 \tag{2.3}$$

Using the above equations:

The surface area of the smaller sphere is 12.56 m², and that of the larger is 1256 m².

The volume of the smaller sphere is 4.19 m³, and that of the larger is 4190 m³.

$$\text{Surface area to volume ratio} = \frac{\text{surface area}}{\text{volume}} \tag{2.4}$$

For the smaller sphere, surface area to volume ratio $= \dfrac{12.56}{4.19} = 3.0$ (to 2 sig. figs)

For the larger sphere, surface area to volume ratio $= \dfrac{1256}{4190} = 0.30$ (to 2 sig. figs)

To summarize, the surface area to volume ratio of the smaller sphere is ten times that of the larger sphere. Therefore, because there is more surface area per unit mass for the internal heat to escape from, cooling through surface heat loss will be much more efficient for the smaller sphere (since there is more surface area, per unit mass, for the heat to escape through).

The elements U and Th and their radiogenic isotopes are particularly concentrated into the silicate-dominated outer layers of the Earth, and in particular within the continental crust. They are thought to be virtually absent in the core. As a result, the radiogenic heat produced per unit mass of the continental crust is, on average, over one hundred times greater than that of the underlying mantle. In effect, this means that the overall radiogenic heating budget is roughly split equally between mantle and crust despite the much greater mass of mantle material. It is the decay of these long-lived isotopes that provides sufficient heat energy to keep the Earth geologically active. This is achieved through constant convective and conductive heat loss, and is manifest at the surface as radiative heat loss into space. Therefore, the surface heat flux is not simply the slow cooling of a once molten body, as originally envisaged by Kelvin.

■　Using the value determined in Question 2.6, what proportion of Earth's surface heat flux loss is due to radioactive decay, compared with the 1.5×10^{-13} W kg^{-1} created by tidal heating effects?

❏　By summing the four values quoted in Table 2.6, the present rate of radiogenic heating is about 5×10^{-12} W kg^{-1}, which is about 30 times greater than that attributable to estimates of tidal heating, and so represents a much more important source of Earth's present-day internal heating.

We can see, therefore, that radiogenic heating is a very important source of heat generation. This heat is slowly lost through the surface, so cooling the planet. However, it is important to remember that the planet is also cooling by the loss of the primordial heat (Section 2.4.1) and, it turns out, that radiogenic and primordial heating make approximately equal contributions to the total present-day heat loss from the surface of the Earth. Geological activity on Venus may also be the result of internal radiogenic heating, because Venus is of a similar size to Earth and is thought to have a crust, mantle and core of comparable size. The smaller terrestrial planets Mercury and Mars may also have been geologically active in the past through heat loss supplemented by radiogenic heating. However, their smaller size, coupled with their greater surface area to volume ratio (Box 2.8), means that whilst they may have the same proportion of radiogenic isotopes as Earth, these would have been present in smaller amounts, and the heat they produce would be more efficiently radiated to space. These factors are compounded in the case of Mercury which contains a large and presumably unradiogenic core (Figure 2.8). By contrast, the small core and proportionally thicker mantle of Mars may have enabled it to have remained geologically active for much longer than its small size might otherwise dictate. The geological history of these planets is discussed further in following sections.

2.4.4 Heat transfer and its effects within the terrestrial planets

Layering in the other terrestrial planets is known to exist, and must have arisen by processes similar to those that created layering within Earth. The presence of this layering has been inferred by careful observation of data collected from space probe fly-bys. In the case of the Moon these data have been augmented by seismic monitoring experiments deployed during the Apollo missions of the late-1960s and early-1970s. The nature of surface materials and topography of the terrestrial planets can also provide important clues about the nature and development of planetary layering and subsequent geological history.

Having already discussed how planets came to differentiate into separate layers through the effects of various heating processes, and the processes by which the heating was (and continues to be) generated, it is now necessary to discuss how this heating might affect a planet once the core, mantle and crust have been formed. In the previous section it was stated that heat generated within a planet by tidal effects or radioactive decay, as well as the primordial heat, was eventually radiated into space from its surface. But how does this internal heat get to the surface? Three main mechanisms operate within a planet such as Earth; these are conduction, convection and advection.

Conduction

This is perhaps the most familiar since it is the process of heat transfer experienced when the handle of a pan becomes hot. The heat is conducted from the stove to the pan, and then to its handle. Different materials, such as rocks of various compositions, will conduct heat at different rates, and the efficiency of heat transfer in this manner is known as heat conductivity. This method of heat transfer is the most important in the outermost layer (i.e. crust).

Convection

This process relies on the fact that most materials expand when heated and, as a result, become less dense (i.e. more buoyant), allowing them to rise up through more dense surrounding material. In the case of a planet, this causes hot, deep material to rise up through the cooler outer regions during which the material gives up its heat, thus transferring heat to the cooler regions. Convection is a particularly efficient method of heat transfer, but the medium through which transfer takes place must be fluid. It should be noted, however, that the term 'fluid' describes any substance capable of flowing, and is not just restricted to liquids and gases. Under the correct conditions, solids, and even rocks can flow, albeit at a very slow rate. Thus, over long periods of geological time, the effects of such solid-state flow become a highly significant way of transferring heat toward a planet's surface.

In Section 2.2.2 it was stated that the Earth's outer core must be liquid since it does not conduct S-waves, and because the convection currents within it are thought to be responsible for generating our planet's magnetic field. We know that the overlying mantle is solid because it conducts both P-waves and S-waves. However, the silicate rock that comprises the mantle does flow when pressure and temperature become sufficiently high. In fact, some peridotites contain highly deformed crystals which are thought to be indicative of this flow process. The process itself is known as **solid-state convection** and, whilst rates may be no more than a few centimetres per year, it is the most efficient form of heat transfer within all but the outermost part of the mantle. Near the Earth's surface the rocks are too cold and rigid to permit convection, so conduction is the most significant process. The zone within the mantle in which both temperature and pressure are sufficient to permit flow is known as the **asthenosphere** (from the Greek *astheno* meaning weak), and this comprises much of the mantle thickness down to the core (Figure 2.14). Above the asthenosphere the uppermost mantle and all of the overlying crust is solid and rigid and is given the name **lithosphere**. It is important to realize that the difference between the asthenosphere and lithosphere is not the same as the boundary that defines the difference between the mantle and the crust (i.e. the Moho) which is primarily a compositional change (Section 2.2.3). Instead, the division between asthenosphere and lithosphere is based upon their physical properties.

> The lithosphere is an outer, rigid layer through which heat is transferred by conduction. The asthenosphere is an underlying, mechanically weaker layer that is capable of flow, and in which the principal process of heat transfer is convection.

The top of the convecting mantle can be identified using seismic data since the velocity profiles of both P-waves and S-waves briefly decrease through a thin layer known as the **low velocity zone** (LVZ), or more strictly low speed zone, created due to its partially liquid nature (Figure 2.4). Here, the drop in velocities is thought to be the result of tiny amounts (<0.1%) of partial melting of peridotite in the upper mantle. This melt collects in the interstices between crystals and, through this partial fluidity, aids in augmenting mobility of the mantle material. Importantly, at depths below Earth's LVZ, the pressure created by self-compression and the associated changes in crystal structure or mineralogical composition (Section 2.2.1) prevent further melting of the mantle. Thus, seismic velocities rise, and mobility is entirely due to solid-state flow.

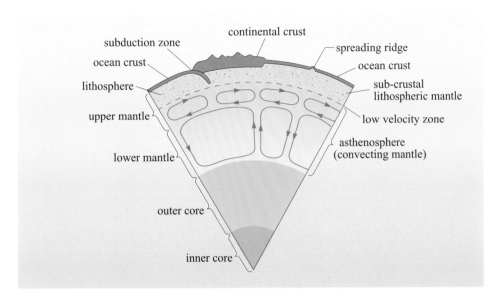

Figure 2.14 Section through the Earth showing the division of the mantle into the uppermost rigid lithosphere, and the mobile, convecting asthenosphere (not to scale).

Advection

The final process of transferring heat is by physically moving molten material (magma) up through fractures in the lithosphere. This process is termed advection and operates whether the magma spreads out at the surface as a lava flow or, if it is injected, whether it cools and crystallizes within the lithosphere itself. The effect is the same in both cases since heat is transferred by the molten rock from deeper levels where melting is taking place, to shallower levels where it solidifies, losing its heat by conduction into the overlying crust. Any planetary body which exhibits, or has exhibited, volcanic activity must have lost some of its internal heat in this manner.

2.4.5 Heat loss and plate tectonics

The importance of the difference between the properties of the lithosphere and asthenosphere became apparent during the 1960s with the development of the theory of plate tectonics. This theory recognizes that the different parts of Earth's lithosphere, comprising the non-convecting upper mantle together with the overlying oceanic or continental crust, can move relative to each other as a series of rigid plates (Figure 2.15). Current rates of movement are mostly between 50 mm and 100 mm per year, and the motion is thought to be a response to convection and heat loss in the mantle below the lithosphere. The movement of Earth's tectonic plates is enabled by the fact that new crustal material is added incrementally along **mid-ocean ridge** systems, and re-absorbed into the mantle at **subduction zones** that are associated with deep trenches at the edges of ocean basins. The main features of the outer layers of the Earth that give rise to plate tectonic motion are summarized in Figure 2.16. A consequence of this generation and destruction of ocean lithosphere is that it recycles material within the upper mantle. **Plate recycling** in this manner effectively adds hot material to the lithosphere, and removes and re-absorbs cold, solidified material, and in doing so assists in the process of the surfaceward transfer of heat. A further consequence of tectonic recycling is that it can resurface large areas of a planet's lithospheric crust and, in the process, destroy the record of previous impact cratering or other surface-modifying processes.

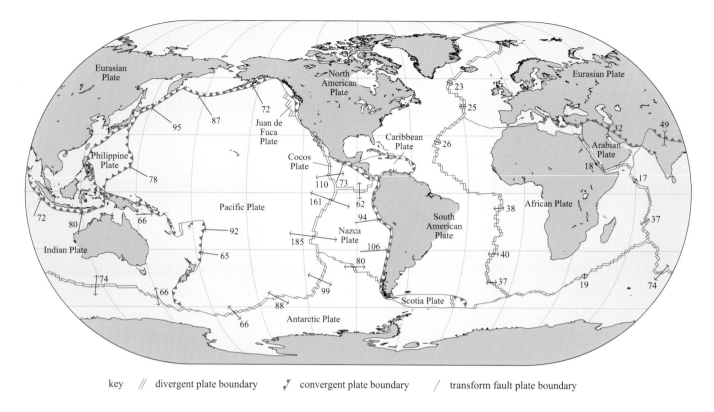

key // divergent plate boundary   ⌇ convergent plate boundary   / transform fault plate boundary

Figure 2.15 Map showing the global distribution of plates and plate boundaries. The black arrows and numbers give the direction and speed of relative motion between plates. Speeds of motion are given in mm yr^{-1}.

Another topographic effect observed on Earth is the generation of broad surface uplifts up to 1000 km across, which produce bulges of hundreds of metres elevation in the overlying lithosphere. These are apparently unrelated to plate boundaries and are thought to be created by pipe-like zones of anomalously hot material upwelling from deep within the mantle. The exact cause of these upwellings, or **mantle plumes** as they are more commonly called, is not yet known but they clearly represent another process by which heat may be transferred from deep within the Earth. Surface uplift occurs not only because there is a surfaceward (i.e. dynamic) movement of mantle material, but also because this rising mass of hotter mantle material expands due to the release of pressure in its upward journey. In addition, the release of pressure can result in the melting of the upwelling mantle (i.e. decompression melting), so such bulges are often associated with **hot spots** because the escape of this melted material produces volcanic activity. The best examples of present-day hot-spot volcanism related to mantle plumes are the volcanic islands of Hawaii and Iceland. Similar uplifted regions associated with volcanic constructions have been recognized on the surface of Mars in the form of the Tharsis 'bulge' (Section 2.5.5).

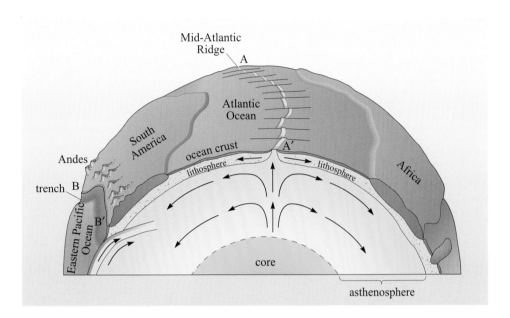

Figure 2.16 A vertically exaggerated model showing the main elements leading to plate tectonic movement on Earth, and showing the relationships between three plates: the African plate, the South American plate and the Pacific Ocean plate, called the Nazcan plate (beneath the eastern Pacific Ocean). Since the lithosphere is rigid and comprises the crust and uppermost mantle, heat transfer is primarily by conduction, with a component of volcanically driven advection. The asthenosphere is weak and comprises the underlying convecting mantle. Lithospheric plates move apart at divergent plate boundaries (mid-ocean ridge systems), where material is continually added incrementally by volcanism (A–A'), and is destroyed at convergent plate boundaries where one plate is forced below another and reabsorbed into the asthenospheric mantle at depth (B–B').

QUESTION 2.8

(a) By completing Table 2.7, indicate which of the following list of Earth's layers are defined seismically, and which by other means. What properties do the differences between these layers represent?

(b) Briefly outline the properties that the differences between these layers represent.

Table 2.7 How the layers within Earth are defined (to be completed).

Layer	How layer is defined
lithosphere	
asthenosphere	
crust	
mantle	
core	

Figure 2.17 compares the relative thicknesses of the lithosphere, asthenosphere and core for each of the terrestrial planets, and the Moon.

These diagrams are mainly based on model calculations. We only have well-constrained interior data (based on seismic studies) for the Earth and Moon.

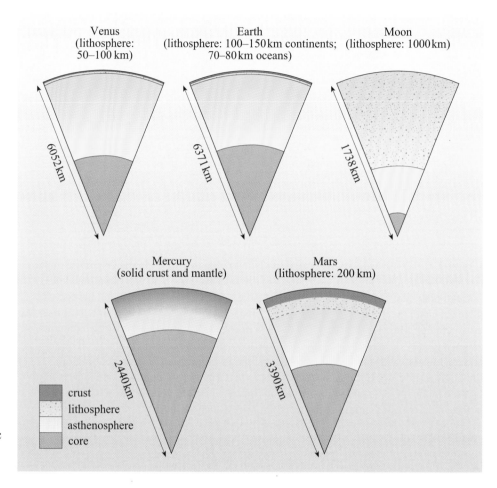

Figure 2.17 A comparison of lithospheric, and/or asthenospheric thicknesses, and core in the terrestrial planets and the Moon.

2.5 Dead or alive? Plate tectonics and resurfacing

The convection in Earth's asthenosphere is a response to internally generated heat sources and remaining primordial heat, and enables transferral of this heat to the lithosphere, after which it is conducted and eventually lost into space by radiative processes. These processes of heat loss are the main controls upon volcanic and tectonic activity on Earth. On Earth, resurfacing of the planet occurs largely in response to tectonism or volcanism. For instance, the effusion of lava flows or the deposition of volcanic ash are effective resurfacing processes. Resurfacing also occurs where the convergence of two lithospheric plates carrying continental crust cause buckling and deformation. The regions deformed in this manner form mountain belts such as the Alps or Himalayas, which are then eroded and supply sediments which cover lowlands and the floors of ocean basins. If plate tectonics were to occur on other planets then similar styles of deformation at the plate boundaries may be expected. However, unlike Earth, these will not necessarily be modified by the action of weathering, erosion and sedimentation. To examine the possibility of planetary resurfacing having operated elsewhere, it is useful to consider how the heat sources may differ both in their total amount, and in their relative contribution to the heat budget in other planets. Clearly, the efficiency of heat loss may also vary from planet to planet due to differences in their size, or because of differences in the relative thicknesses of their asthenosphere and lithosphere. Differences such as these can lead to substantial variation in the degree and nature of volcanic activity and the tectonic features that may be expected.

As a consequence of heat transfer and radiative loss, all planetary bodies will gradually cool over time. This is because the available sources of internal heat cannot be renewed, a fact which is perhaps most apparent in the case of primordial or radiogenic heat sources (Section 2.4). However, any tidal heating effects must similarly diminish over time (Section 2.4.2). As this cooling progresses, the lithosphere will thicken, and the top of the convecting asthenosphere will retreat inwards. In a body which generates only modest amounts of internal heat, perhaps as a consequence of its small size or because it contains lower concentrations of radiogenic isotopes, this process will progress more rapidly. Correspondingly, any associated volcanism or tectonic movement will diminish. If the process continues to the point where asthenospheric convection ceases altogether, all the outer layers of the planet will, by definition, become lithosphere (Figure 2.17; as for Mercury). In such cases, the planet will become tectonically 'dead' and the associated resurfacing processes effectively terminated.

Whilst a diminishing degree and eventual cessation of tectonic activity may be the fate for all terrestrial planetary bodies, the timing of when this inactivity will take place depends upon the planet's composition and size (Section 2.4.3 and Box 2.8). However, even on those bodies where plate tectonic movement has ceased altogether, there may still remain sufficient internal heat to allow partial melting to occur. Consequently, this molten material may continue to generate surface volcanism even on 'tectonically dead' planets, provided the thickened lithosphere can be breached by the rising magma. Under these circumstances volcanic resurfacing, such as blanketing of earlier topography by lava flows or ash deposits, could continue even after surface tectonic processes have apparently ceased altogether (e.g. on Mars). The terrestrial-like bodies within the Solar System display a wide variety of surface features. From studying such features it is possible to determine which of these bodies are tectonically and volcanically dead, and which examples display surface features which are indicative of geological activity.

2.5.1 Earth

Mantle convection, together with lithospheric conduction, are the processes that enable the generation of internal heat (i.e. by radiogenic and tidal processes) to be balanced efficiently over the short term by heat loss at the Earth's surface. Above the convecting part of the mantle, Earth's lithosphere is broken into a number of tectonic plates. Some are surfaced only by an oceanic crust, whilst others carry both oceanic and continental material (Table 2.8). The higher density oceanic crust lies on average 4–5 km below sea-level, whilst many of the continental plains lie just above sea-level. This confers a very distinct bimodal height distribution to the global topography (Figure 2.18).

Figure 2.18 View of the east Pacific Ocean, the Americas and, in the Gulf of Mexico, Hurricane Andrew (25 August 1992). (NASA)

Table 2.8 Size, mass, density and layering within the terrestrial planets and the Moon.

	Mercury	Venus	Earth	(Moon)	Mars
mean radius/km	2440	6052	6371	1738	3390
mass relative to Earth	0.055	0.815	1	0.0123	0.107
density/10^3 kg m^{-3}	5.43	5.20	5.51	3.34	3.93
layer thicknesses					
crust/km	(see below)	20–40	25–90 (continents)	20–100 70 (average)	Thought to be highly variable
			6–11 (ocean basins)		150–200
lithosphere/km (crust + rigid part of upper mantle)	600 (crust + all of the mantle)	50–100	100–150 (continents)	1000	200–300 (?)
			70–80 (ocean basins)		
asthenosphere/km	absent	3000–3100	2750–2850	300–500	1200–1900
core/km	1800	2900	3470	220–450 (if present)	1300–2000
magnetic field	yes (weak, 1% of Earth's magnetic field)	no	yes (active field created by motion in liquid core)	no	very weak (largely remnant, but stronger in past.

The movement of the different plates (plate tectonics) occurs through the generation of new oceanic crust at mid-ocean ridge boundaries where adjacent plates are moving apart, and the destruction of old, cold oceanic crust at boundaries where ocean crust descends either beneath the continental or oceanic edge of an adjacent plate (Figure 2.16) called, respectively, constructive and destructive plate boundaries. The generation of crust in one part of the globe is always balanced by its destruction elsewhere, so that the total surface of the Earth remains constant. The tectonic plates are in continual motion and it is thought that plate tectonic motion on Earth has been occurring for billions of years. However, in the distant past when the rate of radiogenic heat production was greater, tectonic plates may have been smaller, and their motion more rapid.

Much of the volcanic activity is closely associated with the boundaries of the tectonic plates (Figure 2.15), and most of the Earth's best known **volcanoes** occur along destructive plate boundaries. These are located at a position on the over-riding plate above where the descending plate is being destroyed. Submarine volcanism also occurs along the mid-ocean ridges where oceanic crust is being created. Notable exceptions to this tectonically determined distribution of volcanoes do occur, such as the Hawaiian chain of volcanic islands. Here, the volcanic islands are the result of a rising 'plume' of anomalously hot mantle producing hot-spot volcanism characterized by **basalt lava** flows. The motion of a lithospheric plate over such a mantle plume creates a chain of progressively younger volcanic structures (Figure 2.19).

Figure 2.19 View from an aeroplane looking toward the summits of the volcanoes of Mauna Kea (foreground), and Mauna Loa (background) on the island of Hawaii. Gradually rising to more than 4 km above sea-level, Mauna Loa is the largest volcano on our planet, covering half the island of Hawaii. Its submarine flanks descend an additional 5 km to the sea floor, and the sea floor in turn is flexed downwards another 8 km by Mauna Loa's great mass. This makes the entire volcanic construction about 17 km from base to summit. (Mike Widdowson/Open University)

Compared with the other terrestrial planets, our Earth is a very dynamic world and has a relatively young surface. There is very little remaining as a record of Earth's earliest past because evidence has been obliterated by the creation and destruction of crust through the action of plate tectonics. Additionally, surface weathering, erosion and deposition of sediment continually destroy, erase, and bury surface features, including ancient crust. This means that, unlike many of the other terrestrial planets, the preservation of ancient impact craters is very uncommon. By contrast, our nearest neighbour, the Moon, has no plate tectonics or processes of weathering and erosion comparable to those on Earth, and so its cratering record has largely been preserved. The timing of the formation of Earth and the Moon, based upon the available chronological data, is summarized in Figure 2.20.

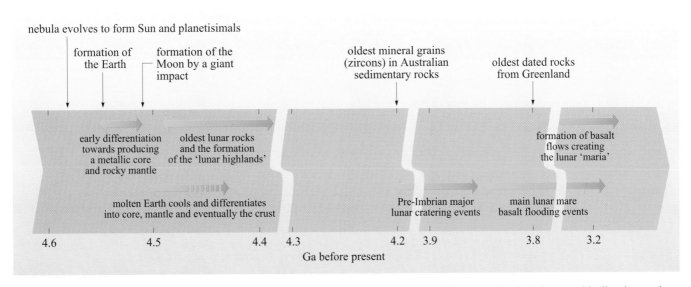

Figure 2.20 The timing of the formation of Earth and the Moon, based upon available chronological data, and indicating major planetary-forming and differentiation events.

2.5.2 The Moon

The Moon is not a planet in the strictest sense, but its composition indicates that it is a terrestrial-like body. The history of its formation, as determined by a wealth of observational data and some sample material, together with the nature of its surface-cratering record, provide a useful analogue, which aids in deciphering other planetary histories. The Moon's formation is thought to be the result of a late-stage collision of large planetary embryos (Section 2.3.3). Energy from this impact melted much of the material that then coalesced into the Moon, giving rise to a deep, all-enveloping magma ocean.

The Moon's surface is clearly composed of two types of surface – the darker **maria** or 'lunar seas' and the brighter, highly cratered, 'highland regions' (Figure 2.21). These two surfaces reflect a fundamental compositional variation. The highlands, standing about 2.75 km above the surrounding maria terrain, are the most ancient crust and are about 20 km thick. Samples from these areas have been dated as 4.4–4.5 Ga (Figure 2.20). This material is thought to have formed as a less dense, feldspar-rich differentiate, or crystal-rich 'scum', that rose to the surface of the magma ocean following accretion. The maria comprise basaltic lava flows dated at 3.8–3.2 Ga, and post-date the major crater-forming impact events that affected the highland regions (Table 2.9). However, these lunar lava flows differ compositionally from terrestrial basalts (Table 2.2) since they typically have a lower SiO_2 content (i.e. <45%). In most cases, these lavas welled up in ancient impact basins simply because the crust was thinner and more fractured in these regions. Internal heat is likely to have been responsible for this volcanism rather than heat generated by later crater-forming impact events. Some evidence suggests that the mare-forming lava flow eruptions may not have ceased completely until as late as 0.8–1.0 Ga ago, after which there was either insufficient internal heat generation or the continued cooling meant the Moon's lithosphere had become too thick for surface volcanism to continue.

Figure 2.21 The Moon. The bright, cratered regions (for instance, at the top of the image) are the lunar highlands, and dark regions are the lunar maria. Dark maria are impact basins which have filled with basalt lavas. These include Oceanus Procellarum (on the left-hand side), Mare Imbrium (centre left), Mare Serenitatis and Mare Tranquillitatis (centre), and Mare Crisium (near the right edge). The distinct, bright ray crater at the bottom of the image is the Tycho impact basin. (NASA)

Table 2.9 The stratigraphy of the Moon, based upon stratigraphical mapping of its surface, its cratering record, and radiometric dating of Moon rocks. Ages refer to the years before the present day.

System	Events
Copernican	Young 'ray' craters (younger than 1 Ga), and some final mare-forming eruptions
Eratosthenian	Older post-mare craters (about 1–3 Ga)
Imbrian	Main mare basalt flooding events, following the formation of the Imbrium and Orientale impact basins (<3.2–3.85 Ga)
Pre-Imbrian:	
Nectarian	Formation of 11 major impact basins preceded by the Nectaris basin (3.85–3.92 Ga)
Pre-Nectarian	Formation of about 30 impact basins preceded by the Procellarum basin (before 3.92–4.2? Ga)

2.5.3 Mercury

Mercury's highly cratered surface appears similar to that of the Moon (Figure 2.22). Inspection reveals that its most densely cratered region contains a similar number of impact sites to analogous regions of the Moon, suggesting it has a similar very ancient surface. Like the Moon, regions interpreted as ancient lava flows occur. Most notable, however, are the presence of sinuous ridges, 2 km in height and 200–500 km in length, which are thought to have been formed by lithospheric compression (Figure 2.23). However, unlike compressional structures in Earth's lithosphere (i.e. mountain belts) which are caused by tectonic convergence, Mercury's ridges are extremely ancient features that are several billions of years old. They have been interpreted as being a consequence of either mantle cooling or solidification of previously molten core material. This cooling and solidification would have caused thermal contraction resulting in an estimated 0.1% reduction of the planet's diameter, and making the lithosphere wrinkle into a series of ridges.

Mercury is too small to have retained much of its accretional heat for long. Moreover, due to its large core (Table 2.8) and relatively thin mantle and crust, the effects of radiogenic heating are likely to have been smaller and less pronounced than those experienced by the larger terrestrial planets. It seems probable that these heat sources were insufficient to drive any long-term mantle convection and, consequently, the planet has been geologically inactive for a considerable part of its history. If this is the case, these ancient sinuous ridges may record the last geological activity to have affected Mercury's surface.

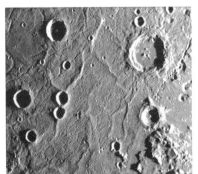

Figure 2.22 A mosaic of images of Mercury's highly cratered surface taken from 125 000 miles away by Mariner 10. (NASA)

Figure 2.23 The ridges and fractures on part of the floor of Mercury's Caloris basin. (Courtesy of Calvin J. Hamilton)

2.5.4 Venus

Venus is similar to Earth in size and density and, consequently, its core dimensions and mantle thickness may be comparable to those of Earth. If this assumption is correct, we may expect to observe similarities in the arrangement of the global terrain of both planets, and perhaps even evidence of Earth-like tectonism on the surface of Venus. However, inspection of elevational data does not reveal a bimodal distribution like that of Earth (Figure 2.24). Instead, although there are lowland and highland regions, much of the topography occurs within 500 m elevation of the average planetary radius. In addition, the cratering record reveals that much of the surface dates from around 500 Ma ago, indicating a catastrophic resurfacing took place, followed by a long period of relative quiescence.

If Venus really is currently geologically inactive, yet otherwise similar to Earth in its dimensions and structure, it prompts the question of how the planet is able to lose its internally generated heat. It is unlikely that the composition of its crust and mantle differ significantly to that of Earth with respect to radiogenic components, so the surface features must reflect a fundamentally different tectonic behaviour in response to radiogenic heat production.

Highland and lowland regions may correspond to the upwelling and downwelling of convection currents within Venus's asthenosphere. However, the apparent absence of any tectonic activity or active volcanoes suggests that conduction through the lithosphere is the only manner in which internal heat can currently be lost. Such lithospheric conduction is likely to be far too slow to balance the internal heating and so it is believed that Venus's interior is gradually heating up, and that the heat is becoming trapped in the upper mantle, below this slowly conducting lithospheric 'lid'. Continuation of this process over the long term will eventually lead to increased partial melting of the sub-lithospheric mantle, a key source of magma and volcanism on Earth, and associated reduction of mantle density. If this heating process continues, this hot mantle material may eventually become less dense than the overlying rigid lithosphere. This is an unstable situation that could lead to a catastrophic 'overturn', causing sinking of large areas of the lithosphere and a surfaceward escape of immense volumes of trapped magma. Such an overturn would produce a massive resurfacing of the planet. After this violent, but geologically short-lived, episode and associated release of trapped heat, the surface would once again cool, thicken and become rigid, and the slow process culminating in lithospheric overturn would begin once again. Such periodic, planet-wide resurfacing may explain the apparently uniform age of much of Venus's surface, and is currently the most widely accepted model for the planet's long-term tectonic behaviour.

Figure 2.24 Two hemispheric views of Venus (a) from the north pole and (b) from the south pole as revealed by radar investigations by the Magellan space probe. These composite images have been colour-coded to provide elevational information across Venus's 15 km topographic range (blue is lowest elevation, and green, brown and white progressively higher elevations). (NASA)

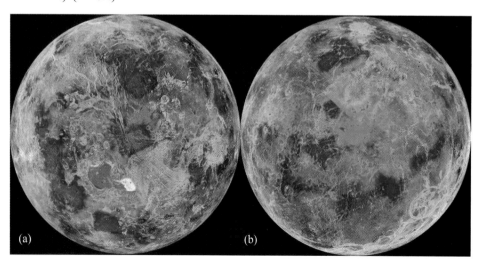

(a)   (b)

2.5.5 Mars

The mass of Mars is only one-tenth that of Earth, and its low density is most similar to that of the Moon (Table 2.8). This lower density may, in part, be due to its smaller mass, having generated less self-compression than that experienced by the larger terrestrial-like bodies. Mars also currently lacks any strong magnetic field.

There are superb examples of volcanoes on Mars but, unlike Earth, they do not occur in linear chains, which suggests that the Martian lithosphere is not divided into mobile plates. Martian volcanoes such as Olympus Mons contribute to the 30 km range in surface elevation (Figure 2.25), thus greatly exceeding both Earth and Venus, which have surface elevation ranges of 20 km and 15 km, respectively. The height of the Martian volcanoes may indicate the presence of a very thick lithosphere. The largest comparable structure on Earth, the volcano comprising the island of Hawaii, is only one-hundredth the volume of Olympus Mons, yet its mass has created a substantial moat-like depression in the surrounding oceanic lithosphere. Despite its size, no similar depression has been observed around Olympus Mons, which is 24 km in height.

Studies of the Martian surface suggest that whilst there are no obvious tectonic boundaries in the lithosphere, there does appear to be two distinct types of crust, which may be broadly analogous to continental and oceanic crust on Earth. The southern hemisphere and part of the northern hemisphere form a huge ancient region of heavily cratered highlands. This elevated, older crust could be compared with Earth's continental areas. By contrast, the northern hemisphere is low-lying, much less heavily cratered, and is apparently covered with lava flows and sedimentary material. Moreover, the southern highlands have regions with remnant magnetic properties that are similar in pattern to those produced by ocean crust spreading ridges on Earth. This has led to speculation that there may have once

Data from the Mars Global Surveyor spacecraft suggests that, despite its weak magnetic field, Mars may have a partially liquid core.

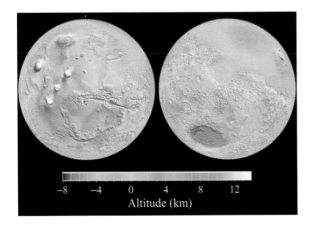

Figure 2.25 Topographic maps of Mars. The left-hand image shows the elevated topography of the Tharsis bulge (in red), a region of thickened and uplifted crust. The huge volcanoes associated with the Tharsis region can easily be seen as three white dots, with two further volcanic structures located in the top left-hand quarter. The youngest and biggest of all is Olympus Mons, positioned on the western flank of the Tharsis bulge. The fissure-like complex of canyons crossing into the Tharsis region from the right (centre of the image) is the Valles Marineris. This may owe its origins to fracturing of the Martian crust during uplift of the Tharsis region. The right-hand image view shows the Hellas impact basin (in purple), which is 2000 km in diameter. (NASA)

been crustal spreading on Mars, and that these regions are thus analogous to oceanic crust. However, the Martian lithosphere must have thickened significantly since, to such a point where any mantle convection must occur at depth, and the only likely surface expression of surfaceward transfer of internal heat is a globalwide distribution of volcanoes, and the presence of elevated surface regions such as the Tharsis bulge, which is 4000 km wide. The elevated Tharsis region, upon which large volcanoes including Olympus Mons are located, is believed to be the expression of a particularly persistent site of mantle upwelling. Such an upwelling is possibly analogous to Earth's mantle plumes, which are similarly associated with the construction of major volcanic edifices (e.g. Hawaii). Nevertheless, volcanic activity on Mars does appear to have waned. This must, in part, be a consequence of its small size since its radiogenic heat production will have undoubtedly significantly diminished over time. This, together with its greater surface to volume ratio, which promotes cooling, may explain why its lithosphere has thickened so much, and in so doing allowed less opportunity for surface volcanism to occur.

From cratering studies, Mars's highland crust, like that of the Moon, appears to have originated about 4.5 Ga ago (Section 3.2.5). However, on Mars this ancient terrain has been substantially modified by later volcanic activity. The volcanic lowlands in the northern hemisphere, and the equatorial Tharsis region probably began to develop 3.8 Ga ago, and some lava flows here may be no older than 10 Ma. Martian volcanic activity cannot yet be considered as entirely extinct.

2.5.6 Io

Io is the innermost of Jupiter's large satellites. Although it is not strictly a terrestrial planet, it ranks as a terrestrial-like body due to its size and density (Appendix A, Table A2). Its density indicates that it is composed largely of silicates and probably also possesses a small core of Fe or FeS. It is worth discussing Io in detail for two reasons. First, its major source of internal heating is tidal. The energy released by tidal heating is estimated to be about 4×10^{13} W, two orders of magnitude greater than that likely to be generated by its own internal radioactive decay (about 5×10^{11} W). In this respect it is significantly different to the dominantly radiogenic heating that is characteristic of the terrestrial planets. Secondly, besides Earth, Io is the only body in the Solar System upon which active hot volcanism has been observed. This was one of the remarkable discoveries made by the space probe Voyager 1 during its Jupiter fly-by in March 1979 when no less than nine active volcanoes were detected. In fact Io's surface appears to be dominated by the products of volcanic eruptions – analysis of all returned images has failed to identify any impact craters which must mean that most of Io's surface is relatively young. The only other method of resurfacing that could be possible on Io would be through some form of tectonic activity and associated crustal deformation. However, other than a small number of minor features that may represent faults, convincing evidence for tectonism is absent.

QUESTION 2.9

(a) What changes will occur over time in the thickness of the lithosphere of a planetary body in which the dominant heat source is radiogenic decay?

(b) Giving reasons, discuss whether the same conclusion can apply to bodies that are dominated by tidal heating.

Io's internal structure is still a matter for conjecture, but the large amount of heat generated by tidal effects, together with its expression as active volcanism, suggest heat production has been sufficient to allow significant internal differentiation (Figure 2.26). Its core may be molten but much of the tidal-energy dissipation is thought to occur in its rigid lower mantle. The overlying asthenosphere must be convecting rapidly in order to transfer heat surfacewards, and here there may even be a thin liquid magma level. It is also likely that as a consequence of this continuing differentiation process the upper part of Io's lithosphere has become geologically very distinct. This is believed to have generated an outer Mg-poor silicate layer and an Fe–Mg-rich layer beneath what is effectively the lithospheric part of the upper mantle. If this interpretation is correct Io may have the thinnest lithosphere of all the terrestrial-like bodies, as well as a capacity to generate both silicate- and sulfur-rich styles of volcanism (Section 3.4.6). In this respect, it is perhaps surprising that there is no obvious expression of surface tectonics as might be expected with such a thin lithosphere overlying a vigorously convecting asthenosphere. Instead, it appears that heat loss is not achieved by convection and associated plate tectonics as on Earth, or by the periodic, catastrophic resurfacing thought to occur on Venus, but instead by conduction and surfaceward advective heat transfer via a network of highly active volcanic vents.

2.5.7 Europa

Europa is both slightly smaller and further away from Jupiter than Io. It is quite a different world, but it is described here because its density is similar to that of the terrestrial-like bodies. Gravity data collected by space probes indicate that below its rocky mantle Europa has a dense, iron-rich core, about 1250 km in diameter. Europa also has its own magnetic field, but it remains uncertain whether this is created by convection within a liquid part of the core. Like Io, it is probable that tidal heating of the rocky mantle occurs. This heat, produced by flexing and relaxation of the mantle, and supplemented by radioactive decay within its rocky mass, will be conducted surfacewards. Such conduction may then cause melting of the overlying ice layer and associated resurfacing through the eruption of 'icy lavas'. The nature of the intriguing phenomenon of icy volcanism will be discussed further in Chapter 3.

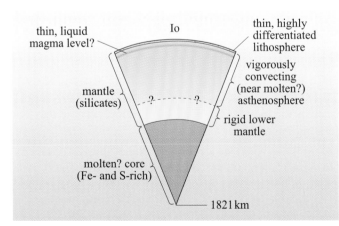

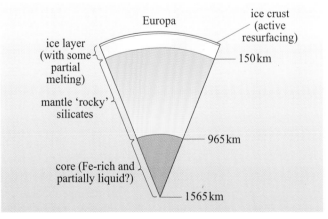

Figure 2.26 Schematic model for the internal structure of Io. The planetary layering within Io is assumed to be highly differentiated because of continual partial melting resulting from the tidally generated high heat flow.

Figure 2.27 Schematic model for the internal structure of Europa.

2.6 Summary of Chapter 2

- Detail of the internal structure of Earth is provided by both direct and indirect evidence. Composition of the lithosphere, that is the crust and upper mantle, can be determined by examination of key rock types. The structure and composition of the deep mantle and core is revealed from the properties and response of seismic waves that pass through the planet.

- The asthenosphere of a planet is the zone where pressure and temperature are sufficiently high to allow a material to flow, even in its solid state. Consequently, in this zone the internal heat is mainly transferred surfacewards by convection.

- The lithosphere of a planet is defined as the rigid outermost layer that cannot convect. Instead, the internal heat is carried through it by conduction. On geologically active worlds, such as Earth and Io, internal heat can also be transported by advection through the lithosphere by volcanic processes. In addition, heat is also transferred to Earth's surface by wholesale recycling of the lithosphere. This results in a phenomenon known as plate tectonics, which helps describe the movement of lithospheric plates. The Earth's lithosphere consists of oceanic and continental crust, and the uppermost part of the mantle. Convective and conductive processes of heat transfer can occur equally well in ice-dominated bodies as well as silicate bodies.

- The accretion and final assembly of Earth and other terrestrial planets is thought to have followed a similar pattern of evolution. As a consequence, their layered structure must have been the result of differentiation and element partitioning that operated during and after their assembly from colliding planetary embryos. The differentiation came about as a result of melting following energy release during these collisions.

- Primordial heat is that retained from processes operating in the early stages of planetary evolution, and represents one of the important heat sources within terrestrial-like bodies. The other two important sources are radiogenic heating and tidal heating. Primordial heat includes that derived from the collision and assembly of planetary embryos, and that delivered to the surface by incoming impactors after the planet had assembled. It also includes heat released by the separation of denser components during core formation.

- Internal heat generation within terrestrial planets such as Earth, is mainly the result of radioactive decay of ^{235}U, ^{238}U, ^{232}Th and ^{40}K in their silicate-rich mantle and crustal layers. The amount of radioactive decay was greater early in a planet's evolution because there would have been considerably more radioactive elements present. This radiogenic heat would have been augmented by the decay of short-lived isotopes such as ^{26}Al in those early stages. In some instances, notably large orbiting satellite bodies such as the moons of Jupiter, tidal effects become the dominant process of generating internal heat.

- Geologically active bodies are those where heat is either more efficiently retained or continually generated in significant amounts. In larger bodies, radiogenic sources are likely to continue to be important because they contain a greater mass of radiogenic elements to begin with, and because cooling is less efficient due to a lower surface area to volume ratio. Tidal heating is a further process whereby internal heat can be generated over the long term.

- The heat retained or generated within a terrestrial-like planetary body represents a key control in shaping the nature of the planetary surface. It also controls the rapidity and extent to which planetary resurfacing has occurred, or continues to occur.

CHAPTER 3
PLANETARY VOLCANISM – *ULTIMA THULE?*

3.1 Introduction

The ancient Greeks and Romans believed that the mythical island of *Thule* existed some six days' sail north of Britain. This island represented the edge of their known world and, consequently, the Latin phrase *ultima Thule* became used to refer to journeys to the utmost limits of exploration. The expression is also particularly apt for this chapter for two reasons. First, the classic volcanic province of the North Atlantic, of which Iceland is the best known locality, is known by geologists as the Thulean province after the land of this ancient myth (Figure 3.1). Secondly, examination of features on the other terrestrial planets and terrestrial-like bodies provides much evidence for volcanism having existed on these distant worlds (Figures 3.2 and 3.3), and it could be argued that spacecraft investigations that allow us to see these features in detail represent the *utmost limits* of modern exploration.

Figure 3.1 Aligned volcanic cinder cones on a basaltic flow field, Laki, Iceland. These cones were constructed as part of a line of over 100 vents that formed along a fissure, 27 km in length, from which 14 km^3 of basalt lava erupted during 1783–84. Historical records indicate that toxic gas and ash from this eruption caused widespread crop failure in northern Europe and a short-term climatic cooling. (Steve Self/Open University)

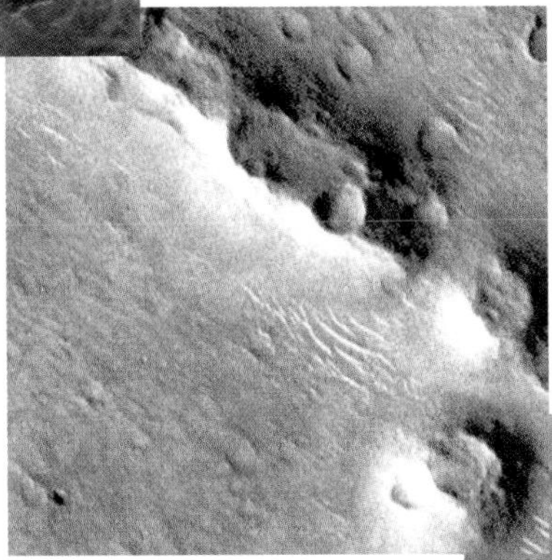

Figure 3.2 Aligned cone features on Mars. The image shows an area of 1.5 km^2 located within the 1100 km-diameter Isidis Planitia impact basin. Many such cones have been identified within the basin, often arranged in clusters or aligned in chains. Although no associated lava flows have been identified, a volcanic origin is favoured. It is thought that the cones were constructed from the eruption of ash and cinders at points along fractures and fissures in the Martian crust. (NASA)

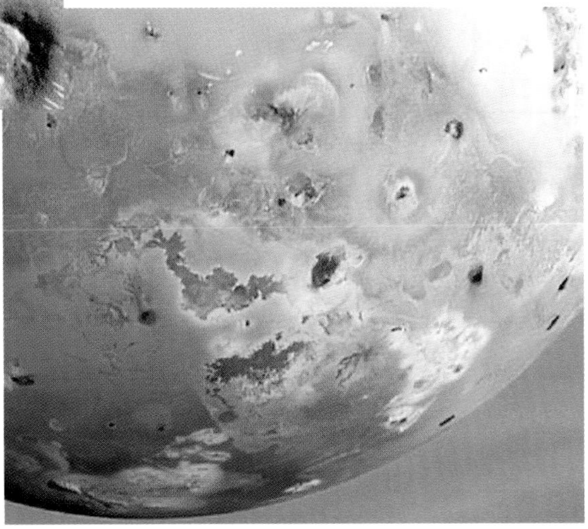

Figure 3.3 The surface of Io (a satellite of Jupiter). The dark spots are volcanic centres from which a variety of effusive (lava flows) and explosive (ash deposits) volcanic products have been erupted. A dark, sinuous area of lava flows can be seen extending from the volcano in the centre towards the left of the image. Red and yellow-coloured halos around the volcanic centres near the upper region of the image are ash and sulfur products deposited by eruption plumes during phases of explosive activity. (NASA)

Exploration missions, including unmanned fly-bys, orbiters and landers, have returned a wealth of data that details the surfaces of the terrestrial planets and satellites of the giant planets. For geologists, and volcanologists in particular, one of the most exhilarating discoveries was that volcanism is a process that occurs throughout the Solar System. Its constructions and its effects can be observed from the alternately freezing and baking surface of Mercury located relatively near to the Sun, to the icy surface of Triton, the largest of Neptune's satellites that lies in the outer reaches of our Solar System. The products and style of the volcanism on these very contrasting worlds vary according to differences in their structure, layering and composition, since these factors control the type of materials erupted. Volcanic products also vary in response to the different planetary environments, including properties such as gravity, surface temperature and atmospheric pressure. These factors control the form and extent of lava flows and the pattern of volcanic ash deposition.

One of the most amazing manifestations of volcanism yet found is the eruption of **ices**.

> The eruption of ices has occurred on numerous moons of the giant planets. Here, volcanism is not the result of rock melting, as is the case on Earth, but involves the melting of mixtures of water and other compounds formed from solidified gases. The term cryovolcanism (from the Greek *kryos*, meaning frost) is used for these icy examples.

Fortunately, despite these fundamental differences in lava composition and planetary surface conditions, understanding and interpreting extraterrestrial volcanism does not present as much of a problem as might first be anticipated. The laws of physics are universal and the nature of volcanism, whether it is the eruption of ices on a distant satellite (Figure 3.4), or a red-hot magma flowing down the side of Mount Etna, Sicily,

Figure 3.4 A high-resolution image, about 500 km across, of part of the equatorial region of Triton, the largest satellite of Neptune. Most of Triton's surface comprises water-ice mixed with frozen methane and probably ammonia. When melted, this mixed icy material behaves just as molten rock does on Earth. The various ridges and dimples appear to be the result of icy volcanic processes (cryovolcanism). Cryovolcanic icy lavas have probably flooded earlier impact craters and covered fractures to produce smooth terrains. For instance, note how the linear feature extending from the lower left-hand corner (inside the white box) becomes obscured by the smooth terrain located towards the centre of the image. The outline of a large irregular crater can be made out at the top edge of the white box. This has a smooth floor, interrupted only by a single small crater, suggesting that it has been flooded more recently by icy lavas. (NASA)

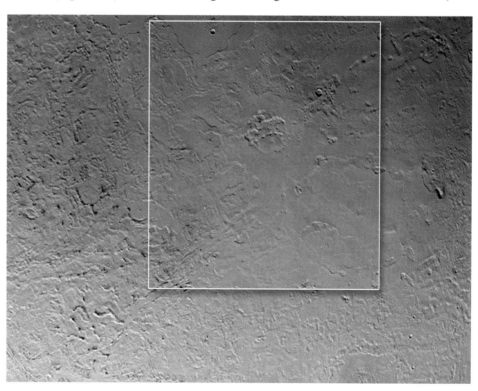

or Kilauea, Hawaii (Figure 3.5), or any other of Earth's spectacular volcanoes, can be understood by employing some basic principles. The object of this chapter is to provide you with an understanding of the underlying causes of volcanism and the physical processes that govern the style and extent of volcanic eruptions. You should then be able to make a prediction about the nature and style of volcanic phenomena likely to occur on even the most obscure planetary body, given only basic information about its size, mass, density and orbital parameters.

Figure 3.5 A red-hot, blocky lava flow, given the Hawaiian name of a'a, which is advancing slowly across an earlier lava field consisting of toe-like flows of pahoehoe (another Hawaiian term) on the lower flanks of Kilauea volcano, Hawaii. (Steve Self/Open University)

3.1.1 What is volcanism?

Volcanoes are the surface expression of the melting processes that occur deep within a body. The volcanic 'plumbing' systems beneath the volcano link these deep processes to the surface, and to the atmosphere, if present.

> Volcanism can be defined as the processes associated with the transfer of molten material (magma), associated volatiles and any suspended load of crystallized material, from the interior of a planetary body to its surface.

On Earth, these processes are most obviously demonstrated by the presence of volcanoes that erupt lavas and produce ash clouds and toxic gases. There are many types of volcanoes on Earth – different styles of volcanic activity are largely controlled by the chemical and physical properties of the magma. As a result, volcanic activity ranges from relatively quiescent effusions of lava, such as might be seen on Hawaii (Figure 3.6) or Iceland, to massively explosive eruptions (Figure 3.7) such as those of Mount St Helens (USA, 1980) and Mount Pinatubo (Philippines, 1991). Yet, for any type of volcanic eruption there are two prerequisites: a process that causes melting, and material that can be melted.

Figure 3.7 A volcanic explosion. The eruption of Mount St Helens, Washington State, USA. At 8.32 am on 18 May 1980, an earthquake of magnitude 5.1 shook the volcano. Part of its flank slid away in a gigantic rockslide and debris avalanche, releasing pressure within the volcano. This triggered a major explosion, a pumice and ash eruption, and the generation of a huge eruption column. (USGS: Photo by Austin Post)

Figure 3.6 Quiescent lava eruption. Lobes of lava advance quietly across an earlier lava flow. This basalt eruption took place during 1999 on the lower flanks of Kilauea volcano, and was accompanied by only minor ash and gas release (see also Figure 3.5). (Steve Self/Open University)

You have already encountered the different sources of heat that are likely to exist in planetary bodies in Sections 2.4.1–2.4.4.

QUESTION 3.1

(a) Given what you have already learned in Chapter 2, discuss the sources of heat that are likely to generate heat within:

(i) a young Earth-like planet with a stabilized crust, fully formed core, and convecting mantle.

(ii) an ancient, rocky planet half the size of Earth with a large solid core, and greatly thickened lithosphere.

(iii) a young satellite 3000 km in diameter, with a very small rocky core and a mantle and crust consisting predominantly of water-ice. This satellite is in a close elliptical orbit around a giant planet and displays recent evidence of widespread resurfacing.

(b) Which example (i, ii or iii) is least likely to exhibit volcanic activity, and which is most likely to exhibit volcanic activity? Give reasons for your answers.

3.1.2 Melting by other methods

In Question 3.1 we considered different sources of heat and the manner by which they could lead to melting and volcanism. You may be surprised to learn that two quite different processes, other than temperature rise, are largely responsible for magma generation within the Earth. These two processes are **decompression melting** and **hydration-induced melting** and occur within the Earth's mantle (Box 3.1).

BOX 3.1 WHAT CAUSES EARTH'S MANTLE TO MELT?

Interestingly, an increase in temperature is not the only, or even the main, cause of melting in Earth's upper mantle. There are two other key processes: decompression melting, and hydration-induced melting.

You may recall from Section 2.2.1 that the pressure on rocks becomes greater with increasing depth. As a result of this, the constituent minerals adjust structurally and compositionally towards more compact configurations that are stable at these increased pressures. Consequently, even greater temperatures are required to melt these high-pressure minerals and increasingly higher temperatures are required to initiate melting at increasingly greater depths. This is the reason why the Earth's mantle is solid even though the temperature increases with depth. The converse of this relationship is that any solid mantle rising upwards sufficiently fast enough to avoid significant heat loss can begin to melt simply because it reaches a depth where the pressure is insufficient to keep it solid. This process is known as decompression melting.

Decompression melting is the major process by which melts of a basaltic composition are generated in the Earth's mantle. It is important because it does not require any additional heating of the rock to create a magma. But how does decompression occur in the mantle? It occurs in two circumstances due to 'rising' mantle masses. First, it occurs as a consequence of solid-state convection within the mantle (Figure 3.8a), or as a result of mantle plumes that carry anomalously hot material surfacewards, creating hot-spot volcanism (Section 2.4.5). Second, it can occur as a result of thinning of the overlying crust which, for example, takes place due to lithospheric stretching during continental break-up (Figure 3.8b). In both instances it is the ascent and associated decompression of the mantle material and *not* the addition of heat that triggers melting, the generation of magma, and associated volcanism.

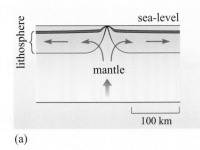

(a)

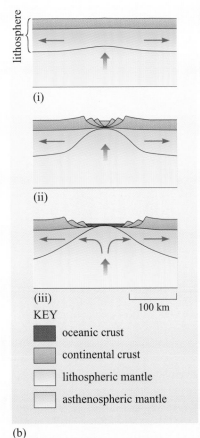

(b)

Figure 3.8 (a) Cross-section through a constructive plate boundary where new material is being added to oceanic crust. Upwelling of mantle is associated with solid-state convection within the asthenosphere. Decompression of upwelling asthenospheric mantle produces a basaltic melt which is erupted at the mid-ocean ridge as **mid-ocean ridge basalt (MORB)**, adding to the oceanic crust. (Thick arrows in a and b depict mantle upwelling, thinner arrows indicate divergent plate motion.) (b) A time-series of cross-sections depicting divergent motion within a continent, and the associated crustal thinning that occurs due to lithospheric stretching during continental break-up. In (i) and (ii), stretching and thinning of the continental crust results in mantle upwelling within the asthenosphere. This mantle upwelling causes decompression of the upper mantle, the generation of melt, and associated volcanic eruptions. By stage (iii) the continent has rifted apart and a new ocean (such as the Red Sea between Arabia and Africa) begins to open between the two rifted fragments.

The second key process that can lead to melting of the mantle occurs when volatiles such as hydrous (water-rich) or gas-rich fluids are added to the mantle and bring about reactions that alter the constituent minerals. When hydration or similar volatile-induced alteration of the mantle mineral assemblage occurs, it reduces the melting point of that assemblage. In other words, mantle altered in this manner begins to melt at lower temperatures and pressures than unaffected, or anhydrous, mantle. This is the process of hydration-induced melting.

Hydration of mantle peridotite occurs as a consequence of plate tectonic recycling (Sections 2.4.5 and 2.5.1). It takes place at depth within subduction zones where old, water-rich oceanic crust is reabsorbed into the mantle (Figure 3.9a). The resulting generation and surfaceward migration of magma produces arcuate belts of volcanoes, known as **volcanic arcs** (Figure 3.9b).

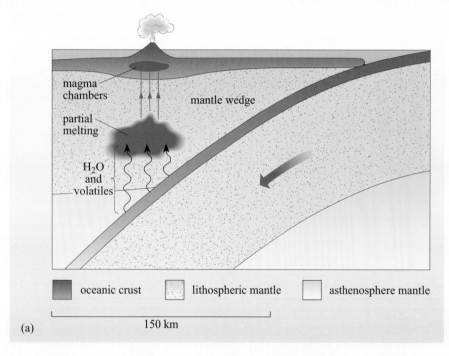

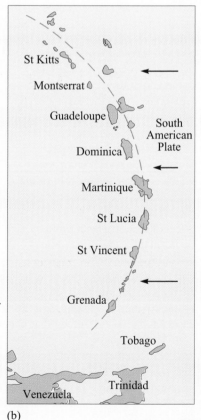

Figure 3.9 (a) Cross-section through a destructive plate boundary where old oceanic-crust material is being recycled back into the mantle during subduction. Heating of subducted material during its descent releases hydrous or gas-rich fluids which then rise and are added to the mantle wedge of the over-riding plate. These released fluids cause hydration reactions that reduce the melting point of the mantle-wedge mineral assemblage, so producing magmas. The volcanoes associated with this magmatism are typically arranged in arcuate belts (volcanic arcs) such as those of the Andes, Central America, and the Lesser Antilles of the Caribbean. (b) Sketch map of the volcanically generated islands of the Lesser Antilles that together form the volcanic arc of the eastern Caribbean. (Arrows show the relative movement of the South American Plate, which is presently being subducted beneath the Caribbean Plate.)

It is very important to realize that lithospheric recycling involving the creation of material at spreading ridges (constructive plate boundaries), and its re-assimilation at subduction zones (destructive plate boundaries), has *not* been identified on the other terrestrial planets or terrestrial-like bodies. Therefore, the generation of magma by hydration reactions, and associated volcanic arcs, may be a feature unique to Earth. However, by contrast, the rise of mantle material in response to plumes initiated by internal heat loss may have been, and in some instances may continue to be, a major process by which basaltic volcanism is generated on terrestrial planets.

3.1.3 The concepts of partial melting

Following the brief discussion of melting processes, we now turn to the second fundamental prerequisite: the nature of the solid materials that can be melted to produce molten magmas. Remember that it is the surfaceward movement of materials and the eruption of magma and associated volatiles that eventually create surface volcanism.

Before we can investigate the different forms of volcanism, it is first necessary to examine the types of raw materials likely to be involved in the melting process, since these exert fundamental controls on the type of volcanism eventually produced. Clearly, the bulk composition of a planetary body and the nature of its layering will play a key role in determining the materials available for melting (Section 2.2.3). In the case of the terrestrial planets, their bulk composition is approximately chondritic since they represent condensation products of the more refractory elements and compounds in the inner part of the solar nebula. By contrast, some of the satellites of the giant planets consist of a significant proportion of water-ice and, with increasing remoteness from the Sun, progressively more volatile compounds such as ammonia and methane. Thus volcanism within the Solar System involves the melting of two different types of raw material: the high-temperature melting of silicates characteristic of the terrestrial planets; and the low-temperature melting of ices, resulting in the cryovolcanism characteristic of some of the satellites of Jupiter, Saturn, Uranus and Neptune. The concept of partial melting was outlined in Section 2.3.2.

> Partial melting is the incomplete melting of a parent material that characteristically produces a melt whose chemical composition is different to that of its parent.

When melting begins in rocks consisting of a mixture of silicate minerals, or in ices consisting of combinations of water-ice and other frozen volatiles, the minerals or constituent ices with a lower melting point naturally begin to melt first. This initial melt often occurs and collects at the interstices along the inter-crystalline boundaries between adjacent crystals (Figure 3.10a and b). However, at these initial very low degrees of partial melting, the melted material cannot migrate since it exists as a series of discrete, minute blobs. Only with further partial melting do these blobs begin to join up into a series of interconnecting networks along crystal boundaries, which eventually allow the partial melt to begin to migrate. The point at which interconnection of melt blobs occurs is a critical point of the melting process since only once the network of micro-conduits have been linked together can the

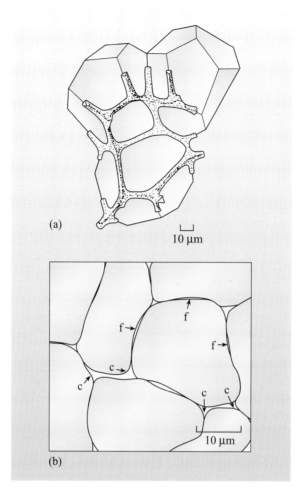

(a)

10 μm

(b)

10 μm

Figure 3.10 (a) Sketch showing a three-dimensional view of adjacent crystals in a rock in which partial melting has begun. The melt initially forms in the pockets (interstices) where three crystals meet and, with further melting, the pockets then become connected by channels of melt running along the grain boundaries. Once interconnected, the melt can then begin to migrate and coalesce into a larger body of magma. (b) Magnified image of a rock beginning to melt (melt: yellow, crystals: blue). Features labelled 'c' are interconnecting melt channels at crystal interstices and boundaries; features labelled 'f' are thinner films of melt coating grain boundaries that have not yet sufficiently developed to form connections.

liberated melt begin to migrate away. Experimentally, this is normally found to occur when the melt-filled grains occupy about 5% of the source rock volume, at which point the pressure of overburden from the layers above causes the melt to move from regions of higher pressure to those of lower pressure conditions. In the case of partial melting within planetary bodies, this high pressure to low pressure migration is typically manifest as a surfaceward movement. Where these migrating magmas breach the surface marks the point at which volcanism occurs.

3.1.4 Magma generation in silicate bodies – peridotite to basalt alchemy

Considering the terrestrial planets as a whole, the most commonly erupted material is broadly basaltic in composition. You may remember from Section 2.2.3 that a typical basalt contains about 50% SiO_2 and that such rocks are the main component of Earth's oceanic crust. Basaltic oceanic crust covers two-thirds of our planet and is continually being produced along the mid-ocean ridges at a global rate of about 20 km^3 per year, as part of Earth's on-going tectonic activity. Similarly, examination of materials and images from the Moon, Mars, Venus and Mercury suggests that basalt lavas are a dominant product of past volcanic eruptions (Table 3.1). So, why should a basaltic composition be such a common volcanic product on these different worlds? To answer this, we need to investigate the composition of the source materials that give rise to basaltic magmas.

Clearly, basaltic compositions are generated from materials that were originally broadly chondritic in composition since this is the bulk composition of terrestrial planets. In other words, it is the chondritic composition of the terrestrial planets that places a fundamental control upon the magmas generated and the volcanic products erupted at the surface. However, if a rock of broadly chondritic composition was completely melted, transferred surfacewards and erupted as lava at a volcano, the bulk composition of the solidified lava would be exactly the same as that of the starting material (i.e. chondritic and with element abundances similar to that found in meteorites). Yet, as shown in Section 2.2.1, basalt is enriched in certain elements (e.g. Si, Al, Ca, Na and K) compared to primitive chondrite composition, and depleted in others (e.g. Fe, Mg, Ni and Cr). To understand how this change of composition takes place, it is necessary to describe the process of partial melting in more detail.

You have already determined that on Earth the mantle rocks (i.e. peridotite) are more chondrite-like than crustal rocks since they show lower degrees of element enrichment and depletion (Question 2.3). However, it has also been argued that Earth's mantle itself differs from primordial chondritic composition largely because some elements, predominantly Fe and Ni, have been extracted from the overlying mantle rocks since they have sunk to form the core. It has also been shown that much of the partial melting that generates Earth's basaltic volcanism occurs in the upper mantle, which is of peridotite composition. Therefore, the fact that the terrestrial planets have similarly experienced core- and lithosphere-forming events early in their history may aid in explaining why basaltic compositions are so widespread on their surfaces. Accordingly, it is logical to use the example of Earth's mantle to examine how materials of broadly peridotite composition give rise to basaltic volcanism at the surface.

When the mixtures of minerals that comprise peridotite begin to melt, some, such as plagioclase, start melting at lower temperatures than olivine and pyroxene (Table 2.1). In this state, the original rocks can be said to be 'partially melted'. If this partial melt is then extracted before any further melting occurs, the extracted melt will have a different composition to that of the bulk composition of the original parent rock. This effect is further reinforced by the fact that some elements within the remaining unmelted material also become mobilized and preferentially escape into the melt. These elements are said to be **incompatible** with the source rock since they 'prefer' to exist within the surrounding melted material. This process is known as element partitioning. With further melting, minerals with an increasingly higher melting point become molten, and progressively less of the incompatible elements become partitioned into the melt. By contrast **compatible** elements tend to remain in the solid residue in minerals that are stable at high temperatures. Much of the mobilization and partitioning into the melt of the more incompatible elements occurs during the initial stages of the melting process. The resulting small degrees of partial melt will, therefore, be the most highly enriched in incompatible elements and so the most different in their composition to that of the starting or 'parent' material. Increasing degrees of partial melting of the same source rock can give rise to quite different magma compositions because, depending upon the degree of incompatibility, different proportions of elements will exist within each increasing increment of the melting process. This is the manner in which the composition of the melt will evolve. Accordingly, those magmas representing very large degrees of partial melting will display an increasing compositional similarity to that of the parent since, by this stage, more of the original parent material will have been liberated to form the melt.

In the case of silicate minerals, like those comprising peridotite, incompatible elements such as K and Na are most readily mobilized into initial melt products (e.g. 1–2% partial melting), whilst compatible elements like Mg tend to remain in the solid residue of minerals that are stable at higher temperatures (e.g. olivine and pyroxene). After 25–30% partial melting of peridotite, the proportions of elements in the melt will have evolved significantly because more elements (e.g. Ca, Al, Ti and to a lesser degree Fe and Si), will have joined the more incompatible elements like K and Na. However, the element abundances in these larger degrees of melting are still very different to those of the parent peridotite and, by the time 25–30% partial melting is reached, will be broadly basaltic in composition. This process can be summarized as follows:

Mantle peridotite (*partial melting*) $\longrightarrow$ basalt magma + solid unmelted residue

Given the relationship between peridotite partial melting and the basaltic magmas outlined above, it was thought for many years that basalts which were erupted at Earth's ocean ridges (i.e. mid-ocean ridge basalts (MORBs)) represented compositions derived *directly* from the partial melting of the mantle. This is now known to be incorrect since other important modifying processes are known to act upon the basaltic magma during its ascent to the surface. These include the crystallization and separation of mineral fractions that form due to cooling during ascent, and the assimilation of minerals and elements from the crustal material through which the basaltic magma passes. Laboratory experiments show that basaltic magmas derived from a peridotite parent, without any such modification (i.e. experimental 'primitive basalt' in Table 3.1) should be characterized by a higher Mg and lower Si content than typical oceanic basalt. In fact, basalts approaching this 'primitive' composition are known to have been erupted on Earth in its distant past. Unfortunately, due to active erosion on our planet, few examples have been preserved except for some rare occurrences on very ancient continental crust. These types of lavas are known as **komatiites** (pronounced komaty-ites), after a locality exposed in the Komati river valley in South Africa. It is thought that komatiites were erupted only during Earth's early history because Earth's internal heat budget was higher then than today (Section 2.4.3), so allowing the generation of larger degrees of partial melting at greater depths and at higher temperatures.

QUESTION 3.2

(a) Does the chemical composition of lunar and Martian basalts (Table 3.1) follow the pattern of element-partioning behaviour expected during the partial melting of peridotite? What does this indicate regarding the source of basaltic magmas on these two bodies? (Hint: to determine the behaviour of the different elements, compare the patterns of element enrichment and depletion shown in the completed peridotite and oceanic basalt spidergrams in the answer to Question 2.3b.)

(b) Which of the analyses listed in Table 3.1 are most like peridotite mantle (i.e. most 'primitive') and which have the most MORB-like composition? Briefly account for these differences.

Table 3.1 Compositions of rocks from the Earth's mantle and other important basalt types.

Weight % oxide	Peridotite (mantle)	Primitive basalt (experimental)	Oceanic basalt (MORB)	Deccan (CFBP[a]) basalt	Komatiite (ancient terrestrial lava)	Lunar mare basalt	Martian basalt[b]
SiO_2	45.4	44.2	48.2	48.8	45.2	42.7	50.1
TiO_2	0.2	3.7	1.4	2.5	0.2	2.6	0.9
Al_2O_3	4.3	12.1	15.4	13.7	3.7	7.7	6.7
Fe_2O_3	8.2	12.1	10.8	14.8	12.2	23.6	20.8
MnO	0.1	0.2	0.2	0.2	0.2	0.3	0.5
MgO	38.2	13.1	9.6	6.2	32.2	14.6	9.4
CaO	3.4	10.1	11.1	10.6	5.3	8.1	10.0
Na_2O	0.3	3.6	3.4	2.4	0.4	0.2	1.3
K_2O	0.03	1.30	0.07	0.30	0.10	0.10	0.20
Cr_2O_3	0.43	0.1	0.05	0.02	0.22	0.25	0.20
Totals	100.6	100.5	100.2	99.5	99.7	100.2	100.1

[a] CFBP = continental flood basalt provinces.

[b] analytical data obtained from a meteorite derived from Mars (Shergotty meteorite).

Basalts with compositions similar to komatiites are also believed to be widespread in the ancient volcanic terrains of the terrestrial planets. The ubiquity of such basaltic compositions arises because they represent the products of melting peridotite mantle, which itself was originally derived from an initial chondritic composition. Since the terrestrial planets were created from the chondritic materials that condensed in the inner region of the solar nebula and, because they are thought to have experienced comparable core-forming events, the bulk composition of their respective silicate mantles is likely to be broadly similar. In their early histories there would have been sufficient remaining primordial heat, together with a much greater input from juvenile radioactive decay, to permit greater large degrees of partial melting at greater depths and higher temperatures. Such circumstances would have provided the necessary conditions for the generation and eruption of komatiitic lavas. Later on, once the primordial and radiogenic heat sources had begun to wane over time and lithospheres had become thicker, the conditions and degree of mantle melting would have also altered giving rise to less primitive, more MORB-like basaltic compositions.

▨ Why might komatiitic lavas be preserved in the volcanic terrains of other terrestrial planets whilst similar occurrences on Earth are so rare?

❑ Given that eruptions of komatiitic compositions were most probable whilst silicate planetary bodies were still relatively young, it should be no surprise that few have been preserved on Earth due to constant tectonic recycling and the processes of weathering and erosion. By contrast, on the other planets the effectiveness of these processes may be much reduced or even absent, permitting long-term preservation of ancient-surface elements. However, tectonic, volcanic and surface processes affecting the surfaces of Mars and Venus may have reduced their preservation potential.

3.1.5 Non-silicate partial melting and the generation of icy magmas

Partial melting of ices provides the magma source that produces volcanism on icy bodies.

■ Is icy magmatism likely to differ significantly in the manner in which it is generated, and the nature of its eruption products, from the largely silicate-dominated magmatism of the terrestrial planets?

❏ Not really, because although icy volcanism sounds exotic, both silicate and icy magmatism require something to melt and a process to cause melting. Thereafter the same physical processes regarding partial melting, the migration of magma and eruption dynamics will apply.

Ices, which are mixtures of water and more volatile compounds such as ammonia or methane, melt at lower temperatures than pure water-ice, in fact about 100 K lower for a mixture of ammonia and water. This mixing effect is analogous to adding salt to water-ice and, incidentally, is the reason why salt is used to melt ice on road surfaces during freezing weather. Partial melting of a frozen water–ammonia mixture will yield an initial melt that will be much richer in ammonia and so quite different in its composition to that of the original mixture. This can be considered analogous to the manner in which basaltic magmas are generated from a peridotite parent.

Since most icy satellites typically contain little rocky material, internally generated radiogenic heat will be negligible, and therefore tidally heated melting within the outer layers of these bodies is the only conceivable way of generating magma. However, in these instances, the amount of tidally generated heat required to bring the ice to its melting point may be relatively small. For instance, if at depth the internal temperature of a small icy body is only 100 K (−173 °C), but the melting point of ice mixtures at this depth begins at 180 K (−93 °C), then a rise of only 80 K would be enough to initiate partial melting. However, as previously discussed (Box 3.1), decompression melting plays an important role on Earth, and could also occur within the mantle of icy bodies as a result of solid-state convection arising from tidally generated heat.

3.2 Styles of volcanism on Earth

The nature and style of volcanism on Earth vary greatly, and range from relatively quiescent lava eruptions with relatively little or no volcanic ash, to mighty explosions that create immense clouds of hazardous ash and gas that can profoundly affect the environment. Volcanologists recognize these two contrasting styles of activity as **effusive volcanism** and **explosive volcanism** (Figures 3.11 and 3.12).

Effusive volcanism is typically a quiet affair since it is characterized by lava emanating from a vent or fissure and then spreading out over the landscape. By contrast, explosive volcanism typically produces fragmented debris or **pyroclastic materials** (from the Greek, meaning 'fire-broken'), which largely comprise ash and other ejected fragments.

(a)

(b)

Figure 3.11 Effusive eruptions. (a) Basalt lava oozes from a vent on Etna. (b) Fire fountain during the eruption of basaltic lava at Pu'u'O'o, Hawaii. ((a) Steve Self/Open University; (b) USGS: Photo by J. D. Griggs)

Figure 3.12 Explosive eruption. View of the eruption column from Mount St Helens, May 1980. The prevailing wind was blowing from right to left, and slightly away from the viewpoint. Much of the ash was deposited in the region downwind of the volcano. (USGS)

The way in which magma is generated and makes its way to the surface, together with the style of the resulting volcanism, are controlled by a range of physical factors. These factors will be discussed in the following sections.

It is important to note that the relative importance of lava and pyroclastic materials varies according to the degree of explosivity of the eruption. We can quantify the degree of explosivity by assigning a volcanic explosivity index (VEI) (Table 3.2).

Large, extremely violent and voluminous pyroclastic eruptions have been associated with some of the major natural catastrophes in Earth's history. For instance, the eruption of Vesuvius in AD 73 buried the Roman city of Pompeii, whilst the effects of the 1620 BC explosion of the volcanic island of Santorini in the Aegean are thought to have destroyed the flourishing Minoan civilization. More recently, the eruption of Krakatau (Indonesia, 1883) as well as more modern examples such as Mount Pinatubo and Mount St Helens mentioned earlier (Section 3.1.1) have also had wide-ranging effects on communities and measurable effects upon global climate (Table 3.3).

Table 3.2 Relative sizes of terrestrial volcanic eruptions and the volcanic explosivity index (VEI) showing the importance of pyroclastic materials according to the degree of explosivity of the eruption.

| | Volcanic explosivity index (VEI) | | | | | | | | |
	0	1	2	3	4	5	6	7	8
general description	non-explosive	small	moderate	moderate –large	large	very large			
qualitative description	gentle	effusive	← explosive →		← cataclysmic or paroxysmal →		← super-eruptions →		
maximum erupted pyroclastic volume (m³)	10^4	10^6	10^7	10^8	10^9 (1 km³)	10^{10} (10 km³)	10^{11} (100 km³)	10^{12} (1000 km³)	10^{13}
eruption cloud column height (km)	<0.1	0.1–1	1–5	3–15	10–25	>25	>25	>25	>25

Table 3.3 Eruption examples: volcanic explosivity index (VEI) and eruption column height.

VEI rating	Eruption column height (km)	Example
0	<0.1	basalt lava eruptions, e.g. Hawaii and Iceland
1	0.1–1	fire fountains in basaltic lavas, e.g. Hawaii and Iceland
2	1–5	Hekla, Iceland, 2000
3	3–15	El Chichon, Mexico, 1982
4	10–25	Mount St Helens, USA, 1980
5	>25	Krakatau, Indonesia, 1883; Pinatubo, Philippines, 1991
6	>25	Bishop Tuff pyroclastic deposit, Long Valley caldera, California, USA, 700 000 years ago
7	>25	Toba, Lake Toba caldera, Sumatra, Indonesia, 75 000 years ago
8	>25	Yellowstone, USA, 2 Ma, 1.3 Ma and 640 000 years ago

3.2.1 Effusive volcanism and lava flow dynamics

When magma reaches the surface and flows across the surface to form a channel or a sheet it is called a **lava flow**. Erupted lavas are typically liquids derived from partial melting at depth. However, they can also contain crystals, which form during cooling as the magma rises, and dissolved gases, which begin to come out of solution with the reduction of pressure so forming bubbles once the magma nears the surface.

Differences in magma composition and gas content are crucially important in controlling the explosivity of the resulting volcanism, because it is these factors that fundamentally influence magma viscosity.

The viscosity of a substance is its internal resistance to flow when a stress is applied to it.

A simple example is when gravitational force exerts a stress upon a liquid (e.g. a lava) lying on an inclined surface, in which case the liquid begins to flow downslope. In fact, two forces control the rate at which the blob of lava can spread: gravity and viscous resistance. In addition to the viscous resistance of the liquid, molten lavas also possess a **yield strength** partly derived from the chilled crust that forms on their surfaces. Before the blob can spread, the yield strength must be exceeded by forces which are acting to make it flow (e.g. gravity). Since all lavas have a yield strength, they never form thin films and always have an appreciable thickness.

Consequently, whether a lava will flow readily at the surface of a planetary body depends upon factors such as the value of gravity, and surface and eruption temperatures, which will control the cooling rate of the lava. In addition, magmas with a higher SiO_2 content tend to have a greater viscosity. Basalt magmas with a SiO_2 content of 48–52% produce particularly fluid lavas, though even these are less fluid than komatiitic lavas. **Rhyolite** magmas containing in excess of 70% SiO_2 are amongst the most viscous (Figure 3.13). Moreover, the greater the silicate content, the lower the melting point of the magma. In other words, a typical viscous rhyolite can be erupted at temperatures as low as 700 °C, whilst a fluid basalt lava can only be molten at temperatures above 1050 °C. Lavas, in common with many other fluids such as motor oil, become more runny (i.e. their viscosity decreases) with increasing temperature. Conversely, they will only remain runny if they continue to remain at temperatures well above their melting point (Figure 3.14). In other words, a magma erupted well above its melting point is more likely to produce a lava flow than one erupted at only a few degrees above its melting point since the latter will freeze before it can travel far. In the first instance, a thin, laterally extensive lava flow would be formed because the magma can spread outwards from the vent, whilst a taller mound or dome-like construction might be created around the vent in the second instance. Surface temperature will also play an important role since, if the surface conditions of a planet are significantly cooler than the temperature of erupting lava, it will be less likely to flow far due to rapid chilling.

Figure 3.13 Dome-like features within the Mount St Helens summit crater produced by viscous rhyolite lavas erupted at about 700 °C. This type of viscous lava can flow only a short distance before freezing, so creating either thick flows or dome-like constructions in close proximity to the vent. This smouldering lava dome is about 250 m high and the crater wall in the background is 600 m high. (Steve Self/Open University)

Figure 3.14 Low viscosity basalt lava oozing out and advancing as a series of 'toes', which build and coalesce to form laterally extensive lava-flow sheets. (Steve Self/Open University)

Eruption rate is another factor known to affect the pattern and structure of lava accumulations. Where magma rises and oozes slowly onto the surface, it rapidly chills and freezes. By comparison, at high rates of eruption the supply of magma is so rapid that chilling cannot occur efficiently, and the lavas can flow greater distances and cover a wider area. For many years, it was thought that eruption rate was the major control upon the distance lava flows could travel. This simple relationship initially seemed an adequate explanation for the extent and dimensions of the relatively small lava flows of $0.1–10\,km^3$, typical of Hawaii and Iceland, since it was observed that when eruption was more rapid, the resulting flows travelled further. However, much larger flows ($1000–1500\,km^3$) are known to have erupted on Earth during the formation of continental flood basalt provinces (CFBPs) such as the Deccan (Box 3.2), which was erupted in northwest peninsular India between 67 million and 64 million years ago, and the Colombia River basalt province of North America which was erupted 14–15 million years ago (Figures 3.15 and 3.16). Some of the flows that comprise these CFBPs can be traced for hundreds of kilometres, but to achieve flows of these huge dimensions using the eruption-rate argument would require colossal rates of eruption. In fact, the eruption rates calculated for these examples were so large that many volcanologists began to consider them geologically improbable.

An alternative model to explain these huge flows was proposed by the volcanologist Steve Self (Open University). Whilst observing the formation and development of Hawaiian lava flows, Self noted that within a relatively short distance of the eruption vent most basalt lavas quickly formed a flexible, chilled, solid crust. However, the lava beneath this crust continued to flow for considerable distances, and progressed by a series of breakouts at the edges of a crusted-over sheet (Figure 3.14). In effect, the crust had formed an insulating layer between the chilling surface and atmospheric conditions above, and the hot magma supplied beneath. Once the crust had fully formed, it could drastically reduce heat loss – possibly to less than 1.5 K for every kilometre of distance flowed. In this manner, huge flows could be created gradually over decades, their crusted surfaces allowing the molten lava beneath to be transported tens or even hundreds of kilometres from the eruptive source via **lava tubes**. Moreover, this model allows the flexible, crusted-over flow to undergo **lava inflation** from within. A basalt flow may begin at its tip as only a few centimetres thick, but with a continuing supply of lava it can, over time, inflate to a thickness of several tens of metres (Figure 3.17).

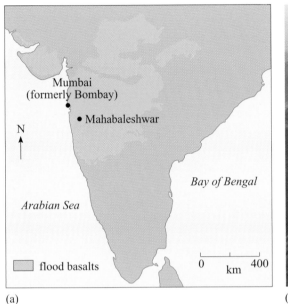

(a)

(b)

Figure 3.15 (a) The areal extent of the Deccan Traps continental flood basalt province (CFBP). Much of northwest peninsular India is covered by these 64–67 million-year-old basalt lavas, which reach a maximum thickness of 2.5 km inland of Mumbai. (b) Panoramic view across approximately 1 km thickness of Deccan lava flows, Elphinstone Point, Western Ghats, Mahabaleshwar. ((b) Mike Widdowson/Open University)

(a)

(b)

Figure 3.16 (a) The Columbia River continental flood basalt province, covering parts of Washington, Oregon and Idaho. Much of this province was erupted between 14 million and 17 million years ago, and contains some of the largest flows and flow fields yet identified on Earth. The deep purple area defines the source and extent of the 14 million-year-old Pomoma flow that can be traced for over 550 km from its source. (b) Layers of stacked lava flows in the Columbia River province of the Columbia River. The flows shown represent just a small thickness of the voluminous and rapidly erupted Grande Ronde Formation part of the succession. ((b) Steve Self/Open University)

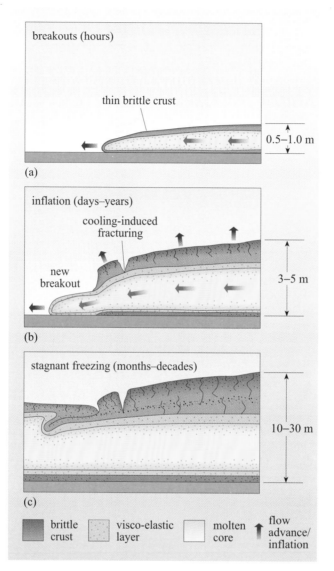

Fig. 3.17 Schematic cross-sections of emplacement and inflation of a CFBP lava flow. (a) Flow arrives as a small, slow-moving lobe of molten lava held inside a stretchable, chilled elastic skin with brittle crust on top. Bubbles become trapped in both the upper and basal crusts. (b) Continued injection of lava into the lobe results in inflation (lifting of the upper crust) and new breakouts around its margin. The growth of the lower crust is much slower. Relatively rapid cooling and motion during inflation results in irregular jointing in the upper crust. (c) Stagnation and slow cooling of the stationary liquid core.

BOX 3.2 FLOOD BASALTS

In addition to basaltic rocks of the ocean crust forming 70% of the Earth's surface area, immense areas of basalt lava can also be found covering large areas of continental crust. Regions of such accumulations are known as **continental flood basalt provinces** (CFBPs), and represent Earth's largest outpourings of lava, producing volumes of up to 2 million cubic kilometres over periods of 1–5 million years. This represents annual eruption rates at least 20 times greater than those observed for a typical modern hot spot such as Hawaii. Particularly well known CFBP examples include the Columbia River basalt province of northwestern USA, and the Deccan Traps of northwest peninsular India.

In a CFBP, individual lava flows can be 20–50 m thick and extend tens, or even hundreds, of kilometres. Many such flows merge to form a single 'flow field' that can cover thousands of square kilometres (Figure 3.18). Each lava field represents a major volcanic episode and, if accompanied by ash and gas release, could have had environmentally devastating effects. In a typical CFBP tens or hundreds of such lava fields are stacked one upon another to a thickness of 1–3 km. Fortunately, the eruption of CFBPs are rare events because only 10 or so such episodes have occurred in the last 250 million

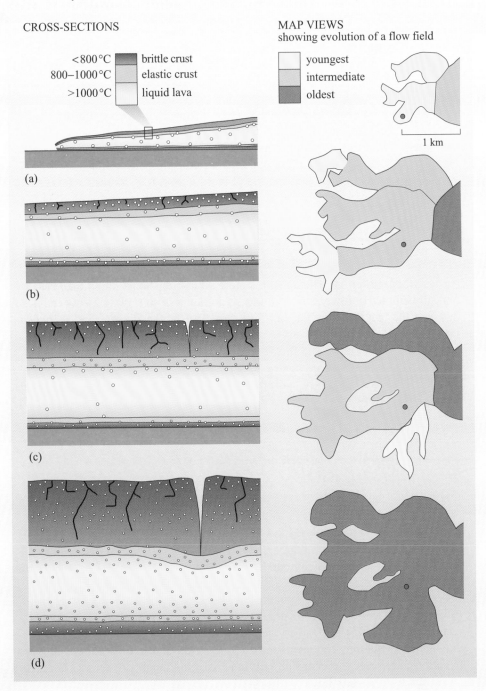

Figure 3.18 Development of an inflated basaltic sheet flow and the evolution of a flow field. On the left are cross-sections at a fixed location, and on the right are concurrent map views of the development of the flow field. The map location of the cross-section is indicated by the red dot. The emplacement of a sheet lobe can take anywhere between days and years. In the cross-sections the shading darkens with cooling, and in the map view it darkens with age. (a) A new lobe advances from right to left. Incandescent lava is exposed only at the very tip of the flow. (b) The lobe thickens by inflation as it extends. Bubbles from the moving lava are trapped in the crust. The lower crust grows much more slowly than the upper crust. (c) Inflation continues. Some cracks in the upper crust become major clefts. (d) Flow stagnates, and cooling is enhanced around deep clefts. Note the complexity of a typical flow field.

years. Nevertheless, the effects of larger ones, such as the Deccan Traps and Siberian Traps, may be linked with major environmental changes 65 million and 248 million years ago, respectively.

These massive outpourings of basalt are thought to be largely the result of decompression melting of a huge mantle plume. Plumes giving rise to CFBPs are much larger than those that currently generate volcanism at hot spots such as Hawaii and Iceland. CFBP plumes are thought to originate as hot and hence unstable masses very deep within the mantle. They rise over millennia towards the surface because of their elevated temperature. However, their development and surfaceward movement appears to be independent of the wider patterns of convection within the mantle. Having risen to upper-mantle levels, the elevated temperatures within the head of these plumes causes the overlying lithosphere to heat up, weaken and thin. In regions where the overlying lithosphere consists of continental crust, this weakening and thinning may eventually result in the crust breaking apart. Where this breaking apart or **rifting** of the overlying crust occurs (Figure 3.8b), it allows the mantle beneath to rise further and so decompress. It is this decompression that produces the vast amounts of basaltic magma which finds its way up through the weakened or rifted lithosphere, and which then erupts to form huge sheets of basaltic lava that comprise the CFBP. This type of paroxysmal basaltic volcanism and associated outpouring of basaltic lavas may provide a possible analogue to major lava resurfacing events that have occurred on the other terrestrial planets.

As described earlier, lava flows with a more silica-rich composition than basalt are typically erupted at lower temperatures so do not cool so rapidly. However, despite this, and the fact that they too can form thick crusts, their greater viscosity usually means they do not flow for very great distances. Instead, rhyolite lavas commonly have clinkery or boulder and rock-strewn surfaces of solidified material. They often build to a thickness of many metres within short distances from the eruption source and, in some instances, these more viscous lava eruptions may form high domes rather than laterally extensive lava flows typical of basaltic eruptions (Figure 3.13).

3.2.2 Explosive volcanism: eruption columns and pyroclastic eruptions

Explosive volcanism is largely the result of two main variables: the viscosity of the magma and the amount of gas it contains. Basaltic lavas have low viscosity (as discussed above) and relatively low gas contents of about 1% by mass. Both factors work together to produce relatively quiescent, effusive basaltic eruptions (Figure 3.11a). Where gas contents are higher in basaltic magmas, **fire fountains** may occur casting lava upwards for several tens to hundreds of metres. Spectacular examples of fire fountains occur from time to time on both Hawaii (Figure 3.11b) and Iceland. Nevertheless, basalt fire fountains are small affairs compared with major explosive volcanoes. In these instances, the expanding magmatic gas not only fragments the erupting lava but also drives it upwards into the atmosphere creating **eruption columns** reaching heights of tens of kilometres above the vent (Figure 3.12).

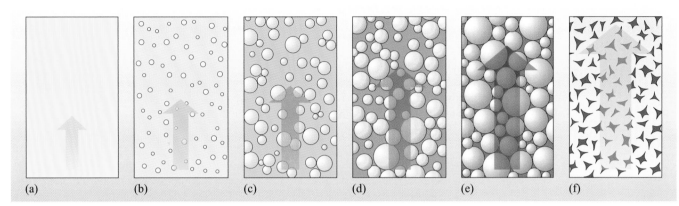

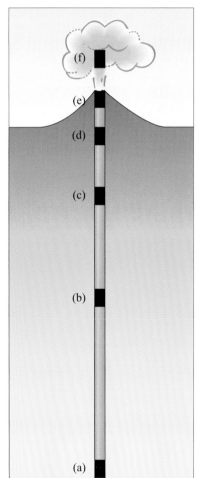

Figure 3.19 Effects of degassing in a volcanic conduit: (a–b) as the magma begins to ascend in the conduit bubbles begin to form, so enhancing the rise of the magma; (b–d) with continued ascent in the conduit additional decompression allows a greater amount of degassing, and so leads to the formation of more bubbles which then coalesce into larger bubbles and further increase the buoyancy; (e) as the rising magma accelerates in the conduit, large amounts of degassing, or else rapid degassing within more viscous magmas, can result in fragmentation and the production of pyroclastic materials (f). Stages (a), (b) and (c) are most typical of the degassing characteristics of less viscous basaltic lavas, or else gas-poor magmas, whereas stages (d), (e) and (f) are more typical of the degassing of gas-rich, or highly viscous magmas such as rhyolite.

So, volcanologically speaking, what are the differences between these styles of eruption? To understand this, it is necessary to consider what happens as the magma approaches the surface. The ascent of magma is driven partly by buoyancy because partial melting typically produces liquids of different composition and lower density than the source, and partly by pressure because these liquids will flow towards the surface away from the higher pressures at depth. Buoyancy is further augmented because the magma will be hotter than the surrounding rocks as it rises surfacewards. Also, during magma ascent, pressure is reduced allowing any gases dissolved in the magma to expand and form bubbles (Figure 3.19). This bubble formation serves to make the magma even less dense and so more buoyant, thus further accelerating its ascent. Bubbles continue to expand as they rise in response to further decrease in confining pressure until they reach the surface where the gases escape into the atmosphere. If the magma is not viscous, then bubble escape may take place in an uninhibited, more gentle fashion, producing lavas with preserved bubbles, or causing fire fountaining in those cases where the gas content is higher.

However, if the lava is viscous, such as that arising from magmas with a higher silica content, this expansion and release of gas cannot take place as easily. As a result, the magma reaches the surface containing highly pressurized bubbles which have been unable to expand fully and escape during magma ascent. Once beyond the confines of the lava conduit, the gases within the erupting lava expand rapidly creating an explosion. An excellent analogue to compare gentle and rapid escape of dissolved gas, is when opening a new bottle of fizzy drink. If opened slowly, the gas has time to escape and the drink can be poured without spillage; if opened quickly, the pressure release causes bubbles to form very rapidly producing a froth which then explodes from the bottle spout (Figure 3.19d–e). In a similar fashion, magma may also froth during gas release producing a low-density material called **pumice** when it freezes.

Pyroclastic eruptions

Violent volcanic eruptions are much more common in volcanoes with more viscous and gas-rich magmas. The explosions they create literally blow apart the ascending magma, creating semi-molten clots of lava known as **volcanic bombs**, together with pumice, **spatter** and ash. These are all pyroclastic materials, since they are produced due to fragmentation of the magma due to gas expansion. The larger fragments thrown into the air cascade surfacewards around the vent, creating volcanic cones, whilst the finer ash material can be carried high into the atmosphere and, depending upon wind direction and the height to which they are thrown, return to the surface many tens, hundreds or even thousands of kilometres from the vent. Where this ash is preserved on land or in ocean sediments, it is given the name **tuff**. Geological evidence indicates that some ancient pyroclastic eruptions were colossal. For instance the explosive eruption of Toba, Sumatra, Indonesia, 75 000 years ago, is the largest single eruption widely known in the geological record (Figure 3.20a). It is estimated to have produced 7.5×10^{15} kg of magma, most of which was

The term 'super-eruptions' has been coined for the most powerful eruptions experienced on Earth.

largely deposited as pyroclastic materials. Such **super-eruptions** can distribute thick pyroclastic deposits over huge regions of the globe, and may even generate tuffs that may be identified within the sedimentary record of entire oceans.

Moreover, since such huge volumes of material are ejected in these instances, large crater-like depressions, termed **calderas**, are left behind marking the site from which the material has been removed. In the case of Toba, a lake 100 km in length now fills the caldera, which was produced by the eruption (Figure 3.20b).

Whilst pyroclastic materials are often associated with rapid gas release and explosive eruptions, they may also be created by non-explosive volcanic phenomena (Box 3.3).

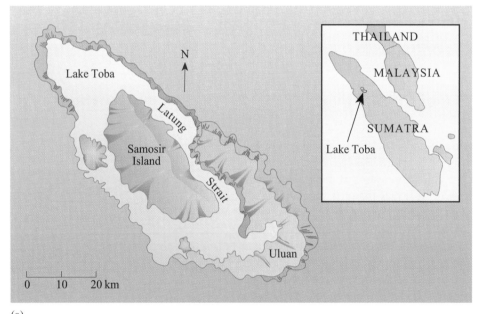

(a)

(b)

Figure 3.20 (a) Topographic sketch and regional setting (inset), of the giant Toba caldera, Sumatra, which is 100 km in length. The caldera consists of a vast, elongate depression, now filled by Lake Toba. Towards the centre, Samosir Island marks a site where later magmatic movements have elevated the caldera floor by about 500 m. (b) View of part of the western caldera rim of Lake Toba. (Steve Blake/Open University)

BOX 3.3 PYROCLASTIC FLOWS

A turbulent, dense mixture of hot air, ash and larger fragments is known as a **pyroclastic flow**, since it behaves as a dense fluid. Once triggered, pyroclastic flows can sweep down hills at speeds in excess of 100 kilometres per hour (about 30 m s^{-1}) leaving behind a blanket of rock and ash fragments covering the landscape (Figure 3.21).

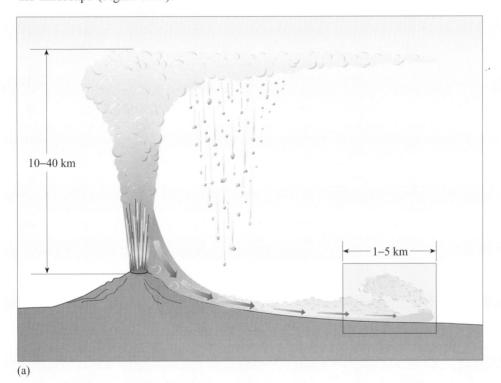

(a)

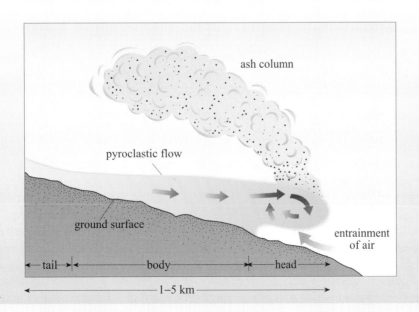

(b)

Figure 3.21 (a) Pyroclastic flows often form by collapse from an eruption column which is so dense that convective rise is inhibited. Descending material acquires enormous velocity and travels over the ground as avalanches of incandescent pumice and dust. Exceptionally, flows on Earth may travel over 100 km. (b) Cross-section through a pyroclastic flow (typically hundreds of metres to tens of kilometres in length), which is moving downslope from left to right.

Figure 3.22 The town of St Pierre, Martinique, one week after the nuée ardente eruption of May 1902 from Mount Pelée. The town was completely destroyed although the resulting pyroclastic deposit amounted to only a few centimetres in thickness. Mount Pelée in the background is partially obscured by cloud. (US Library of Congress)

There are several different processes by which pyroclastic flows may be generated. One is by the collapse of an eruption column. This occurs when part of the column becomes unstable and falls back to the ground, for instance when ash and debris pass outside the gas thrust or convecting region of the column. When this collapsing material hits the ground it creates turbulence, entraining more air, and becomes deflected along the surface topography where it can travel distances of 100 km or more. This type of pyroclastic flow is among the most devastating of all volcanic phenomena, since the fast moving cloud of debris is often extremely hot and can contain huge blocks of lava as well as ash.

By contrast, 'block and ash' pyroclastic flows can be triggered by the collapse of a volcanic edifice such as a cone or a dome. This commonly occurs as a result of the continued growth of such features during a period of volcanic activity. Collapses may be activated either by an explosion of trapped gas within the edifice or simply by the over-steepening and mechanical failure of the growing volcanic structure. Once activated, the falling debris entrains air and travels across the landscape in a similar manner to the collapsing column example. Such pyroclastic eruptions can be 'cold' in instances where the entrained debris does not contain any hot magmatic products, or can form a glowing cloud or **nuée ardente** when hot materials are incorporated. Just such a nuée ardente from Mt Pelée devastated the town of St Pierre on Martinique in 1902 (Figure 3.22) killing over 25 000 inhabitants.

The final example of pyroclastic eruption occurs when the collapse of a volcanic structure results in the sudden removal of confining overburden, thus relieving the pressure upon magma and volatiles held within it. This is what happened during the eruption of Mount St Helens, USA in May 1980 (Figure 3.12) when the bulging mountainside slid away exposing the shallow magma chamber held within. In such a case, the volatiles are expelled explosively in the direction of pressure release, creating a directed blast that entrains both molten magma and avalanching debris from the collapsing segment.

Eruption columns

Expanding magmatic gas drives eruption columns (sometimes termed eruption plumes), but the physics that governs the behaviour and evolution of these columns is complex. Nevertheless, observation shows that eruption columns can be divided into three main parts, each characterized by different processes (Figure 3.23). Immediately above the vent is the **gas thrust region** in which particles of all sizes are propelled upwards, driven by the rapid expansion and release of trapped gases. Larger particles, including rocks and boulders, fall back to the ground along parabolic (ballistic) trajectories. Above the gas thrust region is the **convective ascent region** in which the relative buoyancy of the hot volcanic gases is higher than the cooler surrounding atmosphere and so carries smaller fragments and particles further upward. This convection-driven process is further augmented by heat transfer from the hot particles entrained within the eruption column to the volcanic gases and air surrounding them. Huge volumes of the surrounding cooler atmosphere are sucked into, and heated within, the column in this manner, further

A parabolic or ballistic trajectory describes the path of a projectile after the initial propulsive force stops and it is acted upon only by gravity.

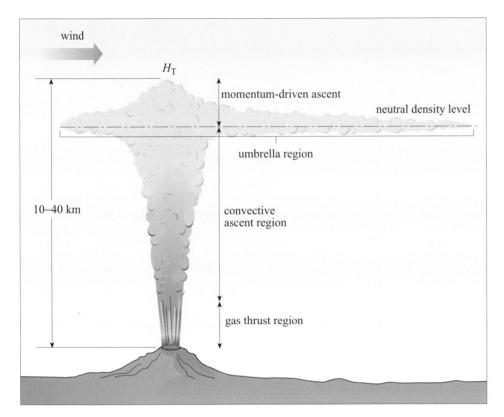

Figure 3.23 Components of a volcanic eruption column rising through an atmosphere (see also Equation 3.1). In large eruptions on Earth, the convective ascent region may reach a height of more than 40 km, while the umbrella region may extend hundreds of kilometres downwind. A small amount of upwind spreading also takes place.

reinforcing the buoyancy effect. These are the main processes that allow the column to rise to such incredible heights. Finally, at high altitude, there is the **umbrella region** where cooling in the higher atmospheric levels eventually causes the plume to become neutrally buoyant, and so it spreads out laterally, much like the infamous 'mushroom cloud'.

The maximum height (H_T) a column can reach is largely dependent upon two variables: the mass eruption rate and the difference between the temperature of the erupted lava and the surrounding atmosphere at ground level.

The mass eruption rate is critical because it provides a measure of the amount of thermal energy that is supplied into the column during eruption (Equation 3.1).

$$H_T = k(M\Delta T)^{\frac{1}{4}}$$

(3.1)

where:

k is a constant of proportionality

M is the mass eruption rate of magma

ΔT is the difference in temperature between the erupting magma and surrounding atmosphere.

■ According to Equation 3.1, what change in eruption rate would be required to double the height of a convecting eruption column?

❏ Since height (H_T) depends upon the ¼ power of eruption rate (M), a sixteen-fold increase in the eruption rate would be needed.

The constant of proportionality, k, in Equation 3.1, depends on various properties, such as atmospheric density. Thus, the height, H_T, will vary between planets with different atmospheric conditions (for the same values of M and ΔT). Eruption columns can reach greater heights in lower density atmospheres. So eruption columns on Mars would typically reach greater heights than on Earth, whereas on Venus eruption columns would reach lesser heights than on Earth.

Mass eruption rate, and hence the density and stability of the eruption column, are controlled by two independent parameters. These are the gas content of the magma and the diameter of the volcanic vent. Basically, the higher the gas content, the more gas will be released, and so higher eruption velocities will be generated. This relationship is corroborated by model calculations which show that, for a vent of given diameter, eruption velocity increases sharply with increasing gas content. Similarly, if the vent diameter is increased, more magma can be delivered resulting in a faster eruption rate. Since eruption velocity in both cases translates to mass eruption rate, and this is directly related to column height (Equation 3.1), either increasing the gas content or increasing the vent diameter can lead to the development of higher eruption columns. However, in reality, after the initial eruption has peaked the mass eruption rate falls because the supply of magma becomes exhausted. When this occurs the column will begin to collapse.

Dispersal of volcanic ejecta

Pyroclastic material ejected from a volcano is also known as **ejecta**. When an eruption column entrains and propels material high into the atmosphere, its return to the surface is controlled by the **terminal fall velocity** of the particles. The pattern of distribution during this fall will be profoundly influenced by particle size and prevailing wind directions (Figure 3.24). The terminal fall velocity is not a difficult concept. It describes the maximum velocity a particle can achieve when falling through an atmosphere. In effect, since **aerodynamic drag** increases with velocity, a point will arise during the fall of an object where its increase in velocity due to the gravitational force will be balanced by the force of aerodynamic drag that slows its descent. At this point, the particle can fall no faster and so is said to have reached its terminal fall velocity. The main controlling factors are summarized in Equation 3.2.

$$\text{Terminal fall velocity} = C_d \sqrt{\frac{dg\rho}{\beta}} \qquad (3.2)$$

where

C_d is the aerodynamic drag coefficient (which is 1.054 for volcanic ash)

d is the particle diameter

g is the acceleration due to gravity

ρ is the particle density

β is the atmospheric density.

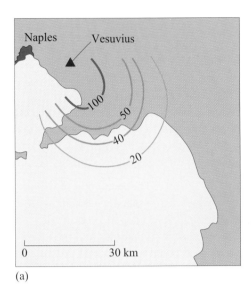

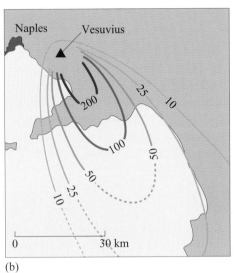

Figure 3.24 Examples of *isopleth* and *isopach* maps for the ash deposited by the famous AD 79 eruption of Vesuvius, which obliterated the city of Pompeii. (a) An isopleth map shows the distribution of maximum sizes (in millimetres) of pumice fragments in the deposit. (b) An isopach map shows the thickness of the deposit in centimetres. The effect of the northwesterly wind is evident from the asymmetrical distribution of ash. (Lirer *et al.*, 1973)

The aerodynamic drag coefficient, C_d, is a dimensionless number that accounts for differences between shapes and surface smoothness. Motor cars and aircraft have a parameter similar to C_d. Volcanic ash fragments are, of course, far removed from the smoothness of luxury cars – they are irregularly shaped and have rough, drag-inducing surfaces. Values of C_d of about 1.054 have been found in experiments.

Terminal fall velocity is important in volcanic eruptions since it is key in controlling the degree of dispersal of pyroclastic materials such as ash. A particle will only continue to be carried upwards in the column and into the umbrella region if the upward velocity within the rising convecting part of the column exceeds its terminal fall velocity. Large eruptions on Earth can create convective velocities great enough to carry fist-sized pumice particles to the tops of columns at altitudes of 30–40 kilometres. However, because larger particles have higher terminal fall velocities, they naturally tend to fall out nearer to the vent (Fig. 3.24a). Thus by examining the particle size and geographical distribution of pyroclastic deposits, volcanologists can gain vital information about the nature and magnitude of ancient eruptions and even about the prevailing atmospheric conditions during the eruptive phase.

3.3 Factors affecting extraterrestrial volcanic eruptions

The same physical laws that govern eruption dynamics on Earth also govern the nature of eruptions on other planetary bodies. However, the physical parameters influencing extraterrestrial volcanism will be significantly different to those of our planet. Therefore, important differences might be expected in the manner in which lava flows evolve, the way in which gas escapes, or the manner in which volcanic products are distributed. You have already been introduced to some key equations that predict the behaviour of volcanic processes. These types of equations often contain terms relating to temperature, pressure and gravity. Clearly, on planetary bodies other than Earth, these parameters can differ significantly. This section briefly explores how these differences can influence the nature and style of extraterrestrial volcanism.

3.3.1 Gravity

Gravity is one of the most obvious variables that differs between planetary bodies. For instance, surface gravity on Mercury and Mars is only about one-third, and on the Moon only one-sixth, of that experienced on Earth (for exact values, see Appendix A, Table A1). Consider again the influence of gas on eruption dynamics discussed in Section 3.2.2. Under low-gravity conditions, the bubbles will begin to form at greater depths than on Earth. This is because for a given depth beneath the surface, the magma (of a given mass) in the conduit above will *weigh* less so the confining pressure will be lower, thus encouraging earlier gas release during ascent.

> Remember that weight is given by *mg*, where *m* is mass, and *g* is the gravitational acceleration.

Other volcanic processes are also affected by low gravity. These include buoyancy and ascent because gravity gives rise to the buoyancy force that allows magma to rise. For a magma body, the magnitude of the buoyant force depends upon its depth, size, how much less dense it is than the surrounding material, and the value of gravity. Therefore, given basaltic magma bodies of equal volume held at comparable depths within similarly sized planets of different masses, the buoyancy effect upon the magma chamber will be greater within the more massive planet because of its greater gravity. Thus, in general, a magma will tend to rise more readily in planets with greater gravity.

Once the magma reaches the surface and begins to form a lava flow, its behaviour is also highly dependent upon gravity. For instance, the critical thickness (h) that a body of lava resting on a slope must achieve before it begins to flow is given in Equation 3.3.

$$h = \frac{S_y}{g\rho \tan\alpha} \qquad (3.3)$$

where:

S_y is the lava yield strength

g is the acceleration due to gravity

ρ is the lava density

α is the angle of slope from the horizontal.

Changing the gravity to values more consistent with that experienced on Mars, Mercury or the Moon, whilst keeping all the other variables the same (for instance basaltic magmas erupted onto slopes of similar angles) will require greater thickness values before magma starts to flow. In other words, lavas of similar composition extruded on to similar topographic slopes would be thicker on Mars and Mercury than on Earth.

Gravity also affects the manner in which any pyroclastic products will be distributed since it is one of the variables that controls the terminal fall velocity of particles in an eruption cloud (Equation 3.2).

■ For bodies with a lower gravity, are particles in an eruption column likely to be taken higher or lower than would be the case in a similar eruption column on Earth?

❑ Assuming none of the other variables alter, the terminal fall velocity should be lower on bodies with lower gravity, so for volcanic eruptions with a similar explosivity the particles should be taken higher.

In fact, the above answer is a rather simplistic view of the manner in which eruption cloud heights might be affected. This is because Equation 3.1 contains the constant k, which depends on (among other things) atmospheric density, which will fundamentally affect the nature, size and style of eruption columns that might occur on other planetary bodies.

3.3.2 Atmospheric density

An eruption column can rise only if the gas within it is less dense than the surrounding atmosphere (i.e. this is the requirement for buoyancy). If it is more dense, it will simply collapse under gravity once the momentum acquired in the gas thrust region ceases to carry the cloud particles upwards.

Column collapse occurs within terrestrial eruption columns once the heat within the convecting part begins to dissipate, or the rate of magma supply begins to wane. When this occurs, the materials held aloft plummet to the ground and spread outwards around the volcano. As these fall, the gravitational potential energy gained in the eruption column is transformed to kinetic energy that drives the fallen materials across the ground at high velocity. This effect creates the highly destructive pyroclastic surges around the volcano as the column material settles out. However, on bodies with an extremely tenuous atmosphere, the ejected pyroclastic materials and gases entrained in the eruption column are unlikely to achieve densities lower than that of the surrounding atmosphere. Therefore, unlike on Earth, eruption columns cannot achieve a convection phase. Indeed, on bodies with almost negligible atmospheres, such as Mercury and the Moon, eruptive columns cannot exist because convection is not possible, and so there is nothing to provide buoyant support to the column (and no atmospheric gases to be entrained and heated). Erupted gases therefore expand into the vacuum of space, driving ejected fragments along ballistic trajectories. Moreover, reference to Equation 3.2 also shows that the terminal fall velocity will be extremely high for ejecta in low-density atmospheres because the value of β is extremely small.

Finally, explosive eruption can only take place if the magma contains sufficient gas to expand, create bubbles and disrupt the ascending magma. Importantly, the degree of magmatic gas expansion is greatest when the atmospheric pressure is least. As a result, on Mars, where the surface atmospheric pressure is only 0.6% that experienced at the Earth's surface, a magma gas content of only 0.01% by mass would be required to drive explosive eruptions. On Earth, the minimum is 0.07% by mass, whilst on Venus, where the dense atmosphere creates a surface atmospheric pressure ninety times that of Earth, a gas content of 3% would be required to create an explosive eruption. Therefore, whilst it seems likely that explosive eruptions could have taken place on Mars, they are much less likely to have played a significant part in the volcanism of Venus.

3.3.3 Surface and atmospheric temperatures

Surface and atmospheric temperatures vary significantly upon and between the terrestrial planets. For instance, Mars's surface experiences extremes of about 290 K (20 °C) during the Martian day to about 130 K (−140 °C) during its night, whilst the surface of Venus is persistently hot at around 730 K (460 °C) due to its all-enveloping cloud cover. Also, the atmospheric temperature profile (the way in which temperature varies with height) is significantly different in the cases of Venus, Earth and Mars. Both atmospheric and surface temperatures will profoundly affect the way in which erupted lavas will cool, the dynamics of any eruption columns, and the associated dispersal of pyroclastic materials.

In Section 3.2.1, it was argued that the eruption temperature and the formation of a chilled crust were important factors that controlled the distances travelled, and hence the lateral extent of basaltic lava flows. In addition, factors such as eruption rate and degree of crust formation can also influence the distances flowed by basaltic and other lava types. As a consequence, it seems reasonable to assume that in environments characterized by lower ambient temperatures, lava flows of similar composition should travel shorter distances due to more rapid cooling. To some extent this will be true, but if extensive solid crusts are formed, these cooling effects will be significantly reduced and flow inflation and breakout may become the dominant process. Therefore, since solid crusts provide such effective insulation against heat loss, differences in style and extent of basaltic lava flows on planetary bodies with cooler ambient surface temperatures may be smaller than expected. Accordingly, whether laterally extensive lava flows or tall volcanic edifices are generated is probably influenced more by magma composition than by planetary surface temperatures.

Nevertheless, in cases such as Venus, the rate of lava cooling will be slower than that experienced on Earth because the differential between surface and eruption temperatures, ΔT, is significantly smaller. For instance, ΔT for a basalt erupted on Earth, is over 1000 K (i.e. 1325 − 288 K), whilst on Venus, ΔT would be nearer 600 K (i.e. 1325 − 730 K).

More bizarre is the idea that unusually fluid lavas similar to **carbonatites** may erupt on Venus and flow immense distances in a manner similar to river waters on Earth. Carbonatite eruptions generate non-silicate lavas that are rich in calcium carbonate ($CaCO_3$), and are known from rare instances in east Africa. Such eruptions produce particularly fluid flows and, unlike silicate magmas, have a particularly low eruption temperature of only about 500 °C. If such lavas were erupted on the surface of Venus, ΔT would be so small that freezing would take place only very slowly, thus providing opportunity for the fluid flows to travel great distances and cut 'channels' in much the same manner as rivers do on Earth (Figure 3.25).

Interestingly, the possibility of slow cooling rates also arises on some icy bodies since the temperature difference, ΔT, between erupting cryovolcanic lavas and surface conditions may be relatively small compared with those characteristic of most terrestrial lava eruptions (ΔT between 700 K and 1200 K). For instance, if an icy lava is erupted at 180 K and the surface temperature of the body is 140 K, then ΔT is only 40 K.

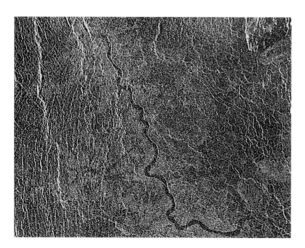

Figure 3.25 A sinuous channel, 2 km in width, crossing fractured lava plains on Venus. The image shows a region 130 km across. This is one of about 200 sinuous channels on the plains of Venus, the longest of which has been traced for 6800 km. Superficially, these resemble meandering river channels, but, unlike drainage patterns, they lack tributaries. Given Venus's high surface temperature (about 460 °C) they are most likely to be channels cut by flowing lava rather than water. However, since it is unlikely that basaltic lava would remain molten long enough to travel such distances, it is more probable that they were carved by flows of carbonatite lava. Carbonatite does not solidify until it cools below 500 °C, and so could remain liquid for a long time after its eruption on to the hot surface of Venus. (NASA)

■ If the ΔT is only 40 K, would the resulting slow cooling of a cryovolcanic lava
flow on a small icy body necessarily result in laterally extensive lava flows?
(Hint: consider Equation 3.3.)

❏ Not necessarily. Although ΔT is comparatively small (only 40 K), permitting
slower rates of cooling and hence the opportunity to flow further before
freezing, the production of laterally extensive lava flows is also dependent upon
gravity (Equation 3.3). Therefore, even if a lava were to cool relatively slowly, on
smaller bodies with low gravity such a lava could not flow and spread easily. Instead,
it would form thick, locally restricted flows and edifices, rather than thin widespread
flow sheets (i.e. the critical thickness, h, would be relatively large).

Following the earlier discussion regarding the development of eruption columns, it
should now also be apparent that in addition to atmospheric density and gravity,
atmospheric temperature will also play a key role. Equation 3.1 indicates that ΔT, the
difference between the temperature of the erupted magma and the surrounding
atmosphere, will control eruption height. Broadly speaking, the lower the atmospheric
temperature, the higher the eruption column can rise for a given magmatic temperature.
However, as previously discussed, the relationship concerning eruption column height
and the velocity of ejected volcanic materials is likely to be more complex on planets
with particularly tenuous atmospheres. Even if such bodies had relatively cold surface and
atmospheric conditions compared with magmatic temperatures, an extremely thin
atmosphere would not allow the development of a convecting region in the column.

From the above arguments, it might be concluded that a dense atmosphere would
promote tall eruption columns since such atmospheres would encourage convection
within the column. However, this is not necessarily the case because the development
of such a column requires the explosive release of gas, and this can only occur if
there is sufficient pressure difference between the released magmatic volatiles and the
ambient planetary atmosphere. Consequently, under particularly dense atmospheres
(e.g. on Venus) the expansion of volatiles would be inhibited, and so explosive
volcanism and associated eruption columns are less likely unless the magma is
particularly volatile-rich (Section 3.3.2).

3.3.4 Cryovolcanism

As outlined earlier (Section 3.1.5), the melting processes that generate icy magmatism
are controlled by factors similar to those which produce the silicate volcanism with which
we are perhaps more familiar. However, given that the eruptive products are themselves
'ices', would it be correct to expect significant differences in the style of cryovolcanism
compared with silicate lava eruptions and associated ash and gas release?

In fact, just because cryovolcanism describes the generation and eruption of magmas
derived from ice rather than molten rock, it does not necessarily follow that the eruptive
behaviour and the style of volcanism associated with these ices must be significantly
different. Partial melting of an ice mixture can result in the release of volatiles and gases in
a similar fashion to silicate melting, and so the resulting surface eruption can be either
effusive or explosive depending on whether this gas is released progressively or
instantaneously. If cryovolcanism becomes explosive, there then arises the possibility of
eruption columns producing the ice equivalent of volcanic ash, which would be deposited
around the vent area and further afield depending upon atmospheric and gravitational
conditions. Similarly, icy lavas may also develop a solidified crust when chilled on
exposure to surface conditions, and so present the possibility of lava tubes and laterally
extensive inflated flows analogous to those of Earth.

Much more observational data is required to further our understanding of cryovolcanism, but the range of cryovolcanic styles is likely to be just as varied as those characteristic of silicate volcanism on Earth and the other terrestrial planets. However, these differences are largely the result of variation in planetary attributes such as gravity, atmospheric density and surface temperatures, rather than major differences in either the cause of melting or the material that is being melted. In other words, although cryovolcanic products may be different to those to which we are more familiar, the same physical laws control the eruption processes.

3.3.5 Summary

Gravity, atmospheric density, and the internal and surface temperature conditions of other planetary bodies place fundamental constraints upon the nature of any volcanic eruption and the manner in which its products are erupted.

The preceding sections have largely dealt with variation in these controlling factors in an unsophisticated manner since we have typically considered only one variable at a time. You may now have begun to realize that any investigation of the nature of volcanism and eruption on bodies elsewhere in the Solar System requires an understanding of the interaction between these controlling variables. Predicting the style and extent of extraterrestrial volcanism therefore requires complex modelling. This is a task that lies at the forefront of modern planetary research.

3.4 Volcanism on the terrestrial planets and planet-like bodies

Having examined and outlined the causes and nature of volcanism, together with the main factors that control volcanic eruptions, the evidence for volcanism on other terrestrial planets and terrestrial-like bodies can now be explored in more detail. In the following sections we briefly examine the evidence for volcanism in the Solar System, and the contrasts between the volcanic histories of different kinds of bodies. Focusing on the observable evidence for different volcanic phenomena, a short description of the observed volcanic features of each body is provided, together with a short account of their differing volcanic histories.

3.4.1 Earth

Terrestrial volcanism is an on-going process. Over 70% of the Earth's surface is covered by basalt lavas that have erupted within the last 200 million years, although most of them are hidden beneath the oceans. Several hundred active volcanoes around the world provide conclusive evidence of continuing thermal processes within the mantle, and both effusive and explosive eruptions are common. Although they are not the world's highest mountains, the volcanic edifices of Mauna Loa and Mauna Kea in Hawaii are the largest volcanoes, reaching elevations of 9 km above their bases on the ocean floor.

In the Deccan area of northwest India, basalt flows that erupted 67–64 million years ago total more than 2 km in thickness and cover almost 10^6 km². Similar continental flood basalt provinces (CFBPs) are found in Brazil, Namibia, South Africa, Siberia and southern China, and testify to periods of intense basaltic volcanism during the last 250 million years. By contrast, evidence of much smaller but far more explosive volcanic eruptions are widespread in volcanic arc settings such as the Central Andes (Box 3.1) where pyroclastic flow deposits can cover hundreds of thousands of square kilometres.

Super-eruptions are thought to be generated by huge volumes of viscous, gas-rich magma that rise to, and are held at, relatively shallow levels within the Earth's crust. A near-instantaneous release of pressure occurs when the magma eventually finds a route and breaks out at the surface, producing a cataclysmic explosion that sends hot gas and pyroclastic materials high into the atmosphere. Such eruptions would have been tens of thousands of times greater than that of Mount St Helens (USA, 1980). Accordingly, the volume of ash is enormous, and can cover huge areas of continent and ocean with many metres of material within only a few hours or days (Figure 3.26). Such eruptions are known in the geological record from the widespread ash 'horizons' preserved in both sea-floor sediments and deposits on land (e.g. Toba, Indonesia which erupted about 75 000 years ago), and from the immense calderas (surface depressions) that are left behind through surface collapse once all the magma beneath the volcano has been expelled. Such eruptions are known to have profoundly affected the environment since they are thought to have produced massive ash clouds and toxic gases which circled the globe, reflected incoming sunlight, and cooled the climate. The term 'volcanic winter' has been given to the cool twilight conditions that are thought to have been experienced at Earth's surface following such eruptions. Indeed, these conditions have many similarities with those that would occur following a large asteroid or comet impact with the Earth. You will consider the role of impacts on planetary surfaces in detail in Chapter 4.

Figure 3.26 Thick volcanic ash deposits in Yellowstone Park, Wyoming, USA. These deposits were produced during a major eruption of the Yellowstone caldera complex about 1.3 million years ago. Two units can be seen. The lower layer was formed from ash and pumice falling out from the eruption column, and consists of larger fragments near the base, with progressively finer material towards the top (the holes at the top are the entrances to birds' nests). The overlying unit also contains coarse material at its base, but was formed later in the eruption when a pyroclastic flow cascaded across the land surface carrying with it more ash and debris. At the very top, the brownish material is ash that became welded together due to the heat in the pyroclastic flow. (Steve Self/ Open University)

3.4.2 The Moon

As discussed in Section 2.5.2, basaltic-type lava flows form the youngest and most extensive rocks on the Moon. Flows filling the mare impact basins are generally 3.8–3.2 Ga old, although some may be as young as 0.8–1.0 Ga old. Spacecraft images reveal evidence of many flows, some of them extending hundreds of kilometres, but the sources of these flows remain obscure. There are no obvious volcanoes and it is thought that the lavas must have flooded to the surface via relatively narrow cracks and fissures (Figure 3.27).

Since the Moon is extremely deficient in volatiles (Table 2.4), including those elements that would normally form volcanic gases, large-scale pyroclastic activity is not expected to have formed a significant part of the Moon's volcanic history. Nevertheless, some evidence for some pyroclastic activity has been detected on spacecraft images and in surface samples.

■ Given the absence of an atmosphere on the Moon and its low surface gravity, how would you expect pyroclastic materials to be dispersed? What might the deposits look like?

❏ Explosive eruptions taking place into the vacuum of space on a low-gravity planetary body would propel ejecta long distances on ballistic trajectories. The deposited material would form thin sheets around the vent, with larger fragments concentrated towards the vent.

Promising-looking deposits of ejecta have been noted in a few sites around the Moon. Termed **dark halo craters**, they are best seen on the floor of the large crater Alphonsus. There, aprons of dark ejecta surround small craters developed on fissures that cut the floor of the larger crater. They have been described as the lunar counterparts of small terrestrial volcanoes such as Stromboli.

Figure 3.27 An astronaut's view looking southeastwards across the Mare Imbrium region of the Moon. The prominent crater towards the upper left is Aristarchus, which is thought to be one of the youngest large craters on the Moon (about 400 Ma). Closer than Aristarchus are the traces of a much older crater, Prinz, which was flooded by the lavas that spread to form the Mare basins 3.5 Ga ago. The sinuous features are lunar rilles. These are thought to be channels or collapsed lava tubes produced by flowing lava towards the end of the Mare-forming events. (Mr Frederick J. Doyle, National Space Science Data Center)

3.4.3 Mercury

Only one spacecraft, Mariner 10, has visited Mercury and it imaged less than half the planet. Consequently, we know much less of its volcanic history than we do of the Moon, or even Io (Section 3.4.6). Mercury is a small, dense body that lacks an atmosphere and so we might expect it to demonstrate evidence of volcanic activity similar to that of the Moon. Mariner 10 images show large areas of smooth plains on Mercury, notably around a large impact basin called Caloris, which are probably equivalents of the lunar mare lavas (Figure 3.28).

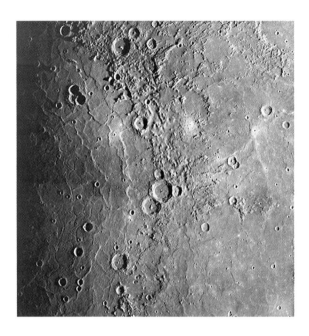

Figure 3.28 Part of the Caloris Basin, which is 1100 km in diameter (left-hand side of the image). This is thought to have been formed by a giant impact early in Mercury's history (about 4 Ga ago), and subsequently filled by lava flows. The nature of the wrinkle ridges on its floor is a matter of debate. They probably represent lava flows escaping from fractures, surface contraction and subsidence effects. Beyond the outer ring of the crater, the right-hand part of the image reveals numerous secondary craters formed when ejecta from the Caloris impact struck the surface. (NASA/JPL/Northwestern University)

3.4.4 Venus

Venus is close in size, mass and density to Earth, but differs in having an atmospheric pressure ninety times greater than Earth's, and a surface temperature of about 730 K. The crushing atmospheric pressure would tend to limit both explosive pyroclastic activity and the formation of convecting eruption columns.

During 1990–94, the Magellan spacecraft mapped the surface of Venus by imaging radar so revealing many details of Venus's volcanic past. It is clear that there are numerous major volcanoes on Venus. Some rise many kilometres above its surface to form broad shields similar to those of Hawaii (Figure 3.29).

Figure 3.29 Three-dimensional, computer-generated, perspective view of the 5 km high shield volcanoes Maat Mons (background), Sapas Mons (foreground), and adjacent region on Venus. The vertical scale has been exaggerated tenfold to show the topography more clearly, so in reality slopes are much gentler than implied here. These low slopes, together with the extensive lava flow fields extending across the adjacent lowland regions, indicate these were probably produced by basaltic eruptions. (Gordon H. Pettengill, The Magellan Project and the National Space Science Data Center)

Figure 3.30 Lava flows on Venus. The flows that appear very bright to the radar have rough surfaces, whereas the darker flows are smoother. This mosaic highlights a system of east-trending, radar-bright and radar-dark lava flows that collide with and breach a north-trending sinuous ridge (running top to bottom, left of centre). Upon breaching the ridge, the lava pooled, forming a radar-bright deposit covering approximately $100\,000\,km^2$ (right-hand side of image). The source of the lava is the Corona Derceto, which lies about 300 km west of the scene. Image shows an area 400 km in width. (NASA)

Figure 3.31 A three-dimensional perspective view of part of the Alpha Regio area on the surface of Venus. This view is of an area containing seven circular dome-like hills, three of which are visible in the centre of the image. Average diameter of the hills is 25 km, with maximum elevations of 750 m. The hills are thought to be the result of the eruption of viscous lava coming from a vent on a relatively level surface, allowing the lava to flow in an even, lateral pattern. These features have been compared to lava domes on Earth (Figure 3.13), but they are an order of magnitude greater in size. The concentric and radial fracture patterns on their surfaces suggest that a chilled outer layer formed, then further intrusion in the interior stretched the surface. The bright margins possibly indicate the presence of rock debris or talus at the slopes of the domes. (NASA)

Volcanoes on Venus probably erupted lavas of basaltic composition. Many lava flows and extensive lava fields have been identified (Figure 3.30). Some long sinuous flows extending nearly 2000 km may be cut by particularly fluid lavas, possibly of carbonatite composition. In addition, there are innumerable small volcanoes a few hundred metres in height, some of which appear to have been the sources of minor pyroclastic eruptions. There are also a number of circular, pancake-like domes which would be consistent with the eruption of more viscous lavas (possibly rhyolitic in composition) (Figure 3.31).

The average age of Venus's surface is thought to be only about 500 million years old, and may indicate a paroxysmal planetary resurfacing event (Section 2.5.4). As yet, there has been no direct evidence for on-going volcanic activity, but given that Venus is so similar to Earth in size, mass and composition, it may be possible that volcanism continues to the present day. However, any current volcanism will be at a much reduced scale compared with that likely to have been associated with a major resurfacing event.

3.4.5 Mars

Much was learned about Mars's geology from the first mapping and lander missions of the 1970s. More recently, high-resolution images defining surface features of only a few metres in scale have been collected. These have provided detailed data for the interpretation of Martian geology and landforms.

QUESTION 3.3

Before reading further, note down what sort of volcanism you might expect to find evidence for on Mars. Think in terms of its sources of heat, thermal evolution, and the likelihood of lava and pyroclastic eruptions.

The volcanic processes described in this chapter should have helped you to formulate reasonable responses to the previous question. However, a complicating factor on Mars is that surface conditions, including its atmospheric pressure, appear to have varied greatly since its origin. These changes have been reflected in its volcanic history, notably the style of pyroclastic eruptions.

Spacecraft images show that Mars has the largest volcanoes in the Solar System. Olympus Mons, and three nearby major volcanoes, lie in an area known as the Tharsis volcanic province. Their morphology is reminiscent of the gently sloping Hawaiian volcanoes, but they are much bigger. Olympus Mons (Figure 3.32) has a base diameter in excess of 600 km, a nested summit caldera complex more than 60 km in diameter and over 2 km deep, and an elevation of nearly 25 km above the surrounding plains, giving it a volume 100 times that of the Hawaiian volcanic edifice. High-resolution images show myriads of fresh-looking flows draping the flanks of Olympus Mons that wind for hundreds of kilometres across the lowland plains. By contrast, some small Martian volcanoes rise abruptly out of the surrounding plains, leading to suggestions that their flanks were partially 'drowned' by extensive flood basalt lavas (the basaltic composition being confirmed by the

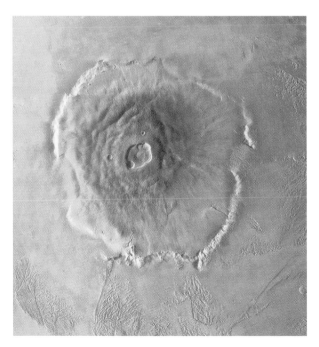

Figure 3.32 A spectacular view of Olympus Mons, the youngest volcano in the planet's volcanic Tharsis region. It is thought that it was last active over 100 million years ago. This volcano is over 24 km high – three times taller than Mount Everest – and as wide as the entire Hawaiian Island chain. The three craters in the summit region are over 2 km deep, and are thought to be caldera collapses formed by the drainage of lava beneath during eruptions from the flanks of the volcano. Despite its large size, the angle of slope is only a few degrees, similar to those of Mauna Loa, Hawaii. Consequently, it has long been inferred that this volcano was built by the repeated eruption of low viscosity basaltic lava flows. The Martian crust is able to support such a large volcanic edifice because, being a smaller planet than Earth, it has lost much of its heat which has enabled its lithosphere to thicken and so become much stronger. (US Geological Survey)

Sojourner rover, part of the Mars Pathfinder lander, Figure 3.33). These lavas are thought to have built up over time and may be up to about 40 km thick in the Tharsis area (Figure 3.34).

■ Plate recycling movements do not take place on Mars. Can you suggest how this might account for the fact that Mars can sustain volcanoes much larger than even Earth's largest volcanoes?

❏ The Hawaiian volcanoes, riding on their lithospheric plate, are rapidly carried away from their thermal sources, thus limiting the maximum size they can achieve. By contrast, a volcano borne above a Martian mantle hot spot remains fixed above it, enabling it to grow indefinitely. Moreover, because the Martian lithosphere is thicker, and hence stronger, than that of Earth, it can support much larger volcanic constructions without flexing downwards, as is the case with the sea-floor around Hawaii.

Typical flows on the plains are about 10 km wide and 20–30 m thick and often display a lack of superimposed impact craters which may indicate that some flows are relatively youthful. However, assigning absolute ages is difficult given only the crater size–frequency statistics (you'll consider this in Chapter 4) to go on, and a relatively scant knowledge of Martian weathering and erosion processes that might otherwise erode and obscure such a record. Consequently, some crater counters suggest it is almost a billion years since the last eruption, whilst others consider that the youngest flows may be only a few hundred million years old.

Figure 3.33 A close-up of the Sojourner rover taken during the Mars Pathfinder mission in 1997 as it placed its alpha proton X-ray spectrometer (APXS) upon the surface of the rock 'Yogi' to determine its composition. (The rover is about the size of a microwave oven.) Results show this rock has a composition similar to common basalts found on Earth. (NASA)

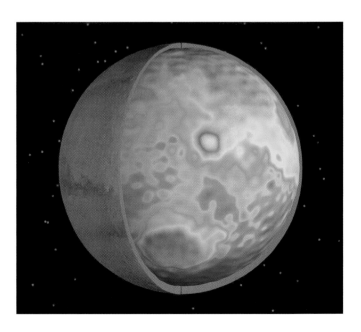

Figure 3.34 Inside Mars. A colour cut-away diagram revealing the thickness of the Martian crust. Red colours correspond to thin areas, and blue to thick crustal areas. The Martian crust typically ranges from about 30 km to 80 km in thickness and shows a dramatic difference between the generally thinner northern hemisphere crust and the thicker southern hemisphere crust. The region of thin crust in the southern hemisphere is the Hellas basin, whilst the other smaller, round area of thin crust just north of the equator is the Isidis Planitia region (see Figure 3.2). The areas of thickest crust associated with the huge volcanoes of the Tharsis region lie beyond the easterly limb of the globe. (NASA)

Some Martian volcanoes may be largely of pyroclastic origin. Two lines of evidence support this suggestion. First, their remarkably flat topographic profiles are consistent with widespread dispersal of pyroclastic fall and flow deposits. Second, the erosion patterns of the deposits on their flanks do not look like those characteristic of lavas (Figure 3.35).

The small cones of Isidis Planitia are similarly thought to be the result of spatter around small vents. Mars's low atmospheric density should allow eruption columns to rise about five times higher than on Earth given the same mass eruption rate (i.e. as a consequence of the constant, k, in Equation 3.1 (Section 3.2.2) having a value larger than it has on Earth). For comparable eruptions, therefore, pyroclastic deposits could be much more widely dispersed on Mars than on Earth. Whenever things become widely dispersed, however, it becomes more difficult to identify them positively on spacecraft images. Thus, while explosive volcanic activity has probably been important on Mars, it will be difficult to resolve its full extent without further detailed investigation.

3.4.6 Io

Io is the most volcanically active body in the Solar System (Figure 3.36). It orbits about 422 000 km away from the colossal mass of Jupiter, and undergoes significant tidal heating due to the large gravitational influence of Jupiter 'kneading' Io as it travels around its orbit (you will consider this effect in more detail in Chapter 7). This perpetual tidal kneading maintains Io's hot interior (Section 2.5.6). It is estimated that the power liberated by this tidal dissipation is two orders of magnitude larger than that released by any internal radioactive decay. Spacecraft and telescopic infrared data suggest that the globally averaged **surface heat flow** on Io is about 30 times that of Earth.

Figure 3.35 Appolinaris Patera, an isolated Martian volcano located near the equator. It is 5 km high, 400 km in diameter, and has a summit caldera that is 70 km in diameter. The dendritic channel-like patterns emanating from the summit caldera are thought to have been produced by erosion and deposition of pyroclastic flows cascading down the flank of the volcano. The pale-blue area is a patch of bright clouds hanging over the summit. (NASA)

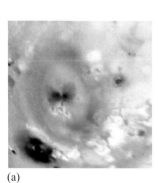

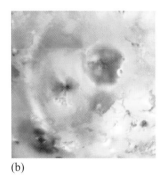

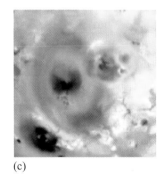

(a)   (b)   (c)

Figure 3.36 Volcanism and the rapidly changing surface of Io. Over a period of three years images from NASA's Galileo spacecraft reveal that dramatic changes have occurred at a volcanically active region of Jupiter's moon Io. (a) Concentric pyroclastic deposits produced by fall-out from the volcano Pele, taken in April 1997. (b) The same area in September 1997 after a huge eruption had occurred producing the large, dark deposit just above and to the right of the centre. These new deposits (400 kilometres in diameter) are the fall-out from a new volcanic centre named Pillan Patera. This has developed on the flank of the older volcano covering part of the bright red ring which form the earlier deposits from Pele's plume. (c) Image acquired in July 1999 showing the changes that have taken place on the surface since the initial Pillan eruption almost two years ago. The red material from a later eruption by Pele has begun to cover, but has not yet entirely obscured, the dark material around Pillan. This image also shows that a small, unnamed volcano to the right of Pillan has erupted, depositing dark material surrounded by a yellow ring, which is most visible where it covers some of the dark material from Pillan's 1997 eruption. (Note: some of the differences in colour between the images are the result of different lighting conditions, for instance, the apparent change in brightness of the dark feature in the lower left-hand corner, and of parts of Pele's red plume deposit.) (NASA)

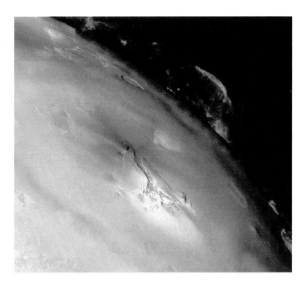

During the Voyager 1 mission, nine distinct eruption columns were observed above the surface of Io. They reached heights of between 70 km and 280 km, implying eruption velocities between 500 m s^{-1} and 1000 m s^{-1}. Some changed visibly over the course of a few hours, whilst others remained apparently unchanged for days. The largest column was 1000 km in diameter, and was fed from a fountain-like vent on a volcano about 35 km across (Figure 3.37). Eight of the columns were re-observed during the Voyager 2 encounter and, in the four months between the two mission encounters, there were considerable changes on Io.

Figure 3.37 Pele in eruption. The vent is in the pale-coloured rugged area in the centre of the image. The plume of sulfur-rich material dispersed by the eruption column shows up against the blackness of space. (NASA)

■ In practical terms, Io lacks an atmosphere. How, then, would you expect the ejecta from its volcanic vents to be distributed?

❏ In the absence of a significant atmosphere, convecting eruption columns cannot be developed. Erupted material should fall back on ballistic trajectories to form sheet-like deposits around the vent areas. Halo-like surface features around volcanoes such as Pele (Figure 3.37) are good evidence for the accumulation of pyroclastic deposits.

Over 500 volcanoes have been identified on Io, although not all are currently active. These cover about 5% of its surface area, and most are low, gently sloping shields of less than 100 km in diameter (Figures 3.36 and 3.37). However, many of the larger shields have shallow summit calderas ranging from 2–200 km in diameter, and radial lava flows snake away from the summit regions, some reaching distances as far as 700 km from the volcano.

Rather than being organized according to obvious plate boundaries, the volcanoes so far identified on Io are randomly distributed, suggesting that Io's tectonic processes are quite different to those of Earth. It is thought that Io loses its internal heat predominantly by advection (Section 2.4.4) via hordes of active volcanoes, rather than by mantle convection and plate recycling as on Earth.

Spectroscopic evidence reveals that the volcanism produces a prodigious amount of sulfur, and Io's distinctive reddish colour is probably the result of the widespread sulfurous deposits ejected by its volcanoes. In addition to sulfur dioxide, lesser amounts of hydrogen sulfide and carbon dioxide, together with atomic oxygen, sodium and potassium also occur as gases in its thin atmosphere, and ionized sulfur has been detected dispersed along the track of Io's orbit about Jupiter. This latter represents the continuous 'leaking away' of Io's atmosphere into space, and is largely maintained through constant replenishment by the on-going volcanism.

The large amounts of sulfur detected in the eruption plumes initially prompted speculation that the volcanism was entirely sulfurous. However, whilst much of Io's surface is covered by sulfur and sulfur oxides, this represents no more than a thin covering compared with the thickness of rocky layers that exist beneath. The fact that no impact craters are visible on Io has been attributed to the rapid deposition of fresh volcanic materials that are thought to accumulate at a global rate of one centimetre per year.

Density and gravitational data suggest Io is dominantly a silicate body with an iron-rich core beneath its rocky mantle. It is assumed that Io's bulk composition is broadly chondritic (i.e. Io is a terrestrial-like body). The huge amount of sulfur in its eruptive products is probably a result of an advanced degree of differentiation that has occurred through prolonged tidal heating and associated partial melting. In fact, the degree of tidal heating is such that it is possible that a large fraction of the mantle may be molten. The lithospheric part of Io's upper mantle is thought to be a Fe–Mg-rich silicate layer, with an overlying Mg-poor layer forming the crust. If this interpretation is correct, melting of the lithospheric mantle would result in Mg-rich lavas of komatiitic composition which, by analogy with examples on Earth (Section 3.1.4), are likely to have very high eruption temperatures (i.e. in excess of 1900 K). Interestingly, thermal images of some of the erupting lava flows on Io's surface indicate very high eruption temperatures, and so seem to support this interpretation (Figure 3.38). These types of lava flows are most probably the result of silicate melting rather than sulfur-based magmatism. Moreover, recent ideas also postulate that the rise of this hot silicate magma through Io's differentiated crust can, in some instances, cause rapid melting of the overlying sulfur-rich layers (Figure 3.39). The resulting release of gases and liquids produced in this manner would be rather like a volcanic 'geyser' on Earth, except that on Io the fountain would consist of liquid and vaporized sulfur, rather than water! If this explanation of Io's spectacular volcanism is correct, these types of eruption are probably more akin to continuous, long-lived 'fountains' (some lasting months or years) than 'eruption plumes' produced by the type of short-lived explosive volcanoes observed on Earth.

Figure 3.38 A volcano on Io photographed during an on-going eruption. This image is colour-enhanced and reveals hot glowing lava, probably of silicate composition, visible on the left-hand side of the image and occurring amongst a landscape of plateau and valleys covered in sulfur. The image shows a region 250 km across. (NASA)

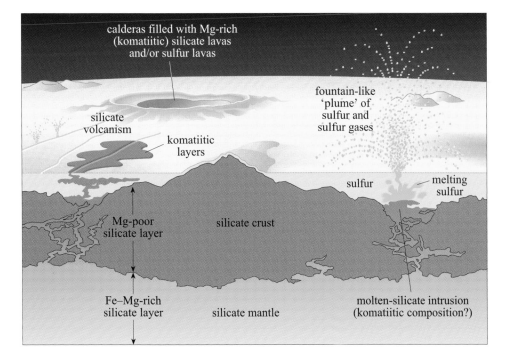

Figure 3.39 Schematic detail (not to scale) of the layers comprising Io's upper mantle and crust (i.e. lithosphere). As described, different types of magmatism and associated volcanism are causing resurfacing of this Jovian satellite, and so aiding in further differentiation of its upper mantle and crustal layers (see also Section 2.5.6).

3.4.7 Cryovolcanism on icy satellites

There is abundant evidence of cryovolcanic eruptions on many of the tidally heated icy satellites of the giant planets. Europa (Figure 3.40), a satellite of Jupiter, and parts of Enceladus (a satellite of Saturn), have very few superimposed impact craters, demonstrating relatively recent cryovolcanic resurfacing has taken place. On other icy satellites such as Ariel and Miranda (satellites of Uranus), the cryovolcanically resurfaced terrains are older but still identifiable, especially where eruptions have been effusive rather than explosive. The most spectacular of these ancient cryovolcanic features occur among the satellites of Uranus (Figures 3.41 and 3.42), whose ice is believed to consist predominantly of a frozen mixture of water and ammonia (NH_3). Interestingly, the partial-melting behaviour of this mixed ice is a particularly close analogue to that occurring during the partial melting of silicate rock, and the resulting icy magma can sometimes produce an especially viscous lava when erupted at the surface.

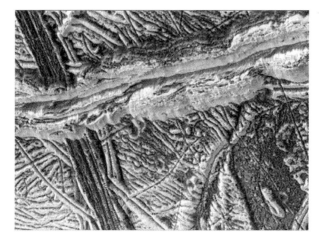

Figure 3.40 A close-up of part of Europa, a satellite of Jupiter. The general ridge-and-groove surface may be a result of cryovolcanic eruptions associated with fractures. The diffuse edges of the white band running across the top of the image may indicate pyroclastic deposits from an explosive fissure eruption. The smooth area in the lower right-hand corner may have been flooded by a low-viscosity cryovolcanic lava. Image shows an area 18 km across. (NASA)

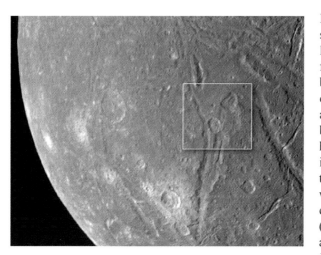

Figure 3.41 Part of Ariel, a satellite of Uranus, whose lithosphere has been widely fractured, creating fault-bounded valleys. The floors of many of these valleys appear to have been flooded by cryovolcanic lavas. A half-flooded pre-existing impact crater can be seen at the edge of one such flow, which also has several later craters superimposed on it (in white box). Image shows an area 500 km wide. (NASA, JPL, Voyager 2)

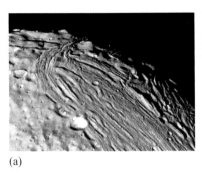
(a)

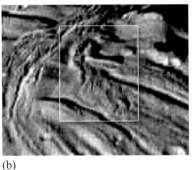
(b)

Figure 3.42 (a) Part of Miranda, a satellite of Uranus, where most of the landscape appears to be cryovolcanic in origin. Image shows an area 120 km across. (b) A 40 km-wide enlargement from the upper left-hand corner of (a), centred on a lobate viscous cryovolcanic lava flow that is 20 km in length (in white box). (NASA)

3.5 Summary of chapter 3

- Volcanism is the surface expression of melting processes occurring within a planetary body. It describes the processes associated with the transfer of magma, volatiles, and suspended crystallized material from the interior to surface and, if present, into the atmosphere. On Earth, these processes are most obviously demonstrated by the presence of volcanoes that erupt lavas and produce ash and gas clouds. Two main types of activity may be identified: effusive volcanism is characterized by the production of lava flows, whilst explosive volcanism is characterized by the production of eruption columns and pyroclastic materials.

- There is abundant evidence for volcanism having affected the surfaces of all the larger silicate bodies in the Solar System, although most do not appear to be currently volcanically active. Io, a satellite of Jupiter is, however, highly active, with numerous volcanoes randomly distributed across its surface.

- Effusion of basaltic lavas has been the most widespread manifestation of volcanism, and is the result of partially melting mantle materials derived from the chondritic composition which is typical of the terrestrial planets and similar bodies. Consequently, basaltic lavas cover 70% of Earth's surface, and large fractions of the surfaces of the Moon, Mercury, Venus and Mars. In Earth's case, the widespread basaltic magmatism that generates ocean floor and continental flood basalt provinces is largely the result of partial melting due to decompression of the upper mantle.

- Partial melting of ices leads to volcanism on icy bodies. The eruption products of icy volcanism are subject to the same physical processes regarding the migration of magma, volatile release, and eruption dynamics that occur on silicate bodies. Both effusive cryovolcanic products in the form of icy lavas and explosive cryovolcanic products generated by volatile release have been recognized on the surfaces of bodies such as Europa and Enceladus.

- Decompression melting may be the source of silicate or icy partial melting on bodies other than Earth. However, Io is unique among silicate bodies in that its volcanism is driven by dissipation of tidal energy, and its dominant volatile is probably sulfur dioxide rather than water. Its lavas are erupted at high temperatures that are indicative of silicate magmas of komatiitic composition.

- The differences in gravity, atmosphere, surface temperature and magmatic composition can modify the style of eruption, and the manner in which the erupted products are distributed at the surface. For instance, Earth and Mars have atmospheric conditions sufficient to support convecting eruption columns, so drastically modifying the dispersal of pyroclastic ejecta.

CHAPTER 4
PLANETARY SURFACE PROCESSES

4.1 Introduction

In this chapter you will be introduced to some of the processes that modify the surfaces of planets. As we have seen in the previous chapter, volcanism plays a part, but cratering – the results of the impact with a planetary surface of objects travelling at many kilometres per second – is probably the most significant factor affecting many planetary surfaces. Evidence for the ubiquity of this process has been provided by images of the heavily cratered surfaces of planetary bodies from Mercury to the satellites of Neptune. Although much of this chapter is given over to a discussion of cratering, it is not the only process that modifies planetary surfaces as erosion and sedimentation are also involved. We conclude with a discussion of the actions of erosion and sedimentation, as illustrated by recent images returned from missions to Mars.

In completing one revolution around the Sun, the Earth travels over 940 million kilometres on a well-defined orbit. But it is not alone. There are probably about 1000 objects over 1 km in diameter and possibly as many as a million over 50 m in diameter whose orbits cross or closely approach that of the Earth. These objects are a mixture of asteroids and comets. In our voyage around the Sun, we travel through a 'shooting gallery', and the same is true for all the planets and satellites in the Solar System. Two to three of those objects, about 1 km in diameter, hit our planet every 1 Ma. From a geological perspective they are a common occurrence – although on the scale of a human lifetime impacts are rare. This is the reason why **impact cratering** is by far the most widespread process shaping the surfaces of solid bodies in the Solar System. Figure 4.1 shows the Earth over the heavily cratered surface of the Moon. The lunar craters are the result of impacts by asteroids and comets. The crater Plaskett (110 km across) is clearly visible in the foreground.

Impacts are intimately associated with the formation and evolution of planets. Although some bodies in the Solar System were formed directly from the dust and gas in the primordial solar nebula (the Sun, and gas giants fall into this category), almost everything else was constructed by impact and accretion of solid objects in the early Solar System. This impact 'cascade' began with sub-millimetre objects that accreted to form bodies that ranged from about a metre to tens of metres in size. These then accreted to form bodies about a kilometre to tens of kilometres in size, which themselves accreted to form planetary embryos. Current models for the formation of the terrestrial planets indicate that the final stages of planetary accretion are characterized by collisions between tens and hundreds of Mercury- or Mars-sized planetary embryos.

Very large impacts appear to be a fundamental part of planet formation. One of these impacts, between a differentiated protoEarth and a Mars-sized body, is the preferred model for the formation of the Moon (Section 2.3.3) and probably occurred 50 Ma after the formation of the solar nebula. A later, heavy bombardment reshaped the lunar surface, and probably frustrated the proliferation of life on Earth until after about 3.9 Ga ago. Impacts with asteroids several hundreds of kilometres in diameter would have vaporized any terrestrial proto-ocean. Impact cratering remains a significant process on Earth, reshaping the surface and possibly causing

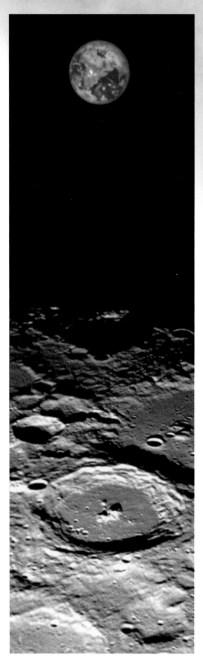

Figure 4.1 View taken by the Clementine spacecraft showing the Earth and Moon, with the crater Plaskett (110 km across), the largest crater on the figure, in the foreground. (NASA)

For craters, except for those specifically described as elongate, the terms 'across' and 'diameter' are interchangeable in this chapter.

occasional mass extinction events, but there are very few obvious craters on the Earth's surface that are visible from space. This is because the surface of the Earth is continually being re-shaped by erosion, volcanism, and crustal deformation and destruction. These processes may be non-existent, less effective or, in some cases, have long since ceased on other planetary bodies, where craters are the dominant landform. Cratering remains *the* significant surface process on other terrestrial planets and many other satellites and asteroids. Figure 4.2 shows a mosaic of images of part of the surface of Mercury.

▣ If impact cratering is a ubiquitous process, why does the Earth show so little evidence of it?

❏ On Earth, several geological processes have been at work, wiping out all traces of the early heavy bombardment and evidence of many other large impacts since then.

Impacts that produce craters on terrestrial planets occur at **hypervelocity**, typically at speeds of several kilometres per second. The average encounter velocity at the top of Earth's atmosphere is about $17\,\mathrm{km\,s^{-1}}$ for asteroidal material, and up to about $70\,\mathrm{km\,s^{-1}}$ for comets.

Figure 4.2 A mosaic of images of Mercury taken $190\,000\,\mathrm{km}$ away from the planet by Mariner 10. The Mariner 10 spacecraft was launched in 1973, and had three encounters with Mercury in March and September 1974, and March 1975. No spacecraft has been to Mercury since Mariner 10, and the majority of the surface remains unmapped. (NASA)

Relative encounter velocities vary throughout the Solar System – asteroidal material at Mars has an average impact velocity of 10–$11\,\mathrm{km\,s^{-1}}$.

Encounter velocities also depend on the gravity of the target planet. Minimum impact velocities (as opposed to average impact velocities) are equal to the **escape velocity** of the target planet, so if any deceleration by atmospheric drag were discounted, the lowest velocity impact on Earth would be $11\,\mathrm{km\,s^{-1}}$ and the lowest on Mars approximately $5\,\mathrm{km\,s^{-1}}$. Because of these high velocities, impact cratering is the most rapid geological process known. Asteroids traverse the Earth's atmosphere in seconds and, following impact with the planet's surface, can form craters that are hundreds of kilometres in diameter in minutes (we will discuss this process in more detail in Section 4.3).

The abrupt nature of the impact process has led some geologists to consider what effect large impacts might have had on the biosphere during the history of life on Earth. The geological history of the Earth can be thought of as a record of environmental conditions, with rocks being deposited one on top of another, with the youngest at the top and the oldest at the bottom. It has long been recognized that within this **stratigraphic record** there is evidence of rapid environmental change. At some of the boundaries between different sedimentary rock types, particularly those denoting dramatic changes in the environment, palaeontologists have documented **mass extinctions** where huge numbers of plant and animal species died out. A mass extinction occurred at the boundary between the Cretaceous and the Tertiary Periods, 65 Ma ago. During the 1980s and 1990s it became clear that a major impact had taken place at the time of this extinction event – diagnostic signatures of impact (many of which you will learn about later in this chapter) have been observed all over the world where the boundary layer is exposed. In addition, a buried crater about 200 km in diameter has been found at Chicxulub (pronounced '*chick*-sulub') on the Yucatan peninsula of Mexico, and is also 65 Ma old. Although the crater is buried, many of the distinctive features of a large crater can be seen in the gravity map of this crater shown in Figure 4.3. Variations in gravity and local magnetic field are commonly used to infer the presence of buried structures. Although some scientists question whether this impact was responsible for the Cretaceous–Tertiary mass extinction, the coincidence of one of the world's largest impacts occurring at the same time as one of the largest mass extinctions is compelling evidence to many others.

In a few cases, it has been possible to document directly the effects of an impact. On 30 June 1908, a huge fireball was observed over Europe and Russia. A large detonation was also recorded by seismometers. Eyewitnesses in a remote part of Siberia reported feeling a powerful shock wave. Later, expeditions to the region found a huge area of forest, over 2000 km^2 in extent, flattened by the force of the blast. This is now thought to have been a small asteroid or comet fragment, about 20–30 m across, which exploded in the Earth's atmosphere. The object was small, but the energy released was huge – equivalent to 10 megatonnes of TNT, which is similar to a large nuclear explosion. By comparison, the Chicxulub impact would have released more than 10^8 megatonnes of energy. More recently, in 1994, observatories around the world and the Galileo space probe en route to Jupiter observed as fragments of comet Shoemaker–Levy 9 plunged into Jupiter's atmosphere and exploded. These observations have highlighted the hazard posed by impacts, and added significantly to our understanding of the impact process itself.

The fact that Earth will inevitably experience other substantial impacts in the future means that impactors and the craters they make are of more than purely academic interest. This chapter should give you an understanding of the cratering process, and enable you to describe the impact craters you would expect to find on an unknown planetary body, given only basic information about its size, mass, density and orbital parameters. In addition, you should be able to identify the diagnostic signatures of an impact in the terrestrial geological record, even if the original crater has been eroded. Finally, you should develop an understanding of the principles involved in using the numbers and sizes of craters in an area to estimate the age of a planetary surface.

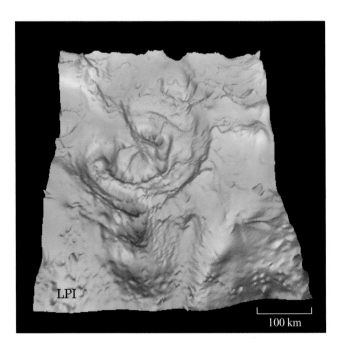

Figure 4.3 A computer-generated three-dimensional gravity anomaly map over the Chicxulub crater. The crater is represented by the near-circular gravity low, shown in blue, at the centre of the figure. This image does not show the shape of the crater; the negative gravity anomaly of the crater corresponds to the relatively low densities of the rocks within the crater. The double-humped central gravity high, shown in green, corresponds to the crater's central uplift now buried deep within the crater. The blue area to the bottom of the figure is part of the regional gravity anomalies of the area. (Courtesy of Buck Sharpton and Lunar and Planetary Institute)

4.2 Historical background

The first specific observation of craters was probably made in 1609 by Thomas Harriot, an Englishman, when he used a telescope to look at the Moon (Figure 4.4). This was several months before Galileo made similar observations (although Galileo was the first to publish, in 1610). Their work triggered a flurry of activity, with progressively more detailed drawings being produced. The engravings of Claud Mellan, made between 1635 and 1637, are remarkably accurate (Figure 4.5). Maps of the Moon appeared throughout the 17th century. Many workers recognized that the features observed were depressions with raised rims and occasional central peaks, but made no attempt to explain their formation. For the majority of the time since then, lunar craters have almost unanimously been thought to be of volcanic origin, even though the topography of the Moon and its craters has been minutely studied by generations of observers. Dissenting voices, such as that of Robert Hooke, who postulated an impact origin in 1665, did little to sway majority opinion. The most serious challenge to the volcanic-formation hypothesis was made in 1893 by the geologist G. K. Gilbert. After analysing the depth-to-diameter ratios of craters, Gilbert suggested that the craters could only have been formed by impact. The central peaks that were commonly observed resulted from rebound of rock, in a similar manner to liquids (think of drops of water hitting a puddle). Bright rays surrounded many craters that were formed from material flung out during the

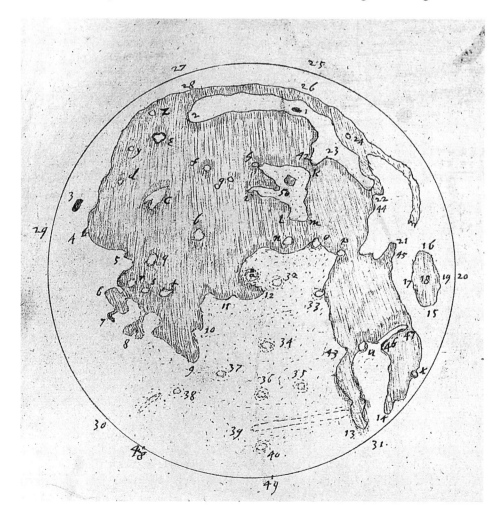

Figure 4.4 Thomas Harriott's map of the Moon, published in 1611, showing the letters and numbers that were used to identify lunar features. (Whittaker, E. A., Lunar and Planetary Laboratory)

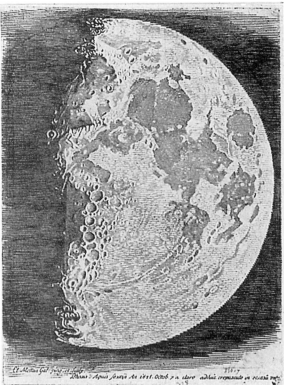

Figure 4.5 Three engravings showing the Moon in different phases, made by Claud Mellan between 1635 and 1637. (Whittaker, E. A., Lunar and Planetary Laboratory)

impact – **impact ejecta**. In Section 4.3 you will see how accurate many of Gilbert's ideas were. The only proviso that Gilbert suggested was that the impact had to be vertical with respect to the surface.

Although impacts with a surface can occur at all angles, from <5° to 90°, early experiments suggested that only vertical (90°) impacts were capable of producing circular craters. So the main argument against an impact mechanism for lunar crater formation was that only circular craters could be seen on the Moon. Gilbert himself showed that the average impact angle was 45° – in fact very few impacts are close to vertical.

In 1916, E. J. Öpik published work that recognized that the impact of meteoroids at very high velocities would be fundamentally different to low-velocity experiments that had previously been carried out. Such impacts would have a similar effect to an explosion, and craters would be circular even at low impact angles. In later years, some of the most persuasive arguments for the impact origin of lunar craters came not from professional scientists, but from R. B. Baldwin, a talented American businessman. He studied the Moon in his spare time and published an influential book entitled *The Face of the Moon* in 1949.

In spite of this and other work, the volcanic hypothesis remained intact, and there was significant unwillingness amongst the scientific community about accepting impact cratering. This began to change in the 1960s with the work of the American geologist Gene Shoemaker, who made a lifetime study of cratering.

■ Terrestrial volcanic craters are poor analogues of lunar craters, as they only occasionally possess central peaks, and have very different ejecta patterns. Suggest one reason why the volcanic hypothesis for lunar crater formation was popular.

❏ Craters produced by early impact experiments at low velocity showed a range of shapes, suggesting that circular craters were formed by a different method.

Figure 4.6 Oblique aerial view of Meteor Crater (or Barringer Crater, as it is sometimes called), Arizona, USA, showing its raised rim and the subdued, surrounding hummocky ejected deposits (foreground). Meteor Crater is 1.2 km in diameter and 200 m deep, with a raised rim 50–60 m high. It was formed about 50 000 years ago. (NASA)

Of the 170 known impact craters on the Earth, **Meteor Crater**, Arizona (Figure 4.6), was the first to be recognized as being formed from an impact and remains the best-known impact crater to this day. In 1906, D. M. Barringer provided good evidence for impact formation, but Meteor Crater was not universally accepted as an impact crater until the 1960s. Its impact origin was finally demonstrated in papers published in 1960 and 1963 by Shoemaker. Three lines of evidence show that an impact, and not a volcanic eruption, formed Meteor Crater.

1   As Barringer had recognized, many large fragments of an iron meteorite have been found on the desert plains surrounding Meteor Crater. Impact structures that are larger than Meteor Crater do not usually preserve fragments of the original projectiles because they vaporize on impact.

2   Most volcanic craters are formed by sustained compression of volcanic gases, which blast out large volumes of fragments. Volcanic ejecta form simple aprons around craters, the first-erupted materials at the bottom, and the last at the top. Shoemaker's field studies showed that a completely different situation exists at Meteor Crater, where the ejecta form an inverted 'flap', flung out and overturned in a single blast. Thus, a drill hole through the ejecta on the rim would pass through layers of the same material twice. In Shoemaker's words, the strata appeared to 'have been peeled back from the area of the crater, somewhat like petals of a flower blossoming'. Similar structures are observed in laboratory experiments involving hypervelocity impacts (Figure 4.7).

3   The final piece of evidence comes from an investigation of the effects of the intense pressures involved in an impact. At Meteor Crater, shattered target rocks are cemented together by **glass** at depths of 200–400 m. The glass is formed by melting of the target during the impact. Melting occurs principally as a result of the very high pressures involved during an impact, and because the pressure is applied almost instantaneously, it is referred to as **shock pressure**. Unusual high-pressure minerals (see Section 4.4.2) are often found in the rocks affected by impact. These minerals are produced by shock pressure during the impact.

In 1962 Shoemaker applied similar ideas to the Moon, and made a case for the impact formation of Copernicus crater, one of the most prominent craters on the nearside of the Moon. Before the decade was out, most scientists had come round to the view that lunar craters were formed by impact rather than volcanism. The

(a)   (b)

Figure 4.7   (a) Impact crater made in a NASA laboratory by firing a projectile with a velocity of about $1\,\mathrm{km\,s^{-1}}$ into a target of colour-banded sand. (b) Detail of the crater rim, showing the overturned 'flap' of ejecta. The pale bands are approximately 1 cm thick. (NASA)

recognition that cratering on Earth is a significant part of our planet's geological history, rather than just a few isolated events, and is an ongoing process, took much longer. It was not until the 1980s and 1990s that impact cratering became part of the mainstream of geological science.

Today, work on impact cratering is progressing rapidly in many different areas:

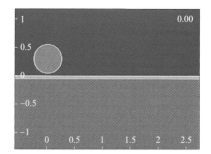

- Analyses of terrestrial-planet surfaces enable the rate at which impacts occur to be constrained.

- Observational astronomy helps in assessing impact hazards by mapping the number and distribution of **Near Earth Objects** (asteroids, and probably some inactive comets, which have orbits that come near to that of the Earth).

- Geological mapping and studies of the target rocks involved in an impact extend our understanding of the forces involved in large impacts, their effects on the environment, and their frequency over geological time.

- Studies of the effects of nuclear tests, and experimental hypervelocity work in which projectiles are accelerated by gas guns to velocities of up to 8 km s^{-1}, have helped to shed light on the physics of the impact process.

- Numerical computation (frequently using so-called **hydrocode** analysis, originally developed to simulate nuclear explosions) has enabled workers to investigate the effects of impacts on scales and at velocities that are impossible to reproduce in the laboratory (Figure 4.8).

4.3 The impact process

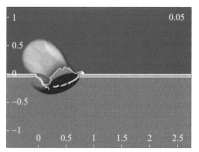

When objects less than 10–20 m in size pass through the Earth's atmosphere they lose most of their initial cosmic velocity, and much of their mass by ablation (i.e. loss of surface layers due to melting and vaporization), and hit the surface at a terminal **free-fall velocity** of a few hundred metres per second. If the object is smaller than about 10–20 m then the atmosphere is sufficient to decelerate it significantly, since smaller objects have a higher surface area (and thus present a relatively larger area to the atmosphere) compared to their mass. Impact of these meteorite-sized objects may excavate a small pit (usually 10 mm to < 10 m across, see Figure 4.9), and in exceptional cases reaching a few tens of metres across.

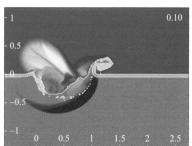

QUESTION 4.1

Study the photograph in Figure 4.9 of the pit produced by a relatively recent British impact. What can you say about the likely velocity of the impactor, and the angle of its approach? How do these account for the appearance of the impact site? *Hint:* think about the size of the object, and the likely effects of the atmosphere.

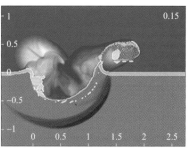

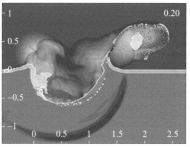

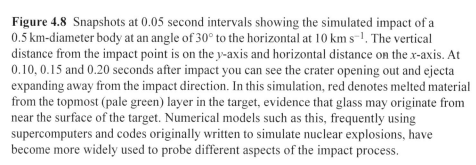

Figure 4.8 Snapshots at 0.05 second intervals showing the simulated impact of a 0.5 km-diameter body at an angle of 30° to the horizontal at 10 km s^{-1}. The vertical distance from the impact point is on the *y*-axis and horizontal distance on the *x*-axis. At 0.10, 0.15 and 0.20 seconds after impact you can see the crater opening out and ejecta expanding away from the impact direction. In this simulation, red denotes melted material from the topmost (pale green) layer in the target, evidence that glass may originate from near the surface of the target. Numerical models such as this, frequently using supercomputers and codes originally written to simulate nuclear explosions, have become more widely used to probe different aspects of the impact process.

Figure 4.9 A suburban setting for one of the smallest terrestrial impact features. This pit was formed by a meteorite that fell at Barwell, Leicestershire, United Kingdom on 24 December 1965. (Institute of Geological Sciences)

In the case of larger objects, a few tens of metres or more in diameter, the situation is very different. These projectiles retain a large portion of their cosmic velocity despite passage through the Earth's atmosphere, and hit the surface at hypervelocity speeds of several kilometres per second or, in some cases, several tens of kilometres per second. Of course, on planets or moons without an atmosphere all objects, whatever their size, strike the surface at hypervelocity.

After a hypervelocity impact, shock waves radiate out from the point of impact, moving huge volumes of target rocks and creating true impact craters. The pressures involved in impacts are much higher than those involved in normal geological processes. The maximum pressures experienced in normal geological processes are typically around 5×10^9 pascals or 5 gigapascals, abbreviated to GPa (Box 2.1).

A complex sequence of events and processes occurs during the formation and immediate modification of a new crater. These may be grouped into three broad stages of impact cratering:

- contact and compression;
- excavation;
- modification.

Figure 4.10 illustrates schematically the stages of the impact process during the formation of a simple crater.

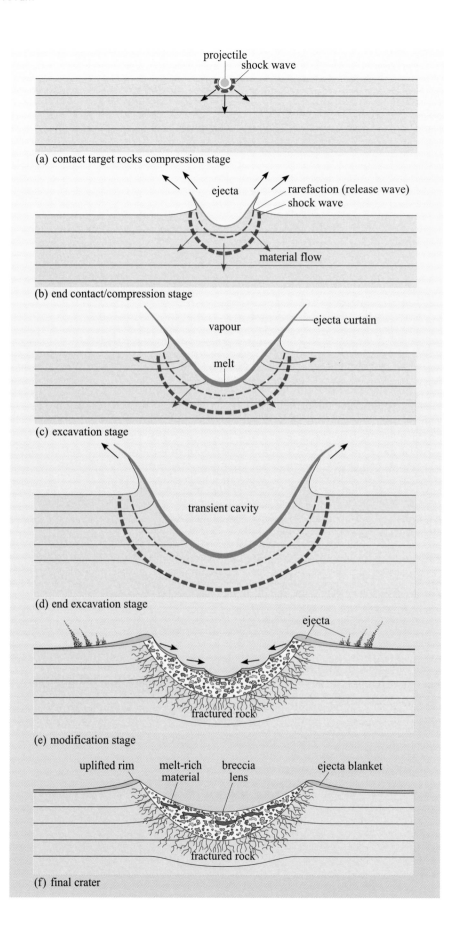

Figure 4.10 Schematic diagram portraying the development of a simple impact structure, through the stages of contact and compression, excavation, and modification. Black arrows show the directions in which target material moves, blue arrows show the directions in which shock waves move at different stages of crater formation. (Adapted from Melosh, 1989)

4.3.1 Contact and compression stage

This first stage of impact cratering begins the moment the projectile makes contact with the ground surface (Figure 4.10a). As the projectile travels into the target, it compresses the target material and accelerates it to high velocities. Simultaneously, the projectile itself decelerates. Shock waves originate at the point where the projectile touches the target surface. Shock pressures during this initial stage may exceed 100 GPa. This pressure represents an extreme environment for rocks at (or near) the surface of the Earth – a rock would have to be buried to > 2000 km to experience this pressure within the Earth. Both target and impactor are vaporized or melted when the pressure is released. The projectile's kinetic energy is largely transferred to the target and heats, deforms and accelerates the target rocks. The contact and compression stage ends when the projectile has unloaded from high pressure (Figure 4.10b). After this point, the projectile itself plays no further part in the formation of the final crater. This stage lasts for less than one second in all but the largest impacts.

4.3.2 Excavation stage

As the contact and compression stage ends, a roughly hemispherical shock wave surrounds the projectile and propagates into the target – the centre of this hemisphere actually lies well below the original ground surface, since the projectile may have penetrated up to twice its own diameter into the target. This initial shock wave, and the following rarefaction, or release waves (waves that are reflected from the original ground surface, and continue downwards) weaken, fracture, and shatter the target rock. These waves also move material, producing an excavation flow around the centre of the incipient structure.

The movement of material upwards and outwards at upper levels, and downwards and outwards at lower levels, eventually opens out the crater, producing a bowl-shaped **transient cavity** that is many times larger than the diameter of the projectile (Figure 4.10c and d). Excavated material is ejected over the surrounding terrain. A point is reached when the shock and release waves can no longer displace rock. The transient cavity reaches its maximum extent at this point. This marks the end of the excavation stage and the beginning of the modification stage.

The excavation stage would have lasted for approximately 6 seconds during the impact that formed Meteor Crater, and approximately 90 seconds for a crater 200 km in diameter, such as Chicxulub.

■ Why is the impact process so different from most other geological processes?

❏ It occurs over much shorter timescales (seconds and minutes, rather than millions of years) and involves much higher peak pressures (> 100 GPa near the impact point, 10–60 GPa in large volumes of the surrounding rock) than any other geological process that exposes rocks at the surface. Note: eclogites, the most highly pressured rock we see naturally exposed at the Earth's surface, may have experienced only 5 GPa in exceptional circumstances.

4.3.3 Modification

Once the expanding shock waves have weakened and moved beyond the crater rim, they play no further part in crater formation. After this point, modification of the transient cavity depends on gravity and the mechanical strength of the target. Small craters, such as Meteor Crater, preserve the approximate original shape of the transient cavity, with only minor modification as debris cascades down the walls to form a layer of fragmented material, called a **breccia** lens, in the base of the crater (Figure 4.10e and f). In larger craters, the transient cavity cannot sustain itself and it frequently collapses under gravity. Slump terraces may form on the walls, as rocks at the edge of the transient cavity collapse inwards along curving, concentric faults. Some craters may contain significant volumes of glassy material, formed from target rock that melted during the impact. Central peaks, or peak rings, form on the floor of the crater as material beneath the impact structure rebounds. This uplift is about one-tenth the final diameter of the crater so, for example, rocks beneath a crater 100 km in diameter will be uplifted vertically by 10 km during the impact event (you can see the central peaks in the crater Plaskett quite clearly in Figure 4.1). Over much longer timescales, as the crust of the planet gradually accommodates the impact, the crater may flatten out until it is defined by little more than slight differences in the material's appearance.

Even for large craters, 200–300 km in diameter, the modification stage will be completed 15 minutes after impact.

4.3.4 Additional effects

In addition to the physical excavation and modification of an impact crater, a variety of other processes occur during an impact. Large impactors may bore a hole in the atmosphere of the target planet, following their passage through it. Vapour and debris from the impact may be drawn upwards into this partial vacuum to high levels in the atmosphere. In addition, ejecta from the impact – glass, and unmelted rock fragments – may be accelerated at high velocities to form an apron of material around the impact site. Some material may be ejected at sufficient velocity to completely escape a planet's atmosphere. Glass that re-enters the planet's atmosphere may fall thousands of kilometres from the impact site, with samples exhibiting aerodynamic shapes. In large impacts, if enough of this material re-enters at the same time, it can significantly heat the atmosphere. The Chicxulub impact probably led to widespread forest fires, even in areas thousands of kilometres from the impact, as re-entering ejecta overheated the atmosphere, causing forests and other plant matter to combust. On Earth, as most of our planet's surface is water, most impacts are more likely to hit an ocean, causing huge tsunamis. Finally, ejecta from an impact may be accelerated with sufficient velocity to escape a planet's gravity, leading to an exchange of material between planets. On Earth, this may happen during impacts that form craters as small as 10 km, so material may be ejected from our planet as often as every few hundred thousand, or a million, years.

QUESTION 4.2

Outline those aspects of the impact process that might have adverse effects on the terrestrial biosphere. In addition, consider the different effects of an impact into continental crust and an impact into an ocean.

4.4 Identifying impacts

4.4.1 Crater morphology

Perhaps contrary to intuition, experiments show that elliptical craters are formed only at angles of incidence that are nearly tangential to the surface, below about 10 degrees. This apparent anomaly arises because hypervelocity impact craters are formed by essentially instantaneous releases of energy rather than mechanical distortion, creating an explosion structure.

QUESTION 4.3

Examine the photograph of the Moon's surface in Figure 4.11. Give a rough estimate of the proportion of the craters that are *obviously* elliptical. You should concentrate on the central part of the photograph – foreshortening will make *all* craters near the edges of the Moon appear elliptical. What inference can you draw from this observation about the process that forms impact craters?

The specific crater morphology that we observe depends on a large number of factors, including the size and velocity of the impactor, composition of impactor and target rock, the strength and porosity of the impactor, angle of impact, and the gravity of the target planet. However, a few broad classes of crater morphology may be recognized.

Microcraters

These are found on lunar samples and man-made objects that have been in orbit for extended periods, such as the Long Duration Exposure Facility (LDEF) and the Hubble Space Telescope (Figure 4.12). These craters may be as small as 10^{-7} m (0.1 μm) across. They are made by the hypervelocity impact of cosmic dust derived from comets, or fine asteroid debris.

Figure 4.11 An Apollo photograph of the Moon. Most of the area shown lies on the far side of the Moon, and consists of heavily cratered highland terrain. (NASA)

LDEF was a NASA satellite designed amongst other things to record the impact of cosmic dust. It remained in orbit for almost 6 years, and was returned to Earth by the Space Shuttle.

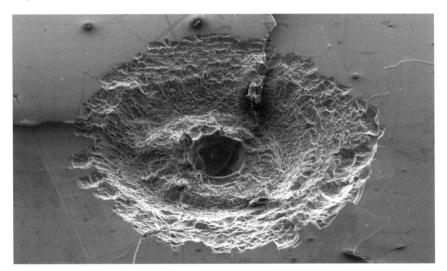

Figure 4.12 Microcrater found on a solar panel recovered from the Hubble Space Telescope (HST) after a servicing mission. The crater is only 0.5 mm across, and was formed when a cosmic dust particle (possibly as small as 0.05 mm in diameter) hit the HST panel at high velocity. Analysis of material in the base of the microcrater suggests that it was made from cosmic dust rather than space debris that originated on Earth. (Giles Graham/Open University)

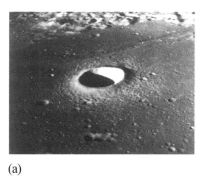
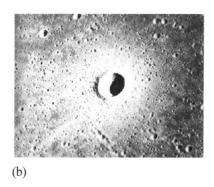
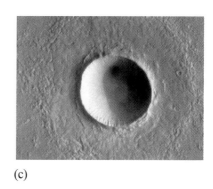

(a)   (b)   (c)

Figure 4.13 Simple craters: (a) and (b) on the Moon, (c) on Mars. The lunar craters are 7 km and 3 km in diameter, respectively; the Martian crater is 2 km in diameter. Note the similarity with Meteor Crater (Figure 4.6). Apart from debris infilling their bases, these bowl-shaped depressions retain the approximate shape of the transient cavity. (NASA)

Simple craters

Meteor Crater is a typical example of a simple crater on Earth. It is similar to others found on all solid surfaces in the Solar System. Simple craters are up to several kilometres in diameter, and are bowl-shaped depressions that lack a central uplift or terracing (Figure 4.13).

The rim-to-floor depth of simple craters is approximately one-fifth of their diameter.

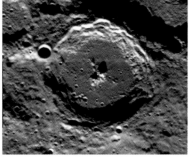

Complex craters

These craters are common in the Solar System. Chicxulub is a terrestrial example. The transition between simple and complex craters depends largely on gravity and the strength of the target material. On Earth the transition point between simple and complex craters may be as low as 2 km in soft sediments (4 km in more competent rock), while on the Moon (Figure 4.14) the transition point occurs in craters between 10 km and 20 km. Mercury shows the transition at about 7 km. So, the transition from simple to complex craters scales approximately inversely with a planet's gravity – Mercury has one-third Earth's gravity and the Moon has one-sixth the gravity of Earth. Complex craters are characterized by terraces of slump blocks, with the terrace width decreasing inwards, and central peaks formed from fluid-like rebound.

(a)

(b)

Figure 4.14 (a) Plaskett crater on the Moon photographed by Apollo 14. This complex crater, 110 km in diameter, shows clearly developed central peaks and slump blocks forming terraces around the rim. (b) A Lunar Orbiter photograph of Copernicus (93 km in diameter), one of the most closely studied lunar craters. Copernicus is located on the south edge of the Mare Imbrium. It is easily visible with binoculars under good lighting conditions. Morphological points to note at increasing distances outwards from the centre are: a cluster of central peaks; the flat crater floor, perhaps filled by impact melt; the multi-terraced crater wall, with terraces formed by inward slumping; a continuous ejecta mantle, extending out to about the same distance as the diameter of the crater – features are diffuse within this zone; a discontinuous ejecta apron, with many secondary craters and crater chains. (NASA)

QUESTION 4.4

Look closely at Figure 4.15. The Clementine data, represented as colours, provides information on the composition of materials in and around the crater. It is clear from the variations in colour that there are significant variations in rock-type over the area. Moving out from the centre of the crater to beyond the rim, what are the broad changes that you observe? Based on what you've learned about cratering, why do you think these differences might arise? What types of rocks do you think the individual colours might correlate with?

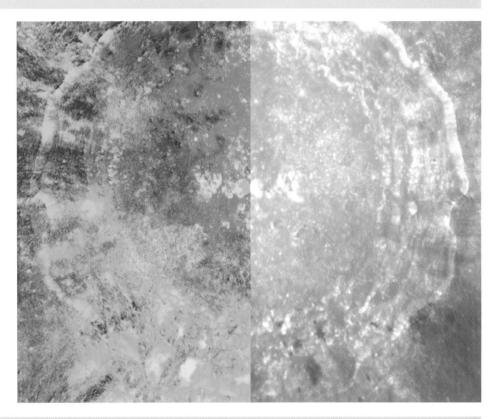

Figure 4.15 Copernicus crater, together with data (the different colours on the left) from the Clementine probe on the composition of materials in and around the crater. (NASA)

Elongate craters

Although almost all craters are circular, if the impact angle falls at an angle below about 10° then elongate craters may be produced (Figure 4.16), such as Messier and Schiller, two elongate lunar craters. The huge Orcus Patera feature on Mars, which is 380 km × 130 km, may also be evidence of a low-angle impact.

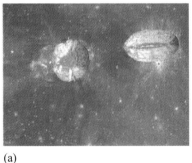

(a)

(c)

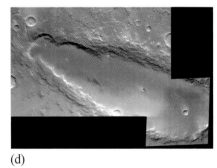

(d)

(b)

Figure 4.16 Three elongate craters. Messier, shown in (a) and (b), comprises two craters. Material may have ricocheted out of the first, more elongate, crater to form the second one. The larger of the two Messier craters is 11 km × 9 km. Schiller (c) is a huge low-angle impact structure, 179 km × 71 km in size, visible in the lower centre of the image. Note that many craters in this image appear elongate due to the perspective, but Schiller is clearly different. (d) Orcus Patera on Mars. This crater, 380 km in length, may also be a low-angle impact structure. (NASA)

Figure 4.17 Apollo photo of the Orientale basin, approximately 900 km in diameter, located on the far side of the Moon. Major concentric ridges are present, but there is no central peak complex. The centre of the basin has been flooded by lavas and there are some small patches of lava in the outer ring. (NASA)

In extreme low-angle impacts, the resulting features cease to be true explosion structures, as the object does not penetrate the target but ricochets away. The scaling laws that apply to elongate craters to get an idea of impactor size based on crater diameter are different from those used for circular craters. If Orcus Patera is a low-angle impact structure, the object that formed it would have been huge, possibly greater than 20 km in diameter.

Multi-ring basins

These are the largest impact structures that have been observed: none are known on Earth, but the Orientale Basin on the Moon (Figure 4.17) and Valhalla Basin on Callisto (Figure 4.18) are good examples from elsewhere in the Solar System. The complex system of scarps that surrounds the central basin of multi-ring basins makes it difficult to define the diameter of the actual crater, but the overall features are huge:

- the basin at the centre of Orientale is encircled by a scarp that is 900 km in diameter and up to 7 km high,
- Valhalla has scarps encircling the basin that are up to 4000 km in diameter.

Multi-ring basins form from large impacts into a lithosphere overlying an asthenosphere. Fluid motions set up in the asthenosphere cause fractures in the lithosphere giving rise to the rings.

Figure 4.18 Voyager image of the vast Valhalla impact site on Callisto. Valhalla is the largest known impact structure in the Solar System, with a diameter of 4000 km, yet its topography is so subdued that each of the numerous visible ring scarps rises only 1–2 km above its surroundings. The impact that formed Valhalla must have taken place at a time when Callisto's lithosphere was thin, and its mantle was hot and mobile enough to smooth out the surface topography. Valhalla's bright central area is about 300 km across: the site of the (now vanished) transient cavity. (NASA)

■ Are complex craters and multi-ring basins merely scaled-up versions of the smaller structures, or are there systematic differences?

❏ Large craters are markedly more complex than smaller ones. In particular, the larger impact structures, such as the 900 km-diameter Orientale Basin on the Moon, exhibit a conspicuous 'bull's eye' pattern of concentric rings.

■ The Orientale Basin on the Moon shows less evidence of **viscous relaxation** than the Valhalla Basin on Callisto (its rings are far more prominent topographic features). Can you think of a reason for this?

❏ The likely answer is that, at the time that Valhalla formed, Callisto had a relatively thin or weak lithosphere, which was unable to support substantial topography.

4.4.2 Diagnostic shock features in rocks

Impact craters are usually clearly visible, and even old craters are well preserved on the terrestrial planets, icy satellites, and asteroids. On Earth, however, geological activity has meant that many craters are partially or completely eroded, or buried. In spite of this, impacts are known to have occurred at given times in the geological record. This is because an impact is a highly unusual geological process, involving materials, and pressure and temperature conditions that are not observed elsewhere on Earth. This gives rise to a set of diagnostic indicators of impact, some of which are local to the impact site, and some of which are observed in more distant settings such as an impact layer in an otherwise normal sequence of rocks. Apart from actual fragments of the impactor, which are only found in the case of some small impacts, these diagnostic indicators take several forms:

- High pressures lead to shock effects on rocks and minerals. Fragments showing shock effects may be carried thousands of miles from the impact site.
- Target rocks may be altered by the impact, but without being completely fragmented or melted. This is **shock metamorphism**, which typically occurs in rocks within, or close to, the impact structure.
- Impact melting of the target rock is another common feature, where the impactor material may contaminate glass.
- Impactor material may be included in the **distal ejecta** carried far from the site, providing a geochemical signature of impact.

Shock effects in minerals

Under the extreme pressures produced during an impact event, various effects are produced in the minerals that make up the target rock. For example, graphite may be converted to diamond; quartz may be converted to stishovite at shock pressures greater than 12 GPa, and coesite at pressures greater than 30 GPa.

Stishovite and coesite are both high-pressure forms of SiO_2.

Shock waves also produce a variety of microstructures in minerals. At comparatively low shock pressures (5–8 GPa), planar fractures can be seen in quartz, but at higher pressures **planar deformation features** are observed. These are sets of narrow planar features, where the original crystalline quartz has been transformed into an amorphous phase (Figure 4.19). These features are distinctive of impact, and are *not* produced by any other known geological process.

Figure 4.19 Planar deformation features in quartz recovered from a drill core into the Woodleigh impact structure, which is a large buried crater close to the coast of Western Australia. The field of view is 0.4 mm. (Robert Hough/Open University)

Shock metamorphosed rocks

Several rock types are produced during impact that reflect the various stages of the impact process of contact and compression, excavation, and modification. **Shatter cones** are a commonly observed feature of many impact rocks. These are distinctive curved, striated fractures that may be present as complete cones. These features can form at quite low shock pressures (2–10 GPa) and have been reproduced in hypervelocity impact experiments in the laboratory.

Breccias, which are rocks composed of angular fragments of pulverized target rock, are found beneath and around the impact site. In some cases this brecciated rock is bound together by melt, and may show flow structures. Such rocks are termed **suevite** (pronounced *sway*-vite) and appear to be characteristic of impacts into crystalline silicate rocks.

Another characteristic impact-produced rock type is **pseudotachylite**. Although not observed in all impact structures, this rock is found extensively at very large impacts. Pseudotachylites typically occur as large veins, sometimes several kilometres long and hundreds of metres wide (Figure 4.20). These veins contain abundant rounded inclusions of target rock, of all grain sizes, set in a black or greenish-black fine-grained material. Its presence provides evidence of local melting.

Figure 4.20 Extremely large pseudotachylite vein in the rocks of the 2 Ga-old Sudbury impact structure in Canada. The pinkish fragments of target rock are set in a dark-coloured finer grained material. (Iain Gilmour/Open University)

Impact glass

During an impact a large volume of rock in the target is heated to well beyond its melting point. When the shock wave has passed through the rock (i.e. the rock has been taken to a very high pressure, and then the pressure released) spontaneous melting occurs throughout this volume. Some of this melt is incorporated into fragmented target rocks, for example suevites, some may remain in the vicinity of the crater as a large melt sheet, and some is expelled from the transient cavity during the early stages of impact at high velocity. This material can travel on ballistic trajectories for thousands of kilometres from the impact site, and will chill rapidly in flight to form a glass. These glasses, termed tektites, often have aerodynamic shapes from their passage through the atmosphere.

Geochemical signatures of impact

Aside from the physical effects of impact, there may also be a chemical effect as impactor material is incorporated in the melt, or fragmented and added to distal ejecta. Knowledge of meteorite chemistry allows this unusual geochemical signature in otherwise normal rocks to be recognized; the occurrence of the Chicxulub impact was inferred from an iridium anomaly in sediments from the Cretaceous–Tertiary boundary, before the crater was discovered. Iridium is a very rare element in rocks in the Earth's crust, but much more abundant in meteorites.

QUESTION 4.5

How could you show that an impact event had occurred where there was no superficial evidence of a crater? Think about things you might analyse in rocks that are distant from the possible impact site and, assuming you have some clue where the buried crater is, what you might do closer to a buried structure.

4.5 Impactors and targets

4.5.1 Types of impactor

Although the size and velocity of the impactor may be the principal control on the size and morphology of the final crater, the composition and density of the object are also important. Impactors fall into two broad populations: asteroids and comets. Apart from having distinct compositions, they also have substantial differences in relative velocities, source regions, and delivery mechanisms (you will consider this in detail in Chapter 7).

Asteroids are relatively dense, composed of silicate rock and metal, and most originate in the asteroid belt between Mars and Jupiter. They are deflected into the inner Solar System following collisions in the asteroid belt and gravitational interactions with Jupiter, and at Earth's orbit (1 AU) they have average impact velocities of $17\,km\,s^{-1}$. Comets are low-density objects made of ices and minor amounts of silicates, which originate in the Kuiper Belt near the orbit of Pluto, or the Oort cloud (see Chapter 1). Occasionally a Kuiper Belt body may be disturbed by the interactions of the giant planets so that its orbit crosses that of Neptune. A subsequent close encounter with Neptune will change its orbit again, and may send it into the inner Solar System where it might be visible from Earth as a comet. Bodies residing in the Oort cloud are much too far out to be disturbed by planets, but as the Solar System orbits the galactic centre it can pass close enough to other stars for their gravity to affect Oort cloud objects. Again, some of these may fall towards the inner Solar System and be observed as comets. At Earth's orbit these objects may have collisional velocities of up to $72\,\mathrm{km\,s^{-1}}$. Given the different source regions for these populations, it is clear that the inner Solar System is subject to impacts from both comets and asteroids, whilst the outer planets rarely experience asteroidal impacts. Instead, planets and satellites in the outer Solar System may be subject to a much higher comet flux.

Little is known about the ratio of asteroid to cometary impactors in the inner Solar System. This is because it is difficult to classify the composition of Near Earth Objects and identify the nature of impactors from geological evidence. It is also likely that the ratio has changed significantly over time. Brief periods of intense cometary bombardment may occur when our Solar System passes close to other stars or giant molecular clouds, scattering comets from the Oort cloud into the inner Solar System.

4.5.2 Nature of the target

On Earth, the comparatively small number of distinct craters is testimony to our geologically active planet, where renewing and wearing away of the surface occurs on a relatively short timescale when compared to the age of the Solar System. Abundant craters on other planets, satellites and asteroids suggest much less geological activity. In some cases, information on the geology of the target can be derived from the types of craters observed.

Many Martian craters look as though they were formed by objects hitting material that had the consistency of wet cement (Figure 4.21). It is thought that the distinctive ejecta patterns of these **rampart craters** result from the mobilization of trapped ground water, or melting of permafrost ice on impact, yielding ejecta with different properties from those formed on 'dry' planets. The observation of **fluidized ejecta** suggests the presence of abundant volatiles, such as CO_2 or water ice, in the subsurface.

The surfaces of most of the satellites of the outer planets are composed of ice. Intuitively, one might suppose that craters on these bodies should be similar to the 'wet cement' craters on Mars, but this is not so – they resemble ordinary lunar craters. One reason for this may be that surface temperatures on the outer-planet satellites are as much as 200 K lower than on Mars. At these frigid temperatures, ice is mechanically much stronger than at higher temperatures and resembles low-density rock. Early in the history of the Solar System, lithosphere temperatures of icy satellites may have been higher than they are today, permitting some glacier-like flow of surface ice structures to take place. Furthermore, the icy lithospheres of large satellites may have been thinner early in their history, when their interiors were warmer. Thus, surface topographic effects would become subdued as the volume of asthenosphere that was deformed on impact gradually returned to its original position by viscous relaxation, so now only the ghostly imprints of former craters are visible. Examples of these imprints are visible on Jupiter's large icy satellites,

Figure 4.21 Fluidized ejecta apron around Arandas, a Martian crater 28 km in diameter. One prominent lobe of 'flow' material extends out from the crater rim to about one crater diameter. Two or three other series of lobes are present, one reaching more than 50 km from the crater rim. (NASA)

Ganymede and Callisto (Chapter 1). Viscous relaxation also affected rocky bodies such as the Moon, when very large impacts deformed both its rigid lithosphere, and its plastic asthenosphere. On Venus, the process continues to play an important role in the evolution of impact structures due to the high heat flow through that planet's lithosphere (Figure 4.22a). By studying the effects of viscous relaxation on craters, we can place constraints on the thickness of the lithosphere. This technique has been used to derive the minimum thickness for the lithosphere of Venus. On Europa very few craters are visible, suggesting a comparatively young surface. The few large craters that are present often have subdued topography (Figure 4.22b–e), suggesting an interior that is still warm, or even a liquid water layer beneath its surface.

In addition, craters can provide information on the specific target rock type. The observation of impact craters in dunes on Mars presents the most unambiguous evidence that Mars has sedimentary rocks. If the dunes were made from loose sand, the craters would be rapidly eroded. Instead, it appears that the sand in these Martian dunes was cemented prior to impact into a form of aeolian sandstone (Figure 4.23).

Finally, studying the cratered surfaces of asteroids can help our understanding of the internal structure of these bodies. Several asteroids show craters of a size that one might expect should have led to the destruction of the asteroid. In addition, very small craters appear to be rare. These data seem to be consistent with impacts into highly porous material – it is predicted that small, porous targets should have

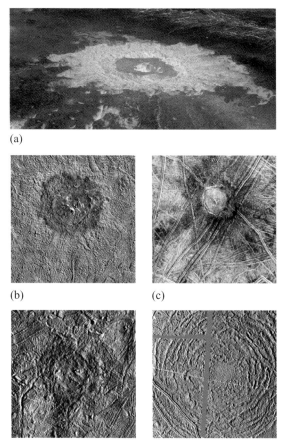

(a)

(b)   (c)

(d)   (e)

Figure 4.22 (a) Craters on Venus often show low relief due to viscous relaxation of the target following impact (false-colour image). (b–e) Examples of craters with subdued relief on Europa, which suggest the possibility of a layer of liquid water several kilometres beneath its icy surface. (NASA)

Figure 4.23 Mars Global Surveyor image of dunes on Mars showing impact cratering. Craters would not be supported in unconsolidated sand, so this image constitutes excellent evidence for the existence of lithified sedimentary rocks on Mars. This image is just over 2 km across. (NASA)

better survivability than non-porous materials and may absorb impacts that would destroy a more solid object. Similarly, small impacts into highly porous targets tend to produce indistinct craters.

■ Would you expect to see *exactly* the same variations in shape and size of craters on other solid bodies? If not, what factors might influence the shapes of craters?

❑ No. Three factors that influence the shapes of craters of varying size are the surface gravity of the body; the thickness of its lithosphere; and the strength of the materials of which it is made.

4.6 Craters as chronometers

In the lead-up to the Apollo Moon landings of the late 1960s and early 1970s, attempts were made to image the surface of the Moon in greater and greater detail. Each refinement in technology revealed more craters, until images from the first spacecraft to land showed that the Moon's surface is cratered down to a scale of millimetres. It became apparent that the longer a surface has been exposed to impacts, the more craters it will exhibit.

The numbers of craters on a surface can be used to estimate its age: older surfaces have been exposed to impacts for longer and show more craters.

If the rate at which bodies impact a surface is known, then the observed density of craters per unit area might be used to estimate the age of the surface (or vice versa – if the age of the surface is known then this could be used to provide information on the rate at which impacts take place). However, although this may sound simple, the technique is far from simple to apply.

4.6.1 Complicating factors

The concept of using populations of craters to estimate surface ages is a relatively simple one to grasp, but there are a number of factors that can complicate its application, and which must be constrained in order to get a meaningful surface age. Important factors to bear in mind are:

- The impact rate and the size distribution of impacting bodies.
- Temporal and spatial variations in impactor population.
- Temporal variation in the target.
- Crater degradation.
- Impacts formed by ejecta from a single large impact, i.e. secondary impacts.
- The need for measured surface ages to calibrate crater counting.

We shall consider each of these factors in more detail.

The number of impacting bodies hitting a given area in a given time is termed the **impact flux**. Whether we measure the observed population of impactors, or look at craters on a surface of a given age, we need to scale from the impactor diameter to the diameter of crater that a given impactor would produce. We also need to know how the size of impactors varies, from the smallest dust to the largest asteroids. Photographs of the Moon show that for every large crater there are many smaller ones. A **size–frequency distribution** can be produced from the measurements of all the crater sizes.

Estimating flux and size–frequency distribution is difficult, but these estimations are made even more difficult by the fact that both may vary in time, and throughout the Solar System. The impact flux was extremely high during the first few 100 Ma of the Solar System as the planets finished accreting – as well as smaller collisions, impacts between Mars-sized objects may have occurred several times. It is possible that the flux then peaked again at around 3.8–4.0 Ga ago, producing the so-called late heavy bombardment that is recorded in the large lunar impact basins. Since approximately 3.8 Ga ago, the flux has declined – probably exponentially – and is now declining at a low rate, or is stable. Mars has a higher flux per unit area than the Earth due to its proximity to the asteroid belt, while bodies in the outer Solar System have a lower flux of asteroids, but a higher flux of comets (Jupiter shields the inner Solar System from many cometary impacts).

Furthermore, the size–frequency distribution varies spatially and temporally. Variation in the ratio of cometary to asteroidal impactors, at different times and in different parts of the Solar System, may change the size–frequency distribution of craters.

Properties of the target that influence cratering statistics may also change over time. For instance, billions of years ago, Mars probably had a much denser atmosphere than it does today, which may have shielded it from small impactors in the same way that the Earth's atmosphere shields us today. We might expect a deficit of ancient small craters in this case. In addition, surface processes, which may have been more intense on Mars during its first 1–2 Ga, may have aided crater degradation.

If projectiles are continuously fired at a surface, early-formed craters will eventually be obliterated by younger ones. When a new crater can form only at the expense of an older one, by overprinting it, the surface is **saturated**. Obviously, the pre-saturation history of a saturated surface is irretrievably lost.

The size–frequency distribution is a statistical term used to describe the relative numbers of objects across a range of sizes.

QUESTION 4.6

Examine Figure 4.24. Does either of the images reveal a saturated surface? What can you say about the relative ages of the two surfaces illustrated?

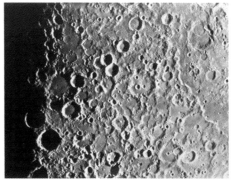

Figure 4.24 Images of planetary surfaces, for use with Question 4.6. (NASA)

Crater-counting statistics become unreliable if the origin of the craters is dubious. Large impacts produce huge numbers of **secondary craters**, so problems can arise if these are included within the population of primary impacts. Statistical techniques have been devised to cope with this issue, such as not counting small craters, many of which may be secondary craters.

Crater counting by itself offers at best only a means of comparing *relative* ages of surfaces. But if we have an absolute age for a surface, and then count the craters on it, we have a calibrated cratering curve (see Box 4.1). Because samples returned from the Moon have been dated by radiometric techniques, as have cratered terrains on Earth, we can use crater statistics in an *absolute* time frame for these bodies. Comparative crater statistics can then be used to estimate reliably the ages of lunar surfaces that have not been dated in the laboratory. But this confidence does not extend far, and so, based on knowledge of the Earth and Moon, and observed impactors, we have to make assumptions about the flux in other parts of the Solar System if we want to use crater numbers to estimate the ages of surfaces.

BOX 4.1 PLOTTING CRATER SIZE–FREQUENCY DISTRIBUTION CURVES

Crater statistics are usually displayed on graphs of crater frequency against diameter. These plots use logarithmic axes, so each interval represents a power of ten. The reason that we use logarithmic axes, instead of the more familiar linear axes, is because the range in diameters is so great, from 1 km to > 4000 km, that a linear graph would be ridiculously long. Furthermore, whereas an increment of 1 km from 2 km to 3 km represents a major *relative* change, an increment of 1 km from 100 km to 101 km is a much smaller relative change. Another peculiarity of these plots is that the vertical axis of the graph

('powers of ten' logarithmic) is not simply 'number of craters per square kilometre', but the number of craters *greater than a given size* per square kilometre of the surface (referred to as the *cumulative* number). Figure 4.25 shows typical crater size–frequency distributions, in this case for different lunar surfaces.

A calibrated cratering curve

One of the striking things about Figure 4.25 is that the cratering curves for the lunar highlands and progressively younger surfaces have the same gradient, suggesting that the impacting bodies had similar size–frequency distributions, even though these data span approximately 4 Ga of Solar System history.

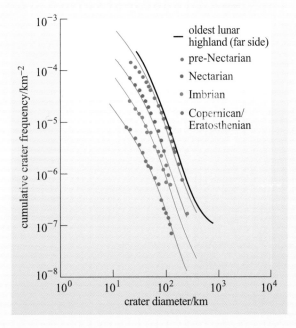

Figure 4.25 Crater size–frequency distributions of the Moon. The curve on the right shows cratering data from the oldest lunar highlands and approximates a saturated surface. The data to the left of this curve are from progressively younger surfaces: pre-Nectarian is the period when some of the oldest surviving lunar impact basins were formed (~ 3.92–4.2 Ga); Nectarian is the period during which most of the major mare basins were formed (3.85–3.92 Ga), and the time of the so-called late heavy bombardment; Imbrian is the period of main mare basalt flooding (3–3.85 Ga) after the Imbrium and Orientale basins were formed; Copernican–Eratosthenian covers the period of cratering after 3 Ga.

Sophisticated statistical manipulations reveal some subtle differences between the populations, and that many of the myriads of small craters (less than 4 km in diameter) on the mare surface may be secondaries. Compilation of crater statistics from all radiometrically dated areas of the Moon has enabled plots like Figure 4.25, and the simplified version in Figure 4.26, to be constructed. These graphs are the key to using craters as chronometers in the inner Solar System. Because there are more accurate ages for the lunar mare surfaces, crater densities for other planetary surfaces may be 'normalized' to the mare value, which is expressed as fractions of the mare value. Alternatively, calculated curves for surfaces of a specific age may be constructed, as in Figure 4.26.

Comparison with Mars

Because no surfaces of Mars have yet been sampled for radiometric dating, crater statistics have been minutely examined to estimate the age of its surface features. Excellent images of Mars are available, so the crater *numbers* are themselves secure, but before we can use them as a chronometer for Martian geological processes you need to understand how the cratering process on Mars compares with that on the Moon. Because Mars has a larger mass than the Moon, and therefore higher gravity, it experiences a higher flux. This is known as gravitational focusing. Mars is also much closer to the orbiting 'scrapyard' of the asteroid belt, so there are more potential impactors. Both these factors indicate that Mars experienced a larger impact flux than the Moon. Exactly how much larger is controversial, but the current 'best estimate' is that it was roughly a factor of 1.6 higher than the Moon. Figure 4.27 is effectively the same plot as Figure 4.26, shifted vertically to take account of the higher overall flux per unit area on Mars. Similar plots can be constructed for the other planets and satellites, but with varying degrees of confidence. For instance, we need to be cautious when applying this method to bodies in the outer Solar System since they experience a very different impactor population (mostly comets) from bodies in the inner Solar System.

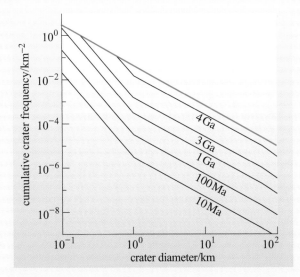

Figure 4.26 'Calibrated' cratering curves for the Moon, based on statistics for surfaces whose ages have been determined radiometrically. The red line shows the crater density on a saturated surface.

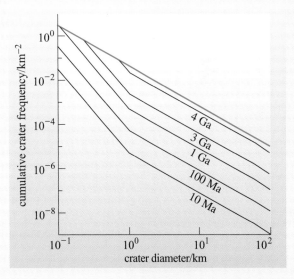

Figure 4.27 Cratering curves for Mars, based on lunar data in Figures 4.25 and 4.26, but adjusted to take account of the slightly higher flux at Mars. Notice it is very similar to Figure 4.26.

4.7 Fluvial and aeolian processes

The processes that have most significantly reshaped planetary surfaces are:

- plate tectonics (on Earth), and
- volcanism and cratering (on other bodies in the Solar System).

On Earth, the effects of water and wind – fluvial and aeolian processes, respectively – are also observed, but these are rarely seen elsewhere in the Solar System. Liquid water is not stable on most planetary surfaces, and most objects do not possess appreciable atmospheres. Mars, however, is a notable exception, and on this planet we see abundant fluvial and aeolian landforms.

Since the Viking missions of the mid-1970s, it has been apparent that Mars has experienced volcanic, aeolian and aqueous activity that has substantially modified its surface. However, it was generally believed that this took place several hundred million years, or even billions of years ago. High-resolution pictures from the Mars Global Surveyor mission have changed this view. As we have seen in the earlier chapters, it seems clear that volcanic activity continues on Mars, and there is cratering evidence for the existence of sedimentary rocks and subsurface volatiles. Additional images from the Mars Orbiting Camera (part of the science payload on board Global Surveyor, abbreviated to MOC) have provided a wealth of information on past and current Martian surface processes.

4.7.1 Fluvial processes

The Viking missions revealed that Mars has experienced a variety of fluvial processes. There is evidence of:

- massive flash floods, extending over hundreds of kilometres (Figure 4.28a), and
- branching river-beds, similar to natural terrestrial drainage systems, suggesting more mature, long-lived river systems (Figure 4.28b).

However, it is clear from crater-counting evidence that neither type of surface drainage has been active for at least 2 Ga. In an extremely exciting development over recent years, MOC images have revealed a third type of Martian drainage feature – small channels, located on the sides of older major valleys or on crater walls (Figure 4.29). These channels are too small to be visible in Viking images. Although dry (no liquid water has been seen on Mars), they appear to be very recent. Many of these surfaces show no discernible craters, and so fall below the limit of resolution of crater counting for determining surface age. This implies that they were made in the last 1 Ma, which is extremely recent on a geological timescale, suggesting that fluvial processes are still occurring on Mars.

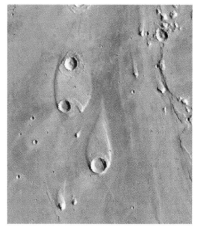

(a)

(b)

Figure 4.28 Evidence for possible flash floods in Mars's distant past. (a) Viking image showing teardrop-shaped craters, formed as floodwaters flowed around them. The craters are 8–10 km in diameter, lying near the mouth of Ares Vallis in Chryse Planitia. The height of the scarp surrounding the upper island is about 400 m, while the scarp surrounding the southern island is about 600 m above the plain. The region pictured is close to the Mars Pathfinder landing site. Branching drainage systems (b) are evidence for more sustained drainage, similar to terrestrial river systems. This type of feature appears to be confined to older terrains, suggesting that sustained river systems on Mars have not been active for several billion years. (NASA)

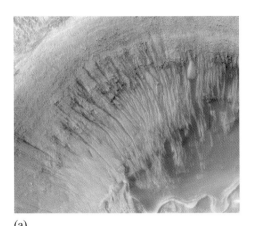

(a)

(b)

Figure 4.29 MOC images showing evidence of recent drainage. Many craters have small channels in the crater walls (a), which appear to be unaffected by subsequent cratering, and have sharp sides, suggesting very recent formation ages. The image is about 5.5 km across. (b) Dunes showing very fresh channels emerging from close to the dune crest. This dune field in Russell crater near the Martian equator is frost covered in the Martian winter so it is possible that the channels have been active during a recent spring thaw. The field of view is 4 km across. (NASA)

4.7.2 Aeolian processes

On Earth a variety of aeolian landforms are recognized in arid and semi-arid environments, with wind-blown material forming dunes of all sizes. A huge variety of dune types may form, depending on the nature and availability of material, and the local wind conditions. Figure 4.30 shows several types of dune that appear to be represented on Mars.

The images provided by Mars Global Surveyor have shown that Mars remains an active planet. There is evidence of recent aeolian activity – new deposition and erosion have taken place in a single Martian year – as well as mobility of volatile phases like CO_2 in the Martian polar regions (Figure 4.31).

4.7.3 Sedimentary rocks

On Earth, fluvial and aeolian processes are fundamental parts of a geological cycle that includes the formation of sedimentary rocks. Eroded material, typically deposited in basins following weathering of upland regions, is cemented by percolating fluids to form a sedimentary rock. As we have seen, cratering of fossilized dunes provides excellent evidence for the presence of sedimentary rocks on Mars. In addition, MOC images provide numerous examples of beautifully layered terrains. Figure 4.32 shows layering in a 64 km-wide impact crater. Horizontal layers have been exposed by erosion, giving the appearance of contour lines inside the crater.

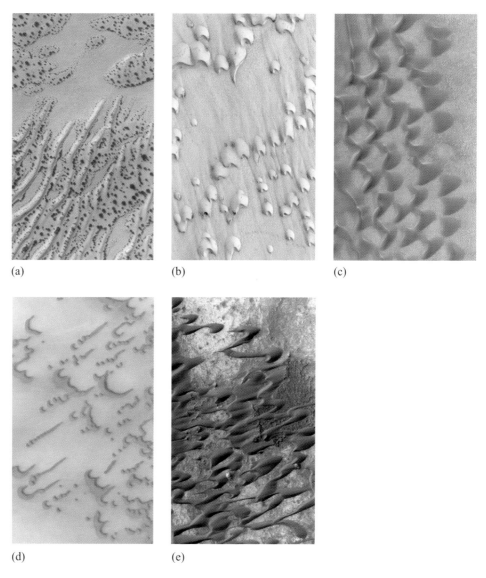

Figure 4.30 A collection of MOC images showing the huge variety of Martian dune forms. The dunes shown in (a) are in the middle of a spring thaw, with dark patches of ice-free sand just beginning to appear. The dunes in (b) are also frost covered – in summer they appear almost black – and are found in Chasma Boreale, a giant trough that almost cuts the north polar cap of Mars in two. The shapes of the dunes suggest that wind is transporting sand from the top of the image towards the bottom. (c) Dunes and smaller ripples. (d) North polar dunes, the morphologies here suggest a reduced sediment supply. (e) A field of dunes in a volcanic depression in northern Syrtis Major. The images in a, c, d and e are 3 km across, while that in b is 1.5 km across. (NASA)

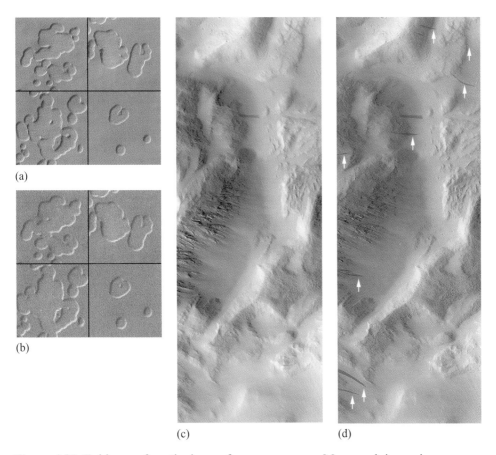

(a)

(b)

(c)

(d)

Figure 4.31 Evidence of continuing surface processes on Mars; each image is approximately 250 m across. (a) and (b) The same area of the Martian south polar cap, the image in (a) taken in 1999, and (b) taken in 2001. When you compare the same area in different years it is clear that small hills have vanished and pit walls have expanded between 1999 and 2001. The pits are formed in frozen carbon dioxide, and a little more sublimates away each year. (c) and (d) are the same portion of a ridged terrain north of the Olympus Mons volcano. Dark streaks are thought to result from avalanching of fine dust. Differences between these two images have occurred over the course of a single Martian year. (NASA)

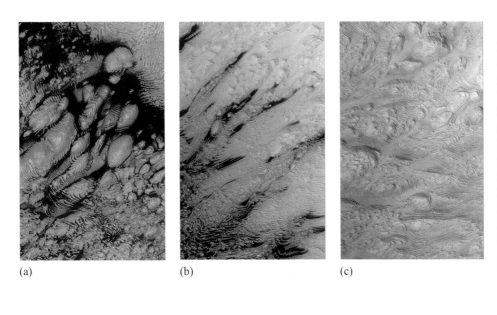

(a)

(b)

(c)

Figure 4.32 Different areas (a–c) of a Martian layered terrain in an impact crater in the western Arabia Terra region. Each image is about 3 km across. Hundreds of layers of similar thickness and texture have been revealed by erosion. This impact crater was a site of repeated sedimentary deposition, possibly related to cyclic changes in climate. It is not known whether the sediments were deposited dry or settled out of water that may have occupied the crater as a lake. (NASA)

4.8 Summary of Chapter 4

- Impact cratering is the most pervasive process affecting the surfaces of solid bodies in the Solar System.

- Unless decelerated by an atmosphere, impacts between different objects in the Solar System occur at hypervelocity, typically $> 10 \, \text{km s}^{-1}$.

- Circular craters form even if the impact angle is quite low. Elliptical craters are formed only at angles of incidence below about 10 degrees.

- The energy released in an impact is huge. On Earth, large impacts may dramatically alter the planet's surface environment, causing mass mortality or mass extinctions.

- The formation of an impact crater may be divided into three distinct phases: contact and compression, excavation, and modification. The first two phases only last for a few seconds and the modification phase lasts only for a few minutes. Even in the largest impacts, cratering is the most rapid geological process known.

- Craters are observed at all scales in a wide range of sizes, from microscopic to thousands of kilometres in diameter. Simple craters of a few kilometres across are bowl-shaped depressions, larger complex craters may have central peaks and terraces, and even larger multi-ring basins may partially disrupt a planet's entire tectonic structure.

- Impacts not only produce craters, they also bring about a variety of changes in the target rocks and minerals – thus the presence of an impact crater can sometimes be inferred from observed diagnostic changes, even if the crater itself has been buried or been eroded.

- There is a variety of impactor types. In the inner Solar System, impactors may be either asteroidal debris derived from the asteroid belt or comets from the Kuiper Belt and the Oort cloud.

- Different types of craters can provide information on the nature of the target, for example, indistinct craters with subdued topography on Europa suggest the presence of a layer of liquid water beneath the surface.

- Impact craters provide a means of dating many planetary surfaces.

- There is abundant evidence for the action of water on Mars in the distant past, for example areas that show the effects of flash floods, or the dry beds of braided river systems similar to those found on Earth. Recent data suggests that small channels may still be active.

- Aeolian processes are ongoing. Changes in the surface of Mars have been observed over times as short as a single Martian year.

- Sedimentary rocks, with detailed, repeated layering have been observed on Mars; some of these rocks may have been laid down in water.

CHAPTER 5
ATMOSPHERES OF TERRESTRIAL PLANETS

5.1 Introduction

In this chapter you will learn about the composition and structure of the atmospheres of the terrestrial planets. So far in this book we have concentrated on the surfaces and interiors of planets and other rocky bodies, but if you view Venus, Earth or Mars from space then you will mainly see the clouds in its atmosphere. Figure 5.1 is an ultraviolet image of Venus showing complete cloud cover. Cloud features are discernible in ultraviolet light, but in visible light Venus appears featureless, resembling a sphere of fog.

It has been known for a long time that Mars and Venus have atmospheres. Drawings of telescopic observations from the 17th century onwards suggest the existence of the polar ice-caps on Mars and complete cloud cover on Venus. By the 1960s, when the first spacecraft missions flew by these planets, the presence of carbon dioxide and water had already been detected in both atmospheres. Nevertheless, little was known at that time of the detailed composition of the atmospheres of planets other than the Earth, and even less was known about their physical properties, such as temperature and pressure. Estimates of the atmospheric pressure, for example, greatly overestimated that of Mars and underestimated that of Venus. Since the early 1960s our knowledge of the planets' atmospheres has expanded enormously due to information from spacecraft and more sophisticated Earth and space-based telescopes.

In Section 5.3, we compare the compositions of the atmospheres of the terrestrial planets. First inspection suggests that they have little in common – Earth has a far greater proportion of nitrogen in its atmosphere than Mars and Venus and is unique in containing a high proportion of oxygen upon which we depend to live. Mercury has such a sparse atmosphere that it would be regarded as an almost total vacuum

Figure 5.1 Ultraviolet image of Venus, taken from the Pioneer Venus Orbiter, showing complete cloud cover. (NASA)

on Earth. A more careful comparison of the atmospheres of Earth, Venus and Mars hints at a common origin and similar amounts of common volatile materials, which prompts the question: could Mars or Venus support or have supported life?

We then look at how the atmosphere changes with altitude and the processes that cause temperature variations for Venus, Earth and Mars. The Earth and Venus are subject to the 'greenhouse effect', which raises the surface temperature. One consequence of this is that water is liquid on much of the Earth's surface. The effect on Venus is much greater, giving a surface temperature some 450 K higher than that of the Earth. At higher altitude, the ozone layer, which protects us from solar ultraviolet (UV) radiation, causes a unique temperature variation in the Earth's atmosphere.

Figure 5.2 Image of Earth taken from Apollo 17 en route for the Moon. (NASA)

If you look up at the sky you will probably see clouds. These are an obvious feature on most days in many parts of the world – and clouds feature prominently in the view of Earth from space (Figure 5.2). In Section 5.5, we consider the nature of clouds in the atmospheres of the terrestrial planets and the conditions required for their formation.

On Earth we are all familiar with wind and weather. Not surprisingly, winds also occur on other planets. In Section 5.6 we examine the causes of atmospheric motion and compare their effects on different planets.

Beyond what is normally considered as the atmosphere lies a region of charged particles (ions and electrons) that is under the influence of the planet. This region, called the ionosphere, can interact with the Sun's magnetic field. Further interactions occur if the planet itself has a magnetic field. The region affected by the Sun's magnetic field is known as the magnetosphere. In Section 5.7 we explore the magnetospheres of the terrestrial planets. Figure 5.3 shows an aurora, which is a spectacular light display produced in the Earth's atmosphere by the disturbance of the magnetosphere. Aurorae have also been observed in Jupiter's atmosphere. This chapter describes the atmospheres of the terrestrial planets in detail – but why do some planetary bodies have atmospheres whilst others do not? There are two parts to this question:

- What are the origins of planetary atmospheres?
- How do planets retain their atmospheres?

Figure 5.3 Aurora borealis as seen from Sweden. Aurorae are generally only seen at high latitudes.

5.1.1 Origins and retention of planetary atmospheres

The origins of the atmospheres include material from the solar nebula (as is the case for the giant planets), outgassing of volatile materials from planetary interiors (the case for Mars, Earth, and Venus) and the solar wind (the case for Mercury).

An important factor dictating whether or not an object in the Solar System can retain an atmosphere is the strength of the gravitational field at its surface – the stronger the field, the stronger the gravitational forces acting on the molecules in the atmosphere. Atmospheric molecules are in perpetual motion in all directions. Without the gravitational field, those moving away from the planet would be lost – and even with the gravitational field, those molecules with particularly high speeds can still escape. This leads to the notion of *escape velocity*, which is defined as: the minimum speed needed before a body has enough *kinetic energy* to escape from the surface of a planet (i.e. overcome its gravitational field). It can be shown that the escape velocity, v_{esc}, for a body of mass M and radius R is given by:

$$v_{esc} = \sqrt{\frac{2GM}{R}} \qquad\qquad (5.1)$$

where G is the gravitational constant.

Whether atmospheric molecules have sufficient speed or not to overcome the gravitational forces depends on the temperature. As the temperature of a gas increases, its molecules move around more quickly, i.e. the average speed of its molecules increases. A fraction of the molecules will always be travelling fast enough to overcome gravitational forces, allowing them to escape to space. At low temperatures, this proportion is negligible, but at higher temperatures the proportion becomes progressively more significant, until most molecules exceed the escape velocity for the planetary body. Escape occurs from a level in the upper atmosphere above which the atmosphere is so thin that a molecule moving outwards has little chance of colliding with another and *will* therefore escape if it has sufficient speed. The relevant temperature is thus the temperature at this level in the atmosphere.

The temperatures where the atmosphere can escape are generally higher than the effective cloud top temperatures (as listed in Appendix A, Table A1).

Different gases have different molecular masses, so their average speeds are different at a given temperature. In order for a planetary body to retain a particular gas in its atmosphere for a period of time of the same order as the age of the Solar System, the average speed of the molecules in the gas should be less than about one-sixth of the escape velocity. (If the average speed exceeds one-sixth of the escape velocity, a significant proportion of molecules will be moving faster than the escape velocity, and will be lost.) This condition is achieved on only a few planets and satellites. Figure 5.4 explores these relationships further. For each of the planets (and the Moon), the temperature is plotted along the horizontal axis and one-sixth of the corresponding escape velocity is plotted on the vertical axis. Note that, in order to cover the range of values needed, the scales are not linear – a particular interval along an axis corresponds to a doubling of the quantity. The sloping lines show the average molecular speeds of each named gas at each temperature. Figure 5.4 thus defines the conditions under which a planet would lose or retain that gas over geological timescales (over several Ga). For example, the giant planets plot well above all the lines and can therefore retain any of the named gases whereas the Moon plots below all the lines and cannot retain any of the gases.

Figure 5.4 Graph summarizing conditions of temperature and escape velocity for which planetary bodies can retain the common gases hydrogen H_2, helium He, water vapour H_2O, ammonia NH_3, methane CH_4, nitrogen N_2, oxygen O_2, and carbon dioxide CO_2 in their atmospheres for long periods. For bodies that do not have a substantial atmosphere, approximate surface temperatures are shown.

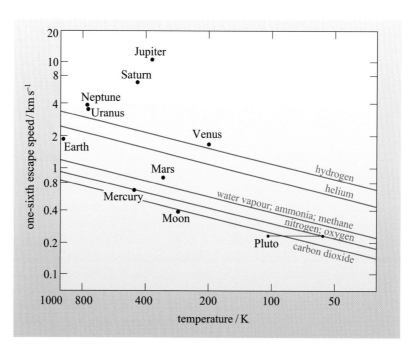

■ Which gases should Mars be able to retain?

❏ Mars plots below the hydrogen and helium lines so cannot retain these gases, but it plots above the lines for water vapour, nitrogen, ammonia, methane, oxygen and carbon dioxide, so these gases could be retained. Water vapour, nitrogen, oxygen and carbon dioxide have indeed been found on Mars. Although Mars plots above the ammonia and methane lines, if these gases were retained they would be converted to nitrogen and carbon dioxide under the conditions found on Mars.

In the image of Mercury in Figure 5.5, there is no discernible atmosphere. The atmosphere of Mercury is an extremely thin layer of gas with a pressure at the surface of 10^{-15} times that of the Earth's atmosphere (see Appendix A, Table A1).

■ Using the information in Figure 5.4, would you expect Mercury to have an atmosphere?

❏ Mercury plots above the CO_2 line so you might have expected it to retain this gas.

Figure 5.4 gives the average surface temperature. The solar-facing side of Mercury is considerably hotter than this (its equatorial temperature can be as high as 740 K). As Mercury rotates relatively slowly, carbon dioxide can escape from the solar-facing side. There are also other mechanisms by which atmospheres can lose gases. Hydrogen and helium may have been present in substantial amounts in early atmospheres but are easily lost through thermal escape, as you have seen. If large amounts of these gases escape rapidly, then other heavier gases can be lost at the same time by being carried along with them. In addition, atmospheric components can be lost through interaction with the solar wind and after impacts.

Retention of gases can be enhanced if the volatile components condense or become incorporated into the planetary surface or deeper layers within the planet. For example carbon dioxide from the Earth's atmosphere has dissolved in the oceans and become incorporated into carbonate rocks.

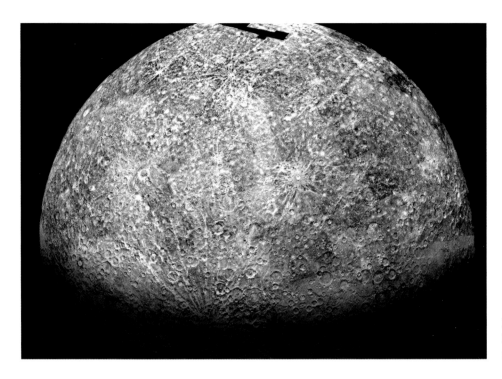

Figure 5.5 Mercury, as seen by Mariner 10 as it sped away from the planet. (NASA)

5.2 Exploration

Direct access to an atmosphere, such as the lower layers of the Earth's atmosphere, enables us to separate and identify the various constituents. In the 18th century, the **diatomic molecules** (molecules consisting of two atoms bonded together) oxygen, O_2, and nitrogen, N_2, were recognized as separate components of Earth's atmosphere. In the last decade of the 19th century, the **noble gases** (Table 5.1) were discovered, which extended our knowledge of the composition of the Earth's atmosphere. The discovery of these unreactive gases was made using spectroscopy, a technique devised some thirty years earlier, which we shall consider shortly. The identification of the other minor components of the Earth's atmosphere depended on the development of other techniques, which have also proved useful in studying the atmospheres of other planets.

Today, spacecraft can be sent to take direct measurements and obtain samples of the atmospheres of other planets. Such missions usually employ the techniques of gas chromatography and mass spectrometry. For our purposes, it is more important to take note of the information that is obtained by these techniques rather than the details of their application.

In principle, **gas chromatography** is a means of separating a complex mixture of gases into its chemically distinct components. This is achieved by pumping a small sample of the mixture along a tube that is filled with a packing material that has been coated with a liquid. As the mixture of gases travels along the tube, those gases that are soluble in the liquid dissolve and are released later by a fresh influx of gas. This solution process slows down the more soluble gases so that the gases that tend to dissolve in the liquid take longer to emerge from the tube than those that are insoluble. Provided the tube is long enough and the liquid suitably chosen, each gas in a mixture of gases will emerge from the gas chromatograph at a

Table 5.1 The noble gases.

Element	Symbol
helium	He
neon	Ne
argon	Ar
krypton	Kr
xenon	Xe
radon	Rn

different time. If the instrument has been calibrated with authentic samples of the component gases then their identification is often possible, but without such a calibration it is difficult to identify the components of a mixture. In planetary exploration it is therefore important to anticipate the results of analyses by this technique, or to couple it with a method of identifying the components independently.

One method of identifying the components is by **mass spectrometry**. This provides a way of measuring the masses of molecules (and of fragments of molecules if they break up on ionization, as they often do) and the relative number of molecules of each mass. Each mass is usually expressed as its **relative molecular mass** (RMM).

The RMM is defined as the mass of the molecule relative to the mass of an atom of an isotope of carbon, ^{12}C, which is given the value of 12 on the RMM scale:

$$\text{RMM (molecule)} = 12 \times \frac{\text{mass of molecule}}{\text{mass of } ^{12}C \text{ atom}} \tag{5.2}$$

The use of the RMM scale conveniently avoids the powers of ten involved in the masses of individual molecules. For example, the mass of a molecule of carbon dioxide, $^{12}C^{16}O_2$, is 7.31×10^{-26} kg, whereas its RMM is 43.99. RMM values can be estimated approximately by summing the mass numbers of the constituent atoms.

◼ Can you envisage any problems in the application of mass spectrometry in identifying molecules such as CO and N_2 composed of their most abundant isotopes: ^{12}C, ^{14}N and ^{16}O (masses 12.000, 14.003 and 15.995, respectively)?

❑ These two molecules have approximately the same RMM of 28, and so will only be distinguishable by mass spectrometry at high resolution.

The mass spectrometers sent on planetary missions so far have been of low resolution and cannot distinguish between molecules with similar RMMs.

Mass spectrometry has, however, a distinct advantage over many other analytical methods. As it measures RMMs, it enables us to distinguish between isotopic variants, for example between $^{12}C^{16}O$ and $^{13}C^{16}O$, the carbon monoxide molecules containing the isotopes ^{12}C and ^{13}C, respectively. These results have been of great value in providing evidence for the origin of planetary atmospheres.

Gas chromatography–mass spectrometry (GCMS) can only be used if a sample of atmosphere can be introduced into the instrument. However, as an alternative, if the amount of light or other electromagnetic radiation that is absorbed or emitted by the sample is studied, then information can be obtained from telescopes and fly-by spacecraft. This technique is called spectroscopy. Some of the noble gases in Table 5.1 were first identified as constituents of the Earth's atmosphere through spectroscopic analysis. Box 5.1 describes the basic principles of spectroscopy.

In planetary atmospheres, the most common atmospheric components tend to occur as molecules rather than free atoms and so they are studied through their molecular spectra.

BOX 5.1 SPECTROSCOPY AND SOURCES OF RADIATION

Sources of radiation may be divided into two broad categories:

- **thermal sources**, which emit light because of their temperature (hot things glow), for example the incandescent filament in a light bulb;
- **non-thermal sources**, which emit light for other reasons, for example the tail of a glow-worm and orange streetlights.

Many natural sources of light are a combination of the two: they are partly thermal and partly non-thermal.

Most light sources, thermal and non-thermal, emit a whole range of wavelengths. (The one common exception to this is the laser, which is a device that produces light of a single wavelength.) Such a range is commonly called a **spectrum** (plural spectra). **Spectroscopy** concerns the production and study of spectra.

Continuous spectra

Light from thermal sources is emitted over an unbroken range of wavelengths. Such sources are therefore said to have **continuous spectra**.

If a narrow beam of light passes through a glass prism, the beam is split up in such a way that different wavelengths travel in different directions, as shown in Figure 5.6. If the original beam contained just a few well-separated wavelengths the result would be a set of quite separate and distinct images, each with its own characteristic colour (wavelength). However, if the beam came from a thermal source it would typically contain all visible wavelengths and the result of passing it through the prism would be a continuous multicoloured band somewhat similar to a rainbow (Figure 5.6).

Spectral information is generally displayed in the form of a graph. When spectra are presented in this way the horizontal axis of the graph usually shows wavelength or frequency (or sometimes photon energy). The vertical axis of the graph normally indicates the intensity of the spectrum at any given wavelength. Such a representation is shown in Figure 5.7a–c.

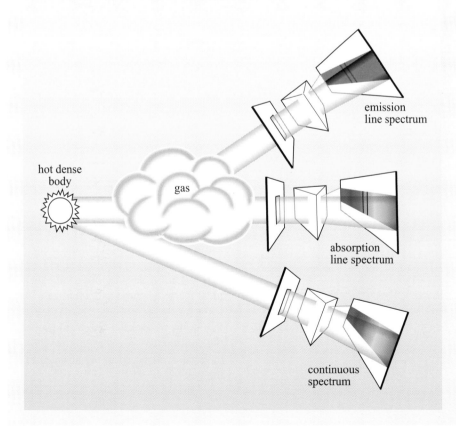

emission line spectrum

hot dense body

gas

absorption line spectrum

continuous spectrum

Figure 5.6 A narrow beam of light split into different wavelengths by a glass prism.

Line spectra: absorption and emission

If a beam of light from an **ideal thermal source** passes through a thin gas of atoms and/or molecules, the spectrum of the emerging beam will generally include a number of narrow *dark* lines. These lines are called absorption lines and correspond to narrow ranges of wavelength that have been wholly or partly absorbed by the gas. This situation is illustrated in Figure 5.7b, which includes a graph of the so-called **absorption spectrum** that arises.

If, instead of the emerging beam, the light emitted by the gas itself is examined, it will be found that its spectrum consists of a number of narrow *bright* lines. These lines are called emission lines, and a spectrum composed of them is called an **emission spectrum** (Figure 5.7c). For many gases, the bright lines emitted cover the same narrow wavelength ranges as the dark lines in the absorption spectrum.

Each type of atom or molecule has its own characteristic set of lines. Figure 5.8a shows the emission spectrum produced by hydrogen atoms; Figure 5.8b shows the emission spectrum produced by helium atoms.

According to quantum theory, each of the electrons belonging to a particular atom or molecule may be in any of a number of allowed states, each of which is associated with some fixed amount of energy. When an electron occupies a particular state in a particular atom or molecule, the atom or molecule has the energy associated with that state. Thus, changes in the pattern of occupied states within an atom or molecule entail changes in the total amount of energy possessed by the atom or molecule. A diagram showing the energy associated with each of the allowed states in a particular kind of atom is called an **energy-level diagram**. The simplest energy-level diagram, that of a hydrogen atom (hydrogen has just one electron), is shown in Figure 5.9.

An absorption spectrum arises when the atom or molecule absorbs light and goes to a higher energy state. An emission spectrum occurs when an atom or molecule that is already in a higher energy state drops back down to a lower energy state. The energy of the photon absorbed or emitted matches the difference in energy between the higher and lower states of the atom or molecule.

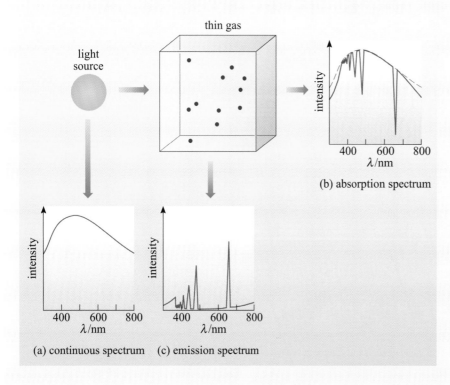

(b) absorption spectrum

(a) continuous spectrum   (c) emission spectrum

Figure 5.7 Three kinds of spectrum: (a) continuous, (b) absorption and (c) emission – seen by observing an ideal thermal source and a thin gas from various directions. The dashed line in the absorption spectrum shows the continuous spectrum that would have been observed in the absence of the gas.

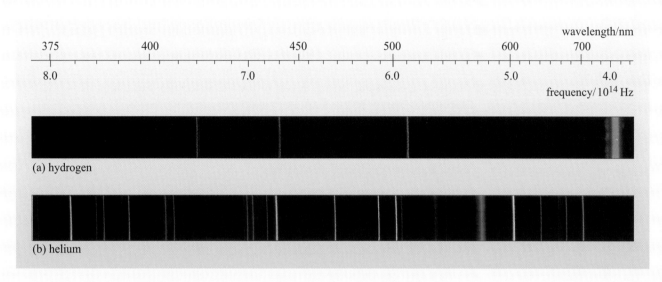

Figure 5.8 Emission spectra produced by (a) hydrogen atoms and (b) helium atoms.

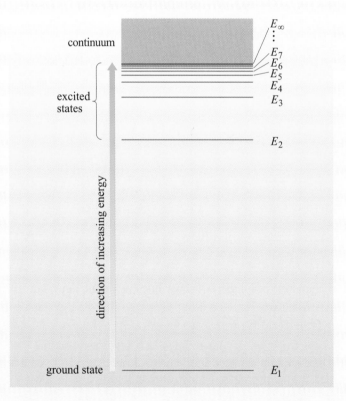

Figure 5.9 The energy-level diagram of a hydrogen atom. Note there is an infinite number of energy levels, but those of highest energy are too closely crowded together to be shown separately on a diagram of this kind.

Molecules, like atoms, have a series of energy levels that give rise to characteristic absorption lines. In principle it is possible to determine which molecules are present by identifying dark lines in the spectrum of sunlight passing through the atmosphere. The energy levels of atoms discussed in Box 5.1 are electronic energy levels; they arise from different states of the electrons in the atom. Figure 5.10 shows the *electronic* energy levels of the CO molecule. For comparison, the electronic energy levels for atomic hydrogen are also shown.

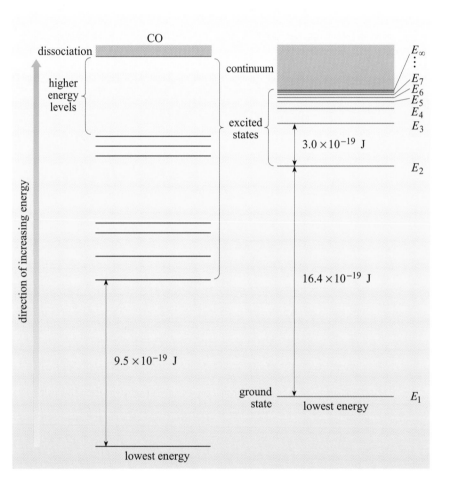

Figure 5.10 Electronic energy levels in the CO molecule and in the H atom.

The emission spectra of H and He in Figure 5.8 show lines in visible light, but spectra are not confined to visible radiation. Other regions of the electromagnetic spectrum are commonly used (see Box 5.2).

CO remains in its lowest electronic energy level unless it is either exposed to photons that are at least as energetic as those in the near-ultraviolet region, or it is at a temperature of the order 10^5 K, or greater. Many other colourless molecules require photons in the UV region to become excited. As you will see later in this chapter, much of the UV radiation from the Sun is absorbed by the Earth's atmosphere so that ground-based telescopes are not of much use in studying the electronic spectra of many molecules.

But as well as electronic transitions, a molecule can also undergo vibrational and rotational transitions. The nuclei in a molecule such as CO are not at a fixed distance apart but act as though joined by a vibrating spring with the bond distance alternately stretching and contracting in a regular manner.

BOX 5.2 THE ELECTROMAGNETIC SPECTRUM

Visible light spans a range of wavelengths, from approximately 400 nm to 700 nm. Electromagnetic waves with wavelengths outside this range cannot, by definition, represent visible light of any colour. However, such waves do provide a useful model of many well-known phenomena that are more or less similar to light. For example, everyone is familiar with radio waves; we all rely on them to deliver radio and TV programmes. Radio waves are known to have wavelengths of about 3 cm or more; their well-established properties include the ability to be reflected by smooth, metal surfaces and to travel through a vacuum at the same speed as light.

The wide range of phenomena that can be modelled by electromagnetic waves is illustrated in Figure 5.11. As you can see, the full **electromagnetic spectrum**, as it is called, ranges from long wavelength **radio waves**, through **microwaves** and **infrared radiation**, across the various colours of **visible light** and on to short wavelength **ultraviolet (UV) radiation**, **X-rays** and **gamma-rays**. These various kinds of **electromagnetic radiation** arise in a wide range of contexts (as illustrated) but fundamentally they differ from one another *only* in the wavelength (and thus frequency) of the electromagnetic waves used to model them.

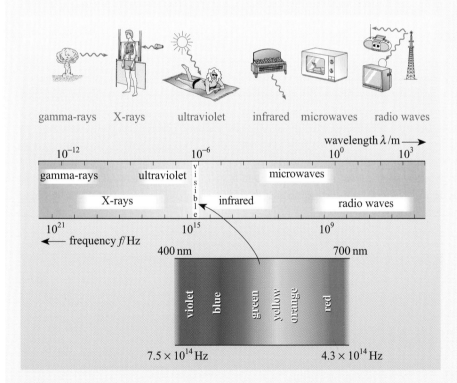

Figure 5.11 The electromagnetic spectrum. Note that the frequency and wavelength scales are logarithmic. Note also that the ultraviolet (meaning 'beyond the violet') adjoins the visible violet, and the infrared (meaning 'below the red') adjoins the visible red. The boundaries of the various regions are deliberately vague; scientists and technologists often draw the divisions somewhat loosely. (NASA)

This vibrational motion gives rise to another set of energy levels corresponding to different vibrational energies of the molecule. A **vibrational transition** of CO is illustrated schematically in Figure 5.12, along with the lowest three vibrational energy levels for the case in which the molecule remains in the electronic state corresponding to the lowest electronic energy level. Note how much smaller the gaps are between the energy levels than is the case for the electronic transitions in Figure 5.10. This means that a vibrational transition typically requires absorption of a photon of lower energy – infrared radiation.

There are also vibrational and rotational energy levels associated with higher energy electronic levels and as a molecule goes to a higher electronic energy level, it can also change its vibrational and/or rotational energy. This leads to the electronic spectra of gaseous molecules being composed of sets of closely spaced lines that are often observed as bands rather than sharp lines.

For vibrational transitions to be recorded as a **vibrational spectrum**, a further condition must be met. In a **heteronuclear** diatomic molecule such as CO where the two atoms are different, the bonding electrons are shared unequally between the two atoms and so the molecule must have an uneven distribution of electric charge, with one end positive and the other negative, as illustrated in Figure 5.13. Here the molecule itself is neutral (it has no overall charge), but one atom retains a small excess of the bonding electrons amounting to 0.18 of an electron and so has a net charge of $-0.18e$, where $-e$ is the charge of an electron. The other atom is slightly deficient in electrons and has a net charge of $+0.18e$. We call this slight separation of charge an **electric dipole**.

For a molecule to have a vibrational absorption spectrum, the electric dipole must change during the vibration. This condition is always fulfilled for heteronuclear diatomic molecules but not for **homonuclear** diatomic molecules (those with two identical atoms) such as O_2. Homonuclear molecules share their electrons equally and so have no electric dipole. They still share the electrons equally when they stretch and compress, and so still have no electric dipole; there is no change in the electric dipole when they vibrate. For molecules with more than two atoms (**polyatomic** molecules, for example carbon dioxide and water) there is always a vibration that leads to a change in electric dipole and so all polyatomic molecules have vibrational spectra.

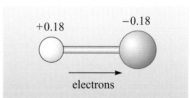

Figure 5.13 Uneven distribution of charge between atoms in a heteronuclear diatomic molecule, giving rise to an electric dipole.

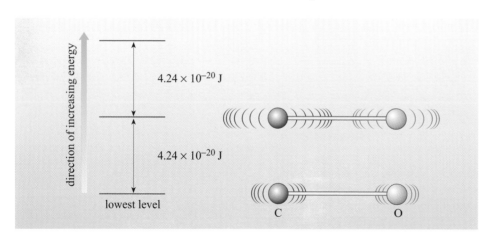

Figure 5.12 Vibrational transitions and vibrational energy levels in CO. To the right are the vibrational states corresponding to the lowest two energy levels.

Homonuclear diatomic molecules, such as H_2 and N_2, will not interact with electromagnetic radiation during their vibration and so will not have a vibrational spectrum. All other molecules do have vibrational spectra.

Infrared radiation emitted by the surface of a planet, which acts as a thermal source of radiation, is used to identify molecules in atmospheres through their vibrational spectra. However, using Earth-based instruments is not easy because the surface of the Earth emits infrared radiation and the atmosphere absorbs it. The use of fly-by spacecraft and space telescopes is preferable.

The application of the techniques described above in fly-by and lander missions to Mars and Venus has provided increasingly detailed descriptions of these planets since the first Mariner fly-by of Venus in 1962. The Soviet Venera and Vega missions to Venus that took place between 1965 and 1986 have provided a lot of information about the physical and chemical properties of the atmosphere of Venus, augmented by the American Pioneer missions of 1978. A series of American missions to Mars, dating from the 1965 Mariner spacecraft to the Viking landers of 1976, the Mars Pathfinder mission of 1997 and the Mars Global Surveyor of 1999–2002, have provided a detailed description of the Martian atmosphere.

The picture that emerges of the three terrestrial planets Mars, Venus and Earth suggests they are bodies with atmospheres that vary considerably in their mass, composition and, as shown in Figures 5.1, 5.2 and 5.14, cloud cover. Earth has an atmosphere of complex structure with much seasonal activity. Clouds of water droplets and ice (H_2O) crystals are extremely variable and cover typically about half of the planet. Venus has an atmosphere that is about 100 times as massive as that of the Earth. It shows no seasonal change and contains an unbroken layer of dense cloud that obscures the planet's surface. The atmosphere of Mars is tenuous and undergoes extreme diurnal (day–night) and seasonal variations, with occasional massive dust storms.

QUESTION 5.1

Comment on the use of mass spectrometry and of gas chromatography in analysing a mixture of CO_2, O_2, N_2, C_2H_2 and C_2H_4. Assume that these molecules are composed of the most abundant isotopes (^{12}C, ^{16}O, ^{14}N and ^1H).

QUESTION 5.2

Which of the following substances could be detected remotely by infrared spectroscopy: CO, H_2 and HCl?

Figure 5.14 Mars as seen by the Hubble Space Telescope in 2001. (NASA)

5.3 Composition of the atmospheres

Although we group Mercury, Venus, Earth and Mars together as terrestrial planets, their atmospheres appear, at least on first inspection, to display more differences than similarities. The major constituents of these atmospheres are listed in Table 5.2 and those of Venus, Earth and Mars are compared in Figure 5.15. Table 5.2 includes the composition of the atmosphere of Mercury although, as we have said, this is very sparse. The compositions given here are those at the surface, where the atmospheres are most dense. The interaction of solar radiation with an atmosphere at higher altitudes leads to the production of reactive species, which undergo chemical reactions that convert some of the molecules found at the surface to different species. For example part of the oxygen component, O_2, of the Earth's atmosphere is converted at higher altitudes to **ozone**, O_3. Consequently, the composition of the atmospheres changes with altitude.

Table 5.2 The major components of the atmospheres at the surfaces of the terrestrial planets.

Mercury		Venus[a]		Earth		Mars	
Gas	Volume ratio[b]	Gas	Volume ratio[b]	Gas	Volume ratio[b]	Gas	Volume ratio[b]
O_2	0.42	CO_2	0.965	N_2	0.781	CO_2	0.953
Na	0.29	N_2	3.5×10^{-2}	O_2	0.209	N_2	2.7×10^{-2}
H_2	0.22	SO_2	1.5×10^{-4}	H_2O[c]	<0.04	Ar	1.6×10^{-2}
He	0.06	H_2O	1×10^{-4}	Ar	9.3×10^{-3}	O_2	1.3×10^{-3}
K	5×10^{-3}	Ar	7×10^{-5}	CO_2	3.4×10^{-4}	CO	7×10^{-4}
		H_2	$<2.5 \times 10^{-5}$	Ne	1.8×10^{-5}	H_2O	3×10^{-4}
		CO	2×10^{-5}				

[a] Some of these values are measured at an altitude of 22 km because surface data are unreliable.

[b] The volume ratio is the fraction by *number* of the atoms or molecules present. Chemists often refer to this as the mole fraction. It is also called the volume-mixing ratio by some atmospheric scientists. When multiplied by the atmospheric pressure, it gives a quantity called the partial pressure (Section 5.5), which may be envisaged as the contribution of a component to the total pressure.

[c] The H_2O in the Earth's atmosphere is highly variable!

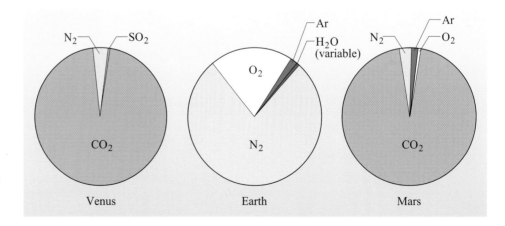

Figure 5.15 The major components of the atmospheres of Venus, Earth and Mars, as measured at the surfaces of the planets. The area of each slice of the pie chart is proportional to the volume ratio of the substance shown.

One major reason for the observed differences in composition of the atmosphere of the Earth compared to those of Mars and Venus is that water exists on Earth in liquid form. The physical conditions on Earth ensure that most of the surface water and atmospheric water exist in the oceans, which act as a reservoir, whereas on Venus water cannot exist in liquid form because of the high surface temperature. The Earth's polar ice-caps form another part of the H_2O reservoir. The ice-caps observed on Mars (Figure 5.14) are composed of both H_2O and CO_2 and there is evidence of ice in the surface rocks. On Earth, CO_2 is largely combined with other elements in carbonate deposits, especially as $CaCO_3$ in sedimentary rocks, whereas on Venus it is the main atmospheric constituent. We can estimate amounts of atmospheric gases transferred to surfaces, such as carbon dioxide lost from the Earth's atmosphere as carbonate deposits; the amount of surface water in the Earth's oceans, seas and lakes; frozen water on Earth and Mars, and frozen carbon dioxide in the polar ice-caps of Mars and use this to compile a **volatile inventory**. The volatile inventory is the total amount of volatile substances in the atmosphere and at or close to the surface of the planet. Figure 5.16 shows estimated lower limits for the inventories of water, carbon dioxide and nitrogen on Venus, Earth and Mars. To enable a direct comparison of the relative amounts of volatiles in each planet, the amount of each volatile is shown as a fraction of the total mass of the planet. The total length of the bars is the lower limit for the volatile inventory. The dark part of each bar represents the amount of that volatile in the atmosphere. The total inventory is still subject to uncertainty as is implied by quoting only the lower limit. Measurements of hydrogen taken in 2002 by the gamma-ray spectrometer on Mars Odyssey indicate that there is more water ice at Mars's poles than previously thought. Thus the lower limit for water on Mars should be raised if this is confirmed and the actual amount estimated.

As you can see in Figure 5.16, the amounts of carbon dioxide are now more evenly balanced for the three planets. You can also see that the predominance of nitrogen in the Earth's atmosphere is not due to it having more N_2 than the other planets but that a larger proportion of the CO_2 has been lost from the atmosphere as carbonate deposits. It does appear, however, that there is a lack of water on Venus and of nitrogen on Mars. This is something that theories of the origin and evolution of planetary atmospheres will need to account for.

Such theories will also need to explain the type of molecules found. If you look at the molecules listed in Table 5.2, you will see that many elements are found as oxides. Atmospheres such as those of Mars and Venus, in which carbon, for example, exists predominantly combined with oxygen as CO_2 rather than combined with hydrogen as CH_4, are called **oxidized atmospheres**. The concepts of **oxidation** and **reduction** are discussed in Box 5.3.

In the atmospheres of the giant planets, which you will study in Chapter 6, the major gas is hydrogen, H_2. Carbon is mainly present combined with hydrogen, as methane for example, and these atmospheres are called **reducing atmospheres**. The atmosphere of the Earth is unusual in that it contains substantial amounts of oxygen, O_2, and is capable of oxidizing surface rocks and other objects on the surface. You are probably aware of the tendency of

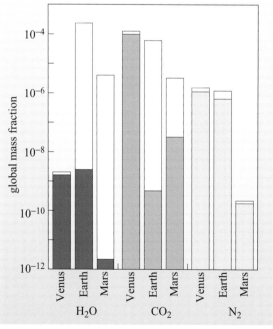

Figure 5.16 Estimated volatile inventories of Venus, Earth and Mars. Amounts of volatiles are shown as a fraction of the total mass of the planets. The total length of the bars is the lower limit for the volatile inventory. The dark part of each bar represents the amount of that volatile in the atmosphere.

iron objects to rust if left out. The Earth's atmosphere is called an **oxidizing atmosphere**. The high proportion of oxygen evolved as a result of photosynthesis by early life on Earth.

The extremely tenuous atmosphere of Mercury is very different in nature and is largely captured from the solar wind. The dissimilarity of this atmosphere from the others suggests a different origin. An important topic for theories of planetary evolution is the origin of the atmospheres of Mars, Earth and Venus.

It is not only the relative amounts of different compounds on these planets that differ greatly but also the total amounts of atmosphere. To compare these amounts we use a quantity called column mass. The pressure at the surface of a planet is not a good measure of the amount of atmosphere because this pressure depends on the gravitational field of the planet. **Column mass**, which is given the symbol m_c, is the mass of atmosphere that sits above $1\ m^2$ of a planet's surface. It may be estimated from measurements of the atmospheric pressure at the surface, P_s, and the gravitational acceleration, g, at the surface, using the expression:

$$m_c = \frac{P_s}{g} \tag{5.3}$$

BOX 5.3 OXIDATION AND REDUCTION

Oxidation was originally defined as the addition of oxygen to a substance. When carbon (as coal or charcoal for example) is burned, it is oxidized to carbon dioxide.

$$C(s) + O_2(g) = CO_2(g) \tag{5.4}$$

where (s) represents the solid phase and (g) represents the gaseous phase. As an extension to this, oxidation can also be defined as the removal of hydrogen. For example when methane, CH_4, is burned in air it is oxidized to carbon dioxide. In the process each carbon atom loses hydrogen and gains oxygen.

$$CH_4(g) + 2O_2(g) = CO_2(g) + 2H_2O(l) \tag{5.5}$$

where (l) represents the liquid phase. If the supply of oxygen is limited, a less oxidized form of carbon, carbon monoxide, is formed.

$$CH_4(g) + \frac{3}{2}O_2(g) = CO(g) + 2H_2O(l) \tag{5.6}$$

Reduction is the removal of oxygen or the addition of hydrogen. For example in the extraction of iron from its ores using carbon, the iron is reduced because it loses oxygen.

$$2Fe_2O_3(s) + 3C(s) = 4Fe(s) + 3CO_2(g) \tag{5.7}$$

With the discovery of electrolysis it was found that similar processes in which elements changed their valence state could occur without the involvement of hydrogen or oxygen. The definition of oxidation was extended to include the removal of electrons. Reduction was similarly extended to include the addition of electrons. Thus adding an electron to an Fe^{3+} ion to form an Fe^{2+} ion would be a reduction. Most of the examples in this book can be understood in terms of the addition/removal of hydrogen or oxygen.

For the Earth the value of m_c is about 10^4 kg m^{-2}. The following question asks you to work out the values for Venus and Mars.

QUESTION 5.3

Estimate the values of the column mass for Venus and Mars, then compare the column masses of the three terrestrial planets, Earth, Venus and Mars.

Venus: P_s = 92 bar; g = 8.90 m s^{-2}

Mars: P_s = 6.3 × 10^{-3} bar; g = 3.72 m s^{-2}

Remember that 1 bar = 10^5 Pa (where Pa is the SI unit for pressure), and that 1 Pa = 1 N m^{-2} = 1 kg m^{-1} s^{-2}.

From the answer to Question 5.3, you can see that the column mass for Venus is about 100 times that for Earth and the column mass for Mars is about one-sixtieth of that for Earth.

Table 5.3 gives the pressures and average temperatures at the surfaces of the terrestrial planets, which also differ markedly.

Table 5.3 The surface temperature T_s, pressure P_s and column mass m_c of the atmospheres of Venus, Earth, Mars and Mercury.

Planet	T_s/K	P_s/bar	m_c/kg m^{-2}
Venus	733	92	1.03 × 10^6
Earth	288	1.0	1.0 × 10^4
Mars	223	6 × 10^{-3}	1.7 × 10^2
Mercury	443	10^{-15}	10^{-11}

5.4 Atmospheric structure

On Earth, the temperature usually decreases with altitude, at least to the tops of the highest mountains. This variation of temperature with altitude is an aspect of **atmospheric structure**, and is determined largely by the absorption of solar radiation by a planet's atmosphere and surface. Together with pressure, which also decreases with altitude, temperature determines the formation of clouds in the atmosphere. The heating of the atmosphere and surface of a planet by solar radiation is also the cause of atmospheric motion.

5.4.1 Temperature profile

Temperatures at the surfaces and at altitudes in the atmospheres have been recorded for Venus, Earth and Mars. Of course, these temperatures also vary with latitude, time of day and the season. To single out the variation with altitude, the temperatures are averaged over time and latitude. These variations of average temperature T with altitude are shown in Figure 5.17.

This figure indicates that several similarities exist between the three planets. In the lowest region of each atmosphere, the temperature drops with increasing altitude.

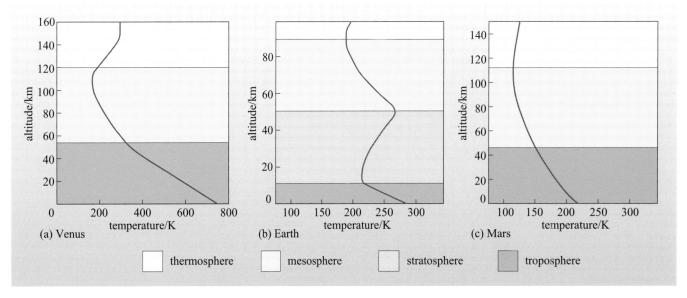

Figure 5.17 Atmospheric structure: the variation of temperature with altitude for (a) Venus, (b) Earth and (c) Mars.

This region is called the **troposphere**, meaning the region of mixing. The rapid decrease of temperature with altitude, which is due to heating of the surface, causes convection (Box 5.4), which leads to vertical mixing. The uppermost region of each atmosphere, called the **thermosphere**, is characterized by an increase of temperature with altitude; this increase shows a large diurnal variation. Between these regions is the **mesosphere** in which the temperature decreases with altitude but more slowly than in the troposphere. The **stratosphere**, which is a region between the troposphere and the mesosphere in which the temperature increases with altitude, is unique to the Earth amongst the terrestrial planets.

We have said that the rapid drop in temperature with altitude leads to convection in the troposphere, but how rapid does that drop need to be for convection to occur?

Convection in the atmospheres of the terrestrial planets occurs when the atmosphere is heated by the surface of the planet, the surface itself being heated by radiation from the Sun. Consider a small parcel of atmosphere close to the surface. This will be heated by the surface and rise. It will now be surrounded by gas at lower pressure since atmospheric pressure decreases with altitude. The parcel of atmosphere expands to equalize its pressure with the surroundings. The temperature of the parcel will drop as a consequence of this expansion even if no heat is transferred to the surrounding gas. If the temperature after expansion is still higher than that of

BOX 5.4 CONVECTION

Convection involves the transfer of energy by the movement of bulk matter. If we take a pan of water and heat the bottom of the pan, a temperature gradient is set up. Some energy is transferred through the water by **conduction**, that is through molecules transferring energy to their neighbours via their motion. However, most energy is transferred by bulk parcels of heated water rising up through the liquid due to their lower density and transferring energy to the adjacent water. These parcels then fall to the bottom of the pan and are reheated so that they rise again and the process is repeated. This cycle of rise and fall of bulk parcels of water produces **convection currents**.

the surrounding gas then the parcel will rise to a higher altitude and expand again. This process will continue until the temperature of the parcel is below that of the surroundings. At this point the parcel will sink back towards the surface.

■ How will the volume and temperature of the parcel change as it sinks?

❑ The pressure of the surrounding atmosphere will increase as the parcel sinks and so the parcel will contract. As a result of this contraction, the temperature will rise.

Convection will take place if the fall in temperature with altitude of the bulk atmosphere is such that a rising parcel of atmosphere is always hotter than the surroundings.

The slowest rate at which the temperature of the rising parcel decreases is that which results solely from the expansion of the gas with no heat lost to the surroundings. This rate is called the **adiabatic lapse rate**. For Earth, the adiabatic lapse rate, assuming a dry atmosphere, is 9.8 K km^{-1}. If the decrease in temperature with altitude of the general atmosphere is greater than the adiabatic lapse rate, then the surrounding atmosphere must be cooler than the rising parcel and the parcel continues to rise (Figure 5.18a). If the decrease in temperature with altitude of the general atmosphere is less than the adiabatic lapse rate then the surrounding atmosphere must be hotter than the rising parcel and the parcel does not continue to rise (Figure 5.18b). Under these circumstances, convection does not occur.

In practice, the parcel of atmosphere will lose some heat to the surroundings so that the temperature of the general atmosphere will have to fall faster than the adiabatic lapse rate for convection to occur.

■ On Venus the adiabatic lapse rate is higher than on the Earth (the temperature drops more quickly with altitude). For a dry atmosphere the rate is 10.5 K km^{-1}. Account for this in terms of the faster rate of pressure decrease with altitude on Venus.

❑ If the pressure drop is greater for the same increase in altitude, then the rising parcel of atmosphere will expand more and its temperature will consequently drop by more.

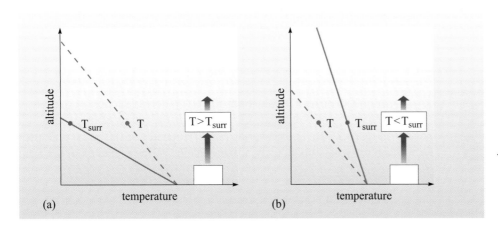

(a)   (b)

Figure 5.18 A rising parcel of gas. In (a) the temperature of the surrounding atmosphere decreases *faster* than the adiabatic lapse rate (dashed line). In (b) the temperature of the surrounding atmosphere decreases *slower* than the adiabatic lapse rate.

Note that Figure 5.17 shows altitude versus temperature, rather than temperature versus altitude. So a *larger* value of the temperature gradient (K km^{-1}) produces a *less steep* curve on the diagram (as in the Earth's troposphere).

In the troposphere, the temperature gradient is close to, but higher than, the adiabatic lapse rate so that convection takes place and bulk mixing of the atmosphere occurs. The lower temperature gradient in the mesosphere rules out convection as a major source of energy transfer. In the stratosphere and thermosphere, the temperature actually increases with altitude so that again we need to look at alternative mechanisms for energy transfer.

In the following subsection we look at how atmospheres are heated and how they cool.

5.4.2 Heating and cooling

Effective temperature

The temperature of the surface and atmosphere of a planet is determined by the balance between the energy that is absorbed and the energy that is emitted.

The amount of energy coming from the Sun is usually given as the solar flux density, which is the amount of energy arriving per unit time on unit area.

For the terrestrial planets, the source of most of the energy reaching the atmosphere and the surface is the Sun; relatively little energy comes from the planet's interior. The solar **flux density** at the top of the Earth's atmosphere is about 1.38×10^3 W m^{-2}. Some of this is reflected back to space, the atmosphere absorbs some, and the rest reaches the surface, where it is either reflected or absorbed. The absorbed radiation heats the surface, which then re-radiates this energy, mainly in the infrared region of the spectrum.

On the assumption that a planet undergoes no net heating or cooling in the short term, it is possible to estimate the temperature necessary for a planet to re-radiate all of the energy absorbed by the atmosphere and the surface. This temperature, called the **effective temperature**, T_e in kelvin, is defined as follows:

$$T_e{}^4 = \frac{L}{4\pi R^2 \times 5.67 \times 10^{-8}} \tag{5.8}$$

where L is the total power radiated by the planet in watts, R is the radius of the planet in metres (its surface area is $4\pi R^2$ and radiation is emitted from the whole surface, Figure 5.19) and 5.67×10^{-8} is a constant which has the units W m^{-2} K^{-4}.

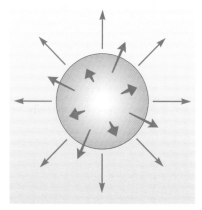

Figure 5.19 The emission of radiation occurs from the whole surface of a planet, which is an area of $4\pi R^2$ for a planet of radius R.

This formula was originally derived for thermal sources or black bodies (see Box 5.5), but now serves to define effective temperature, regardless of the form of the spectrum of the emitted radiation.

It is instructive to estimate T_e for the terrestrial planets and to compare these temperatures with those at the planets' surfaces. The temperature of the surface and the temperature of the atmosphere of a planet are determined by the balance between the radiation absorbed and that emitted by the planet. In order to estimate T_e, the power lost by radiation must be equated with that absorbed from solar radiation. The Sun's radiation arrives from one direction, and so a planet is heated over only half of its surface at any time. The cross-section of the planet is a disc of area πR^2, where R is the radius of the planet (Figure 5.20). The power absorbed depends on this area, and also on the solar flux density at the distance of the planet from the Sun.

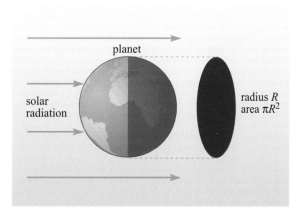

Figure 5.20 The planet presents a disc of area πR^2 to the Sun's radiation and casts a shadow of this area.

BOX 5.5 BLACK-BODY RADIATION

A thermal source emits radiation as a result of its temperature. An ideal thermal source, called a black body, displays a characteristic emission spectrum. This is shown in Figure 5.21.

In any graph that shows how the relative spectral flux density varies with wavelength for an ideal thermal source, the height of the graph will depend on factors such as the size and distance of the source. However, in all such graphs the overall *shape* of the curve is solely determined by the temperature of the source. This means, in particular, that the peak of each curve occurs at a wavelength, λ_{peak}, that characterizes the source's temperature, irrespective of the height of the peak. In fact, there is a simple law, called Wien's displacement law, that relates the value of λ_{peak} (in metres) to the temperature, T (in kelvin), of the source:

$$\lambda_{peak} = \frac{2.90 \times 10^{-3}}{T} \qquad (5.9)$$

where 2.90×10^{-3} is a constant which has the units m K. With the aid of Wien's displacement law it is a simple matter to determine the temperature of any source of light, provided it is an ideal thermal source. Such sources are not common, but many real sources, including the Sun, are reasonable approximations to ideal thermal sources, so their temperatures may be estimated by this spectral technique.

■ If you were to heat a metal ball so that its temperature steadily increased, you would find that above a certain temperature the ball would start to emit a dull red glow. As the temperature increased further the ball would become brighter and the colour would change from red to orange-white to yellowish-white to white. How would you explain these changes in appearance?

❑ Assuming that the heated ball can be treated as an ideal thermal source, it is to be expected that, as the temperature increases, the ball would emit relatively more light with wavelengths towards the blue end of the spectrum rather than the red end (Figure 5.21). At relatively low temperatures red light will predominate and the ball will glow 'red-hot'. As the temperature rises, the proportion of shorter wavelengths will gradually increase making the colours progress from red to orange-white to yellowish-white to white.

The shapes of the curves that describe the spectra of ideal thermal sources are of great importance in science. Such curves are usually referred to as thermal radiation curves, Planck curves or black-body radiation curves. Non-ideal thermal sources and non-thermal sources may also produce continuous spectra, but, from the graphical point of view, those spectra will generally have a different shape from those of ideal thermal sources.

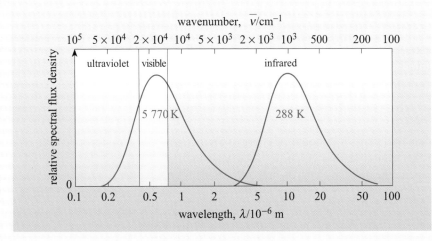

Figure 5.21 Spectra of **black-body sources** at the temperatures of the Sun's photosphere (5770 K) and the Earth's surface (288 K). The vertical scales for the two spectra are not the same; the Sun's radiation is much more intense than that of the Earth. (Note that the spectra extend well beyond those wavelengths that correspond to the various colours of visible light.)

Table 5.4 The albedos of Venus, Earth, Mars, Mercury and the Moon.

Object	Albedo
Venus	0.77
Earth	0.30
Mars	0.25
Mercury	0.10
Moon	0.07

A third factor determines the power absorbed by a planet. Not all the solar radiation reaching the planet is absorbed, some is reflected. On Venus, for example, the clouds reflect a large fraction of the solar radiation, and this strongly affects the heating of the planet. The total fraction of solar radiation that is reflected by a planet is called the **albedo**, a, (where the value of a can be between 0 and 1). Thus the total fraction absorbed is simply $(1 - a)$. Surface reflection also plays a part in determining the overall planetary albedo and this will be the dominant contribution for Mercury. The influence of cloud cover can be seen by a comparison of the albedos for Venus, Earth and Mars in Table 5.4.

A planet that has a light-coloured surface or extensive cloud cover reflects much of the Sun's radiation, whereas a dark surface absorbs most radiation. A planet with a high albedo thus absorbs little of the Sun's radiation. With an average cloud cover of about 50%, the Earth has an overall albedo of 0.30.

▪ From its albedo, does Mercury have a light- or dark-coloured surface?

❑ The albedo is very low, so Mercury absorbs a high proportion of the incident solar radiation and is thus dark.

By equating the power, L, radiated by the Earth with the solar power absorbed, which can be readily estimated independently, an effective temperature of 255 K can be estimated from Equation 5.8.

The effective temperatures of Venus and Mars are 227 K and 217 K, respectively.

▪ How do the effective temperatures of Venus and Mars compare with the value of T_e for the Earth?

❑ Both are significantly lower.

▪ Why are they lower?

❑ Mars is farther from the Sun. Venus, although nearer to the Sun, has a much higher albedo (Table 5.4) because of its total cloud cover.

Notice that this estimate of effective temperature is based entirely on the Sun as a source of energy and ignores a planet's internal heat sources. So, how much higher would the effective temperature be if heat from the planetary interior were taken into account?

QUESTION 5.4

The solar flux density (amount of energy coming from the Sun) at the radius of the Earth's orbit is 1.38×10^3 W m^{-2}. A major internal heat source for Earth today is radiogenic heating and this, as you may recall from Chapter 2, provides a flux of 5×10^{-12} W kg^{-1}. Calculate the total power from the Sun and from radiogenic heating for the Earth and compare the two values. Use the values given in Appendix A, Table A1 for the mass and radius of the Earth.

Radiogenic heating thus accounts for only 0.02% of the power input to the Earth's surface and atmosphere. Similar conclusions are reached for Venus and Mars. So it is valid to compare effective temperatures on these planets as a measure of the power re-radiated by the atmospheres after absorption of solar radiation.

If we compare the mean surface temperatures of the planets with their effective temperatures we find:

- the mean surface temperature of Mars is almost exactly equal to its effective temperature;
- the mean surface temperature of Mercury and its effective temperature are identical.

▨ How do the surface temperature and T_e compare for Earth and for Venus?

❏ For Earth, the mean surface temperature is 288 K, which is 33 K higher than T_e, whereas the mean surface temperature of Venus is 733 K, which is about 500 K higher than T_e.

It is lucky for us that the surface temperature is higher than the effective temperature because at the higher temperature water is a liquid and the presence of liquid water is thought to be necessary for the evolution of life as we know it. But what is the cause of these high surface temperatures? The moderate increase for Earth's surface temperature and dramatic increase for Venus are the result of the so-called greenhouse effect, the origin of which we shall now discuss.

The greenhouse effect

The **greenhouse effect** is the name given to the trapping of heat energy by planetary atmospheres. It gets its name from the higher temperatures observed in an unheated greenhouse than in the air outside. (It was later found that protection from wind also plays a major role in keeping greenhouses warm.)

The greenhouse effect in planetary atmospheres arises in the following way. The surface of a planet is heated by solar radiation, which has a spectrum in which the energy is concentrated in the visible region. This spectrum is approximately that of a black body emitting at a temperature of 5770 K, the Sun's surface temperature. The hot surface of the planet loses heat mainly by radiation from the surface.

QUESTION 5.5

If the Earth, with a surface temperature of 288 K, radiates as a black body, at what wavelength will the maximum energy of the emission spectrum be, using Wien's displacement law (Equation 5.9)? In which region of the spectrum is this radiation most intense?

Because the planets emit radiation mostly in the infrared region and scarcely at all in the visible region, they do not glow like stars or the iron ball in our example of a black body.

The black-body spectra at the Earth's surface temperature and the Sun's surface temperature, 288 K and 5770 K, respectively, are shown in Figure 5.22a. Figure 5.22b shows the absorption spectrum of the Earth's atmosphere. The atmosphere is quite transparent in the visible region, so that if not scattered by clouds, most of the visible radiation reaches the surface where it is absorbed. However, in the infrared region, electromagnetic radiation is absorbed by molecules that, as a result, increase their vibrational energy. During subsequent collisions between molecules, this internal vibrational energy is converted, through chemical reactions and by energy transfer in collisions, into kinetic energy of the molecules (so raising its temperature). So solar (visible) radiation is transmitted by the atmosphere and heats the surface of the planet. The surface radiates at infrared wavelengths, and this energy is trapped by absorption in the atmosphere, thereby raising its temperature. This heating of the lower atmosphere of the planet (and heating of the surface through contact with the atmosphere) is the greenhouse effect.

Notice that most of the absorption at the most intense region of the 288 K spectrum is caused by CO_2 and H_2O, two minor components by volume of the Earth's atmosphere.

■ Why do the major components of the Earth's atmosphere, N_2 and O_2, not contribute to the greenhouse effect?

❏ N_2 and O_2 are homonuclear diatomic molecules and so do not possess an electric dipole. They therefore do not absorb infrared radiation strongly through changes in their vibrational energy (Section 5.2).

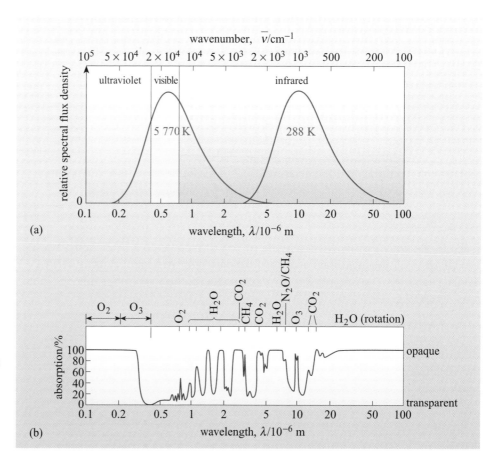

Figure 5.22 (a) Spectra of black-body sources at the temperatures of the Sun's surface (5770 K) and the Earth's surface (288 K). The vertical scales for the two spectra are not the same; the Sun's radiation is much more intense than that of the Earth. (b) The absorption spectrum of the Earth's atmosphere: the wavelengths at which some atmospheric gases absorb energy are indicated.

The high surface temperature of Venus can be attributed to the absorption of radiation emitted by the surface by CO_2 and H_2O molecules in the atmosphere and clouds. Although much visible radiation is scattered and reflected by the cloud layers, enough reaches the surface to heat it and, with a much more dense atmosphere than that of Earth, more of the heat emitted by the surface is trapped by the atmosphere. Thus Venus has a large greenhouse effect, causing its surface temperature, as you have seen, to be about 500 K greater than T_e.

Currently there is much concern about the environmental effect of raising the level of CO_2 in the Earth's atmosphere, and the effect of this on global warming. Human activities have been changing the atmosphere since the Industrial Revolution. In particular, the level of CO_2 is rising steadily, as measurements in several locations over many years have shown. For example, Figure 5.23 gives the variation of atmospheric CO_2 since 1850 as measured by direct sampling of the atmosphere and from air bubbles trapped in polar ice-sheets.

■ What effect will increased CO_2 have on the temperature of the Earth's surface?

❏ CO_2 contributes to the greenhouse effect and so will tend to raise the temperature.

In agreement with this prediction, global mean temperatures over the last 150 years have increased by between 0.3 K and 0.6 K, although studies covering the last few thousand years show that atmospheric CO_2 is not the only influence. Other gases arising from industry and agriculture are also greenhouse gases and contribute to the greenhouse effect. These include the chlorofluorocarbons and their replacements (which are designed to have less effect on the ozone layer), as well as methane (CH_4), which is generated by agriculture, particularly paddy fields. A warmer Earth would affect the H_2O reservoirs, resulting in some melting of polar ice-caps, with release of trapped gases, as well as increased evaporation from the oceans, thereby raising the H_2O level in the atmosphere. Water is itself a greenhouse

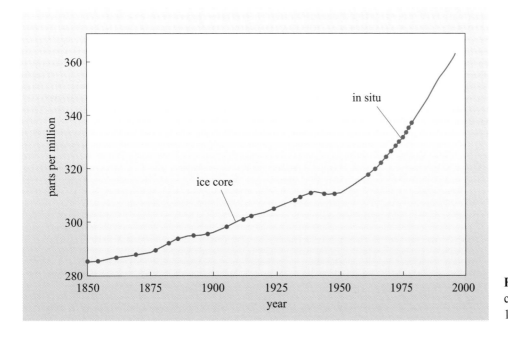

Figure 5.23 The variation of CO_2 content in the atmosphere between 1850 and 2000. (Ledley *et al.*, 1999)

gas and will contribute further to the rise in temperature. The possibility exists that such a positive feedback could result in a runaway greenhouse effect as is believed to have occurred on Venus, although current modelling predicts that this is unlikely.

The high surface temperature of Venus at present is due largely to the greenhouse effect, which might have been operative for most of the lifetime of Venus. Even if the surface temperature of Venus had initially favoured the condensation of H_2O, its closeness to the Sun means that the evaporation of H_2O into the atmosphere would have been sufficiently high for H_2O to make its own contribution to the greenhouse effect. In fact, H_2O absorbs radiation in different parts of the spectrum from CO_2 (Figure 5.22b), so its contribution is important in these circumstances. Any rise in surface temperature resulting from this greenhouse effect would ensure the evaporation of more liquid H_2O. The increased partial pressure of H_2O in the atmosphere would then further contribute to the greenhouse effect.

The stratosphere and the ozone layer

Between the Earth's troposphere (that extends from the Earth's surface up to 10–16 km in altitude depending on latitude) and the thermosphere (which begins at an altitude of approximately 90 km), there is a region in which the temperature first rises and then falls. The lower part of this region (between 10–16 km and 50 km) is called the stratosphere. The stratosphere is unique to Earth – some mechanism must exist for absorbing energy that does not exist on Mars or Venus and is responsible for the temperature rise in the stratosphere. The mechanism responsible for this absorption involves the substance ozone, O_3. Although ozone occurs throughout the atmosphere, it reaches its highest concentration in the stratosphere, as Figure 5.24 shows. It is this peak in concentration of O_3 that is called the *ozone layer*. In fact, the concentration of O_3 is highest in a band between 15 km and 30 km altitude, although it is only a minor atmospheric component. If it were all collected and compressed to 1 atmosphere pressure at the Earth's surface, it would form a layer only 3 mm thick.

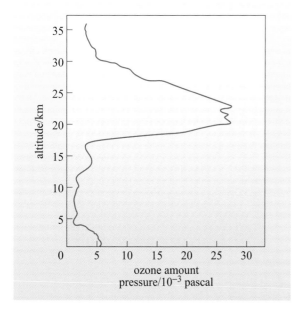

Figure 5.24 The variation of ozone concentration with altitude in the Earth's atmosphere. There is some variation with latitude, but, over the whole of the Earth, ozone occurs mainly between 10 km and 50 km altitude.

As solar radiation enters the Earth's atmosphere, it is absorbed selectively by various components of the atmosphere (Figure 5.22). Ionization of molecules in the thermosphere is responsible for the removal of radiation of wavelength (λ) shorter than about 100 nm. Molecular oxygen (O_2) absorbs radiation of $\lambda < 230$ nm (Figure 5.25), and is split into oxygen atoms by radiation of $\lambda < 140$ nm. Ozone absorbs radiation in the wavelength range from 200 nm to 350 nm. Thus this radiation is transmitted by the thermosphere and largely absorbed in the stratosphere. At wavelengths longer than 230 nm, ozone is the only component of the atmosphere that significantly absorbs solar UV radiation. In spite of its low concentration, it is highly effective at absorbing solar radiation of $\lambda < 350$ nm (Figure 5.25).

Thus, ozone prevents this part of the solar spectrum (230–350 nm) from reaching the Earth's surface. This region of the spectrum is highly damaging to life, causing genetic damage when absorbed by DNA. So life is protected by the ozone layer. The recently observed thinning of the ozone layer (Figure 5.26) is therefore a matter of concern and much research has gone into the mechanisms of formation and destruction of the ozone layer.

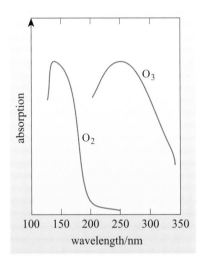

Figure 5.25 The absorption spectra of oxygen (O_2) and ozone (O_3) in the ultraviolet region of the electromagnetic spectrum.

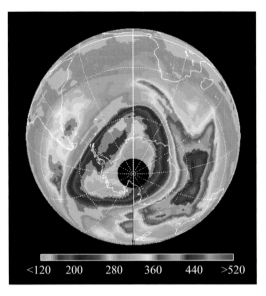

Figure 5.26 Ozone 'hole' over the Antarctic. The contours show the concentration of ozone in Dobson units. (NASA/JLP/Malin Space Science Systems)

One Dobson unit (DU) corresponds to 0.01 mm thickness of ozone under conditions of standard temperature and pressure. So the 'ozone layer' is about 300 DU.

The energy absorbed by ozone is eventually transferred to kinetic energy and this is responsible for the heating of the stratosphere.

Within the stratosphere, ozone is produced from the interaction of solar radiation with molecular oxygen, O_2. The chemistry of the stratosphere consists of a complex series of reactions involving O, O_2 and O_3 as well as a number of other atoms and molecules. One set of reactions involved in this chemistry is called the **Chapman scheme** after the English geophysicist Sydney Chapman (1888–1970), who first proposed it.

The reactions in this scheme are shown schematically in Figure 5.27, and written as:

$$O_2 + photon \longrightarrow O + O \qquad (5.10)$$

$$O_2 + O + M \longrightarrow O_3 + M \qquad (5.11)$$

$$O_3 + photon \longrightarrow O_2 + O \qquad (5.12)$$

$$O + O_3 \longrightarrow 2O_2 \qquad (5.13)$$

In Equation 5.11 the symbol M represents any atom or molecule involved in this collision simultaneously with O and O_2. The presence of M is necessary because a chemical bond is formed in this reaction and the formation of a bond releases energy. This released energy must be transferred to M or it will simply be absorbed by the O_3, which will then dissociate again:

- Equations 5.10 and 5.11 lead to the formation of ozone;
- Equations 5.12 and 5.13 lead to its destruction.

The concentration of ozone at any altitude is dependent on the relative rates of these two pairs of reactions. At higher altitudes than the stratosphere, the atmosphere is too tenuous for Equation 5.11 to lead to a substantial formation of ozone. At lower altitudes, Equation 5.10 is prevented because the required UV radiation has been absorbed and does not reach these layers.

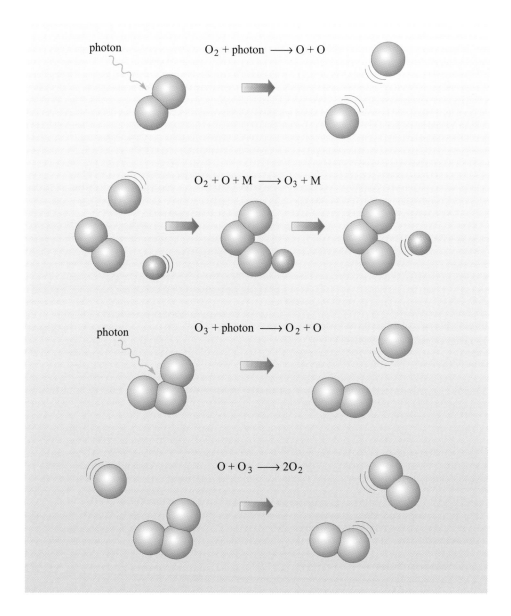

$$O_2 + photon \longrightarrow O + O$$

$$O_2 + O + M \longrightarrow O_3 + M$$

$$O_3 + photon \longrightarrow O_2 + O$$

$$O + O_3 \longrightarrow 2O_2$$

Figure 5.27 A schematic depiction of the four reactions of the Chapman scheme of ozone chemistry in the stratosphere. The body M (shaded brown) can be any atom or molecule in the three-body collision.

Heating and cooling of the mesosphere and thermosphere

In the mesospheres of the planets the temperature falls with altitude, but less steeply than the adiabatic lapse rate. In some cases the temperature is almost constant in this region. Convection does not occur to any major extent and heat transfer is through the absorption and emission of infrared radiation.

Heating of the thermosphere results from the absorption of high frequency (UV) solar radiation, which causes ionization of gases – mainly O_2 for the Earth and CO_2 for Mars and Venus. In the outer reaches of the thermosphere, a substantial fraction of the atmosphere is ionized forming a plasma (separate ions and electrons) that has appreciable electrical conductivity. This region is known as the **ionosphere** and on Earth starts at an altitude of approximately 60 km. Radio waves used for long-range communication, e.g. ship-to-shore, amateur radio and over-the-horizon radar, travel in straight lines, but the Earth is spherical, so in order to cover long distances they have to be sent up into the atmosphere and then reflected back to the receiver. The plasma of the ionosphere makes a suitable reflector. Other effects of the ionosphere will be discussed further in Section 5.7.

Eventually, much of this thermal energy is converted via recombination of ions and electrons, and by chemical reactions, into vibrational and kinetic energy of the constituents of the thermosphere. For this energy to be radiated, the vibrationally excited molecules must relax to a low energy state with the emission of a photon of infrared radiation – the reverse of the absorption process described in Section 5.2. For the components of the Earth's thermosphere, such as the homonuclear molecule O_2, this emission process is very inefficient. So although very little of the solar radiation is absorbed in the thermosphere, it is sufficient to generate a high temperature because the loss of this energy is highly inefficient. By contrast, the thermospheres of Venus and Mars, which consist mainly of CO and CO_2, are able to radiate efficiently, and so we do not observe temperatures as high as those in the Earth's thermosphere, as shown in Figure 5.17.

QUESTION 5.6

Triton, the largest satellite of Neptune, has a high albedo. What does this indicate about its surface? Given that it is a long way from the Sun, is its low surface temperature of 38 K surprising?

QUESTION 5.7

The release into the Earth's atmosphere of chlorofluorocarbons (CFCs), which were once commonly used as aerosol propellants, introduces substances that absorb infrared radiation in a different part of the spectrum from that absorbed by CO_2. What effect would you predict this to have on the temperature of the Earth's atmosphere?

5.5 Cloud formation

A cloud is a region of the atmosphere in which a component has condensed to form small liquid droplets or solid particles. These droplets or particles are so small that they are held suspended by updraughts in the cloud until they grow large enough to fall as rain or snow.

On Venus, the cloud layer is thick and extensive, extending from about 45 km to 65 km above the surface. It covers the entire planet (Figures 5.1 and 5.28). The clouds consist largely of particles, about 2 μm in diameter, the properties of which closely match those of droplets containing a solution of 75% H_2SO_4 (sulfuric acid) and 25% H_2O. Additionally, the clouds contain larger solid crystals of chlorine-containing substances, thought to be $FeCl_2$ and Al_2Cl_6, as well as phosphorus compounds, such as H_3PO_4.

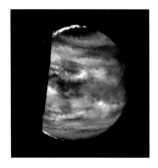

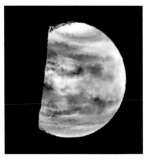

Figure 5.28 Near infrared image of clouds on Venus (Galileo mission, 10 February 1990). These clouds lie 50 km above the surface and about 10 km below the visible cloud tops. The lower image, a modified negative, represents what scientists believe would be the visual appearance of this mid-level cloud deck in daylight. (NASA)

Figure 5.29 Clouds on Mars imaged by the Mars Global Surveyor. (NASA/JPL/Malin Space Science Systems)

On Earth, we are familiar with clouds of H_2O, which may exist as liquid droplets or as particles of snow or ice. These clouds occur up to altitudes of about 10 km. Very thin clouds occur up to altitudes of 50 km.

On Mars, the most extensive clouds are composed of dust (mostly silicate particles) raised by wind. We shall deal with these clouds later as they are not the result of condensation. In addition, clouds of solid particles of H_2O and CO_2 occur, as shown in Figure 5.29, although the total average cloud cover is small.

But how and why do these clouds form? Although it is difficult to forecast when it will rain, it is possible to predict in a model system the conditions when water will exist as vapour, liquid, ice or any combination of these. The two factors that determine the phase in which water will exist are temperature and pressure. To be more precise, it is the temperature and the **partial pressure** that determine the phase rather than pressure; the partial pressure of H_2O is equal to the total pressure of the atmosphere multiplied by that fraction of the molecules in the gas that are H_2O molecules.

The basis of the prediction of the phase in which a substance will exist is the **saturation vapour pressure diagram** (see Box 5.6), which can be constructed from laboratory studies of the substance in question.

BOX 5.6 SATURATION VAPOUR PRESSURE DIAGRAMS

As an example of a saturation vapour pressure diagram we shall consider how rain forms in an atmosphere that contains water vapour. Because the atmosphere of the Earth or any other planet is complicated by the presence of various zones and by atmospheric motion, we shall develop the diagram for a much simplified model system. Our model consists of a closed box that contains water and a gas (for example dry air) that does not interact with the water (Figure 5.30). Because the box is closed, nothing leaves or enters it. However, the lid of the box can move up or down to change the pressure in the box. This movement simulates the effect of changing the total pressure in the Earth's atmosphere. We can also raise or lower the temperature.

Within the box, water is present as a liquid and also as a vapour, these two phases being in equilibrium, a situation represented by Equation 5.14

$$H_2O(l) \rightleftharpoons H_2O(g) \qquad (5.14)$$

At equilibrium, there is no change and so the ratio of the number densities, $n(H_2O, g)/n(H_2O, l)$, of the two

phases is constant. As you might expect, this ratio depends on the temperature and to a lesser extent upon the pressure, and it is this dependence that concerns us.

The number of molecules per unit volume of liquid water is approximately constant. The number density of gaseous H_2O is proportional to the partial pressure and so it is convenient to represent the equilibrium ratio in Equation 5.14 by the partial pressure of $H_2O(g)$ that is in equilibrium with $H_2O(l)$. A plot of the variation of the equilibrium partial pressure with temperature is called a saturation vapour pressure diagram, and that for H_2O is shown in Figure 5.31.

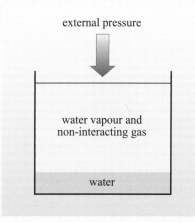

Figure 5.30 A model of an atmosphere above a pool of liquid water. In this model it is possible to vary the pressure and the temperature, factors that determine the amount of water vapour in the atmosphere.

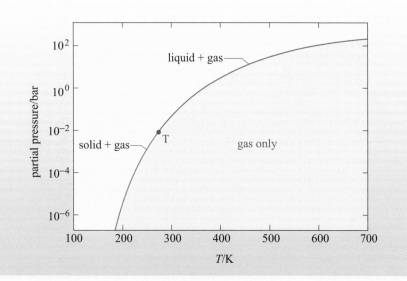

Figure 5.31 The saturation vapour pressure diagram for water. The pressure scale is logarithmic. Note that 1 bar = 10^5 Pa, close to atmospheric pressure at the Earth's surface.

The upper part of the curve (above the point T) represents $H_2O(g)$ in equilibrium with $H_2O(l)$. Below the point T, the curve represents $H_2O(g)$ in equilibrium with solid water, ice, $H_2O(s)$.

So the curve in Figure 5.31 gives the value of the partial pressure of water vapour that is in equilibrium with liquid water or ice in the range of temperatures shown. If the box in Figure 5.30 and its contents are cooled, the partial pressure of water in the gas at equilibrium with liquid water will be lowered, as Figure 5.31 shows. At this lower temperature, the gas will therefore contain initially *more* water than the equilibrium concentration, that is the partial pressure of water in the gas will exceed that at equilibrium.

The result is that some of this water vapour will condense to liquid until the partial pressure of water vapour in the gas reaches the equilibrium value (the value on the saturation vapour pressure curve at the lower temperature).

If the equilibrium partial pressure and temperature are to the *right* of the saturation vapour pressure curve, then the substance is present as a gas only (Figure 5.31). If the pressure or temperature is then changed to lie on this curve, some liquid or solid will form. This condensation of water vapour to liquid or solid in the gas is the same process by which clouds form in the Earth's atmosphere.

From Figure 5.31 we can predict whether water vapour will condense in the atmosphere.

■ Suppose that on a warm, humid day the partial pressure of $H_2O(g)$ at ground level rises to 10^{-2} bar, and then the temperature falls abruptly at night to $-10\,°C$ (263 K). Will water vapour condense in the atmosphere and, if so, in what phase?

❑ As the temperature falls, the point representing the gas moves horizontally across the 'gas only' region of Figure 5.31, until it meets the curve. At a pressure of 10^{-2} bar, this is at a temperature of about 275 K. Thus, condensation will begin at this temperature, and the partial pressure will fall until the point on the curve at 263 K is reached. The point on the curve corresponding to 263 K is below the point T, so water will condense as ice. (*Note*: These are the conditions under which freezing fog forms at ground level on Earth.)

Saturation vapour pressure diagrams, such as Figure 5.31, enable us to predict the equilibrium conditions when clouds will form, but such predictions give only an approximate estimate as they ignore some important effects.

The most important factor that affects cloud formation is the shape of the droplets. Figure 5.31 applies to a flat liquid surface, as in Figure 5.30. At any particular temperature, the partial pressure of vapour in equilibrium with the curved surfaces of small spherical droplets is higher than that for a flat water surface, and it increases as the drop size decreases (as curvature increases). So at the initial formation of very small droplets, a *higher* partial pressure is needed than that predicted by saturation vapour pressure diagrams. However, atmospheres are not simply gaseous mixtures. They contain particles of dust, on which water molecules may condense to begin the process of drop formation. If the surface of these particles is easily wetted, and especially if the particle is soluble in water, the drop will form and grow even at partial vapour pressures *lower* than those predicted by saturation vapour pressure diagrams. This principle is the basis of seeding experiments where particles are introduced into clouds to try and bring about rain. Such experiments have not so far been very successful. Like many aspects of atmospheric science, cloud formation is not highly predictable.

Dust clouds on Mars are raised by wind and occur locally and (occasionally) on a planet-wide scale. Local dust clouds are also seen on Earth. The formation of dust clouds is linked to the motion of the atmosphere, which is the subject of the next section.

QUESTION 5.8

Using Figure 5.32, determine whether CO_2 would condense as a solid or a liquid on Mars at its average surface temperature.

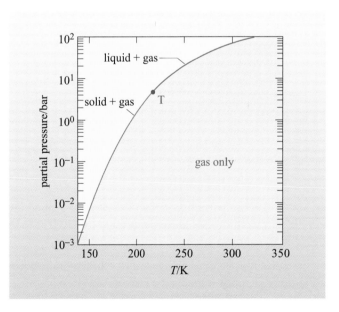

Figure 5.32 The saturation vapour pressure curve for carbon dioxide, CO_2.

5.6 Atmospheric motion

In this section, we concentrate on the major processes that produce a general rather than a local circulation of atmospheres. On the global scale, we recognize several different kinds of major process for transport in the atmosphere. On Earth, Mars and Venus for example, the factors responsible are the heating of the planet by the Sun and the rotation of the planet.

On Earth, the most important atmospheric circulation arises from the differential heating by the Sun due to the spherical shape of the planet. Near the Equator, the Earth's surface receives more energy per unit area than it does near the poles (Figure 5.33). This is because a beam of solar radiation is spread over a larger area near the poles than at the Equator. The higher temperature of the Earth's surface at the Equator causes warm air to rise vertically in the troposphere. It cools as it rises and the cooler air then travels towards the poles. As it moves polewards, this air radiates heat and so cools further, until it eventually sinks back to the surface where it returns to the Equator, completing a cycle that defines what is known as a **Hadley cell**, named after the British scientist, George Hadley (1685–1768). These cells extend to about 30° N and 30° S, as shown in Figure 5.34. The mass transport of heated air tends to even out temperature differences at the surface so that the range of temperatures on Earth is less extreme than that on planets with a thinner atmosphere or no atmosphere at all.

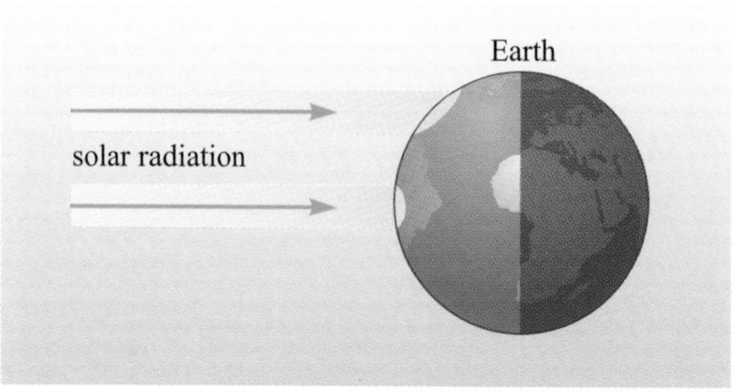

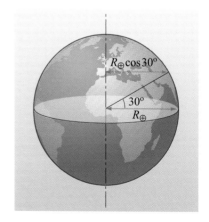

Figure 5.33 The Earth's surface receives more solar energy per unit area of the surface, near the Equator than it receives near the poles.

The fast rotation of the Earth (rotation period of one day) has an additional effect on the atmospheric circulation through the interaction of the rotation with Hadley cell motion. We can understand this effect by following the motion of a piece of atmosphere around a Hadley cell as it circulates on the rotating Earth. Suppose that this piece of atmosphere begins on the surface of the Earth at the Equator, with no initial motion relative to the surface. It is rotating with the planet, with some west-to-east speed, v, relative to a non-rotating observer in space moving at the same speed as the Earth in its orbit. This piece of atmosphere rises without significant change in v (the height of the Hadley cell in Figure 5.34 is exaggerated for clarity). As it moves northwards, the piece of atmosphere at the top of the cell, of mass m, does not change height, but its distance, r, from the Earth's spin axis decreases. At 30° N, its distance from the spin axis is now $R_\oplus \cos 30°$ (Figure 5.35), where $R_\oplus$ is the Earth's radius. During this motion the quantity **angular momentum** must be conserved. Angular momentum is the product of the mass and the **angular speed** (see Box 5.7).

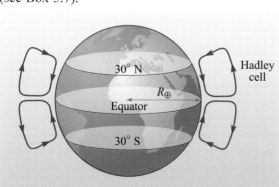

Figure 5.34 Hadley cells in the Earth's tropics extend to about 30° N and 30° S (not to scale).

Figure 5.35 The air in a Hadley cell moves closer to the Earth's spin axis as it travels towards the pole. At 30° N the distance has decreased from $R_\oplus$ to $R_\oplus \cos 30°$.

BOX 5.7 ANGULAR MOMENTUM

If you have ever watched a high diver you may have noticed that, when tight-tucked, the diver somersaults rapidly during the dive, but when stretched out she tumbles more slowly and makes fewer, graceful turns before entering the water. This is an illustration of the conservation of angular momentum.

A pirouetting skater is another example. With the arms outstretched the skater spins slowly; if he pulls them in the rotation speeds up.

Angular momentum is to do with how fast a body is rotating, the angular speed, and how compact or how spread out the mass of the body is around the rotation axis. Conservation of angular momentum means that, unless the body is impeded (for example it hits something), its angular momentum will not change. For an isolated body, left on its own, the angular momentum at the end of a manoeuvre is the same as at the beginning.

In the case of the atmosphere within our Hadley cell, the angular momentum has magnitude mvr, where r is the distance of the mass from the spin axis. We assume the parcel does not lose or gain mass. It follows that in order for the angular momentum to remain constant, the product vr must also stay constant. In the process of moving northwards, the distance, r, decreases. In order to conserve the angular momentum, therefore, the speed, v, must increase.

At the Equator, the west-to-east speed of the atmosphere at the top of the cell, which we assume is equal to the speed of the surface, is $2\pi R_\oplus/T$, where T is the rotation period of the Earth, $R_\oplus$ (at the Equator) is 6378 km and T is 8.617×10^4 s. So this speed relative to an observer in space is 465 m s^{-1}.

■ What will the west-to-east speed of the atmosphere at the top of the Hadley cell be at 30° N relative to the observer?

❑ The distance of the cell from the Earth's axis decreases to $r\cos 30°$, m remains constant and so for the angular momentum, mvr, to remain constant, v must increase to $v/\cos 30°$. So at 30° N, the speed is increased to $(465/\cos 30°)$ m s^{-1} = 537 m s^{-1}.

Because the Earth is a solid body, at 30° N the angular speed of the surface is the same as at the Equator. At this latitude, the west-to-east surface speed is therefore $2\pi R_\oplus \cos 30°/T$ = 403 m s^{-1}. The top of the Hadley cell thus acquires a west-to-east motion *relative to the surface*, as shown in Figure 5.36.

So we conclude that, at 30° N, the top of the Hadley cell has acquired a west-to-east speed relative to the surface of $(537 - 403)$ m s^{-1}, which is 134 m s^{-1}. If, again, no significant change in v occurs as this piece of atmosphere returns to the surface at the northern extremity of the Hadley cell, the calculation predicts a west-to-east wind speed of 134 m s^{-1}, which is very high. The calculation ignores the effects of friction and small eddy currents, which reduces this wind speed, although the prevailing wind direction is accounted for. This wind is an example of the **Coriolis effect** – a displacement arising as matter moves in a rotating system.

Each of the two tropical Hadley cells forms a spiral of ascending and descending air that extends around the Earth (see Figure 5.37).

At higher latitudes, the Hadley circulation becomes unstable, with the result that the tropical cells extend only to 30° N and 30° S. At mid-latitudes (30–60° N and 30–60° S) the north–south transport of energy is dominated by atmospheric waves that give rise to cyclones and anticyclones. These are the areas of high and low pressure familiar from weather maps. The position of these varies from day to day, but if we average the circulation over about a decade, we obtain a pattern that resembles a Hadley cell but in the *reverse* direction, known as a Ferell cell. In the polar regions, Hadley cells, known as polar cells, once more occur but are less prominent than those in the tropics. Figure 5.38 shows the three cells in each hemisphere forming the general pattern of circulation.

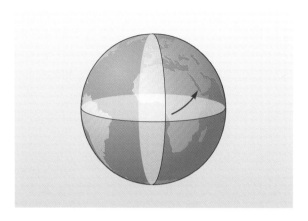

Figure 5.36 As it moves northwards, the top layer of the Hadley cell acquires an easterly motion relative to the surface of the Earth.

The wind patterns established by Hadley cells have been known to travellers for centuries. At the Earth's surface the northern equatorial Hadley cell results in a general northeast to southwest motion of the atmosphere (Figure 5.38) giving the trade winds. Near the Equator, where the two tropical Hadley cells abut, an almost windless region exists. This is known as the doldrums. The top of the most northerly Hadley cell has a strong west-to-east motion known as the jet stream. At high altitudes of over 10 km, at which transatlantic jet aircraft fly, the jet stream ensures that aircraft make the eastbound journey faster than the westbound journey.

The Earth's rate of rotation causes a large Coriolis effect that limits the extent of Hadley cells. We can compare atmospheric motion on Earth with that on Venus and Mars. Venus rotates very slowly, once every 243 days, in a retrograde direction, i.e. in the opposite sense to most Solar System bodies.

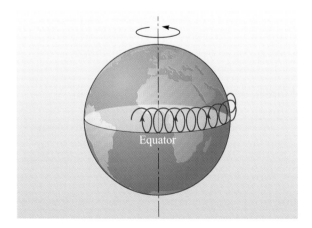

Figure 5.37 The Earth's rotation causes the Hadley cell to spiral. A piece of atmosphere that remains in the Hadley cell follows this flattened and tilted spiral path. This figure shows part of the tropical cell in the Northern Hemisphere; the vertical component is exaggerated.

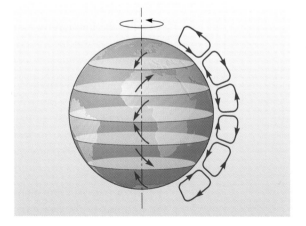

Figure 5.38 On Earth, three atmospheric circulation cells occur in each hemisphere. The tropical cells are the most persistent. The intermediate cells are driven by neighbouring cells in a direction contrary to that expected for a Hadley cell. Arrows indicate the directions of surface winds, i.e. the motion at the bottom of the cell.

◼ In which direction, west to east or east to west, would you predict the atmosphere at the top of the Hadley cells on Venus to move?

❑ As the rotation of Venus is retrograde, this is also the predicted direction of the atmospheric motion (i.e. east to west).

◼ How would you expect this slow rotation of Venus to affect the east-to-west movement of Hadley cell circulation?

❑ The small Coriolis effect should contribute little to this movement, and so east-to-west motion might be expected to be slow.

At visible wavelengths, the cloud tops of Venus are almost featureless, but imaged in UV or infrared radiation (Figures 5.1, 5.28) they show cloud patterns that indicate that Hadley cells extend from the equator to the poles. As might be expected on a slowly rotating planet, wind speeds at the surface are low, only about $1 \, \text{m s}^{-1}$, as measured by entry probes and landers. However, at the cloud tops (altitude 65 km) the wind speed is typically $120 \, \text{m s}^{-1}$ in the retrograde direction of the rotation of the planet. This phenomenon is called **super-rotation** and its cause is not well understood, but is probably related to the fact that most of the solar radiation is absorbed at high altitude in the atmosphere of Venus, with only about 4% reaching the surface. This is in contrast to what happens on Earth and Mars, where most of the solar radiation reaches the surface.

The rotation period of Mars is similar to that of the Earth, so it is not surprising to find a similar pattern of atmospheric circulation. Hadley cells occur near the equator but, owing to the thin atmosphere, the transport of heat from the tropics to the poles is less effective. Surface temperatures of more than 240 K were recorded by the Viking 1 lander at 22° N, with a diurnal variation of up to 50 K. The temperature at the poles drops to less than 150 K, sufficient for CO_2, the major atmospheric component, to solidify. The condensation of CO_2 on Mars, together with the inclination of its spin axis, leads to a further type of atmospheric circulation. Let's consider the inclination first. The angle of inclination of the spin axis, which is about 24°, is similar to that of the Earth. Seasons occur, each hemisphere alternating between summer and winter (see Box 5.8).

With the arrival of summer in the Martian northern hemisphere, the rise in temperature causes the polar CO_2 ice-cap to evaporate. In fact, CO_2 passes directly from solid to gas by *sublimation*. At the same time, the temperature at the south pole decreases and CO_2 solidifies from the atmosphere, increasing the size of the polar ice-cap. This phenomenon causes a flow of atmosphere from one pole to the other and is called the **condensation flow** (Figure 5.41).

A further type of atmospheric motion that is important on Mars can be attributed to the thinness of the Martian atmosphere. On the hemisphere facing the Sun, the surface heats up rapidly because very little solar radiation is absorbed by the overlying atmosphere. When this hemisphere has turned to face away from the Sun, radiation from the surface readily escapes from the planet, again because the atmosphere is so thin. The result is a large difference in temperature between day (220 K) and night (170 K). This temperature difference causes a flow of atmosphere around the planet, referred to as a **thermal tide**, although it is not really a tidal process. Thermal tides also occur on Venus and Earth, especially at higher altitudes, although they are a less significant component of the circulation than on Mars.

BOX 5.8 SEASONS

On Earth, we are familiar with alternating **seasons** of summer and winter. Seasons are caused by the angle that the spin axis makes with the normal to the orbital plane (i.e. a line drawn perpendicular to the orbital plane). This angle is called the axial inclination, i_a (Figure 5.39). It is 23.5° for Earth and 25° for Mars.

In the Northern Hemisphere summer, that hemisphere is tilted towards the Sun (Figure 5.40) with the result that day is longer than night and the Sun is high in the sky. At this time it is winter in the Southern Hemisphere and the opposite conditions prevail. When the Earth has completed a further half-orbit about the Sun, the seasons are reversed.

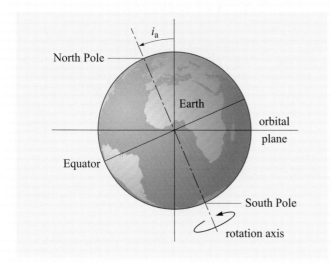

Figure 5.39 The axial inclination, i_a, of a planet.

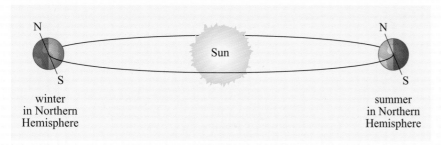

winter
in Northern
Hemisphere

summer
in Northern
Hemisphere

Figure 5.40 The seasons alternate as the Earth orbits the Sun (not to scale).

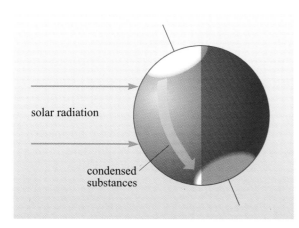

Figure 5.41 During summer in the Martian northern hemisphere, condensed substances (mostly CO_2 with some H_2O) evaporate and migrate to the southern pole, where they condense as ices.

(a)

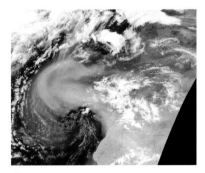

(b)

Figure 5.42 (a) Martian north polar dust storm observed by Mars Orbiter on 29 August 2000, compared with a storm on Earth (b). (NASA/JPL/Malin Space Science Systems)

Although other processes are more significant for the general circulation, the large mass of atmosphere redistributed by this process on Venus leads to tidal interaction with the Sun. There is speculation that this has affected the rotation speed of the planet.

The effects you have been studying in this section determine the general circulation of the planetary atmospheres, that is an average over time and longitude. Localized wind motion also occurs and can be caused, for example, by the need to get around topographical features such as mountains or on Earth by temperature differences between sea and land. The latter leads to offshore breezes on warm days that change to onshore breezes as the Sun sets.

The combination of various atmospheric motions and local eddy currents on Mars can lead to local surface wind speeds as high as $50 \, \text{m s}^{-1}$. These winds raise dust from the surface, creating local dust storms (Figure 5.42a). Within these storms the dust absorbs solar radiation, setting up large temperature differences and hence pressure differences. These lead to further turbulence causing more dust to be raised, which in turn heats up and leads to even more turbulence. In spring these dust storms are promoted by the higher pressure as the CO_2 sublimes, enabling more dust to be suspended for a longer time. Over a Martian year many local dust storms occur on Mars, and once or twice a year the whole planet becomes enveloped in a storm. In 2001, Mars Global Surveyor and the Hubble Space Telescope tracked a storm that resulted in a cloud of dust engulfing the entire planet for about three months. This storm started as several localized dust storms that then spread. The temperature of the upper atmosphere was raised by about 45 K but the surface cooled due to the dust blocking the solar radiation. As the surface cooled, the winds died down and the dust settled.

QUESTION 5.9

What speed, relative to the surface, will a piece of atmosphere gain by the Coriolis effect in travelling on Venus from the equator to 30° N? In which direction will the wind produced by this effect at the top of the Hadley cell be? (Take the surface speed of the rotating planet at its equator to be 0.997/243 that of the Earth, as the two planets are of similar size and Venus rotates once every 243 Earth days in the retrograde direction, east to west.) The Earth's sidereal (with respect to the stars rather than the Sun) rotation period is 0.997 days.

5.7 Ionospheres and magnetospheres

You have seen that solar radiation affects the atmospheres of the terrestrial planets. It provides the energy to heat it and drive the winds. High-energy solar radiation is also responsible for the formation of the charged layer, the ionosphere. The Sun also interacts with the atmospheres via its magnetic field. We can think of the Sun as being like a large bar magnet (Figure 5.43) where the magnetic field lines loop round between the poles. This structure of magnetic field lines is called a magnetosphere (although that does not necessarily mean that the structure is spherical, but implies a sphere of influence).

At large distances from the Sun, the field lines extend so far that they appear to be unconnected, travelling away from the Sun. These field lines are often referred to as the **interplanetary magnetic field** or IMF. The IMF also guides and influences the flow of the *solar wind* away from the Sun. The **solar wind** consists mainly of protons and electrons, along with some heavier ions that have been lost from the outer atmosphere of the Sun, the corona. The charged particles of the solar wind interact with the IMF, so the flow of the solar wind is coupled to the IMF.

As the Sun rotates (it rotates once every 25 days or so) the field lines are carried with it and form a spiral shape rather like the jets of water from a garden sprinkler (Figure 5.44). Charged particles in the solar wind travel along the magnetic field lines.

Now what happens when the interplanetary magnetic field and its entwined solar wind meet a planet? We start with the Moon since it has no atmosphere and no planetary magnetic field.

The Moon is, for some of the time, shielded by the Earth's magnetosphere but when it is not, it is exposed to the solar wind which deposits atoms on its surface. On the side away from the Sun there is a shadow region where there is no solar wind (Figure 5.45).

High-energy particles in the solar wind are potentially lethal to human life. An astronaut standing on the Moon when it is unshielded by the Earth's magnetosphere and when the Sun is particularly active would be in great danger.

Venus has no magnetic field but it does have a substantial atmosphere. The solar wind is prevented from reaching the surface of Venus by its ionosphere. Because the ionosphere consists of charged particles it can exclude the IMF and deflect the solar wind. We find that the IMF field lines tend to wrap around the ionosphere, producing a **bow shock**. This is similar in shape to the shock wave produced by supersonic aircraft or the bow wave of a boat (hence the name). The solar wind then flows

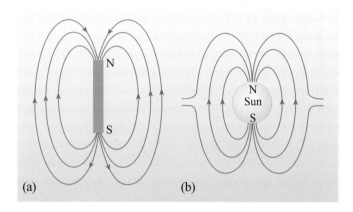

Figure 5.43 The Sun's magnetic field (a) is rather like that of a bar magnet, where magnetic field lines loop round from pole to pole as shown in (b).

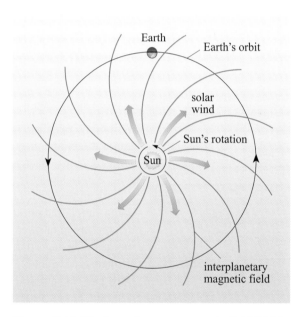

Figure 5.44 The interplanetary magnetic field (IMF) extends away from the Sun. As the Sun rotates, the field lines form a spiral shape, rather like a garden sprinkler.

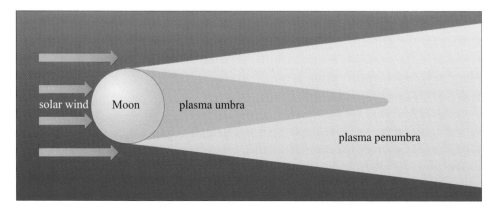

Figure 5.45 Interaction of the IMF and solar wind with the Moon or other body with no atmosphere and little or no magnetic field.

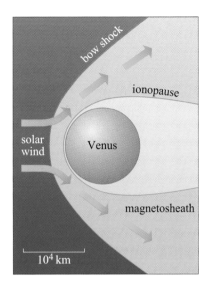

Figure 5.46 As the interplanetary magnetic field (IMF) and solar wind flow over a planetary body with an ionosphere, a bow shock is produced, which decelerates solar wind particles, which then flow around the ionopause. This is the scenario found at Venus.

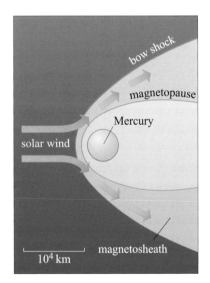

Figure 5.47 Magnetosphere of Mercury. The solar wind is deflected around the planet by the magnetic field.

around the planet leaving a region where few ions from the solar wind enter. The boundary of this region is the **ionopause**. This interaction with the ionosphere gives Venus a modest magnetosphere (Figure 5.46).

The magnetosphere of Mars is less well known but is probably similar to that of Venus.

Mercury has very little atmosphere and thus no ionosphere. It does however have its own magnetic field.

- How might Mercury be affected by the IMF?
- Its magnetic field could interact with the IMF.

In this case a bow shock is produced when the IMF and the solar wind are resisted by the planetary magnetic field. This boundary between a planet's magnetosphere and the external magnetic field or solar wind is known as the **magnetopause** (Figure 5.47).

The Earth has its own strong magnetic field (as you may remember from Chapter 2) and an ionosphere. The magnetic field lines extend further out than the ionosphere, so as the IMF and solar wind flow past the Earth, a bow shock is produced by the Earth's magnetic field lines as on Mercury. The structure of Earth's magnetosphere is more complicated and interesting than those of the other terrestrial planets.

Within the magnetosphere but beyond what is generally thought of as the atmosphere, matter is present in the form of plasma, that is electrons and highly ionized atoms. Because it consists of charged particles, it is affected by the magnetic field and is not uniform throughout the magnetosphere. In 1958 an American physicist James Van Allen (Figure 5.48) and his research students were using simple radiation detectors on the satellites Explorer 1 and Explorer 3 to survey cosmic ray intensities above the atmosphere. They discovered a large population of energetic charged particles that were trapped by the Earth's magnetic field. These formed two distinctive toroidal (doughnut-shaped) regions, the Van Allen radiation belts (Figure 5.49).

Figure 5.48 James Alfred Van Allen, born in 1914 in Iowa. He was a pioneer of space physics starting with high-altitude rockets in 1945. He and his research students discovered the Van Allen belts using instruments on Explorer 1 during International Geophysical Year, 1958. He has since done much more work on planetary magnetospheres. He is emeritus professor at the University of Iowa. (University of Iowa)

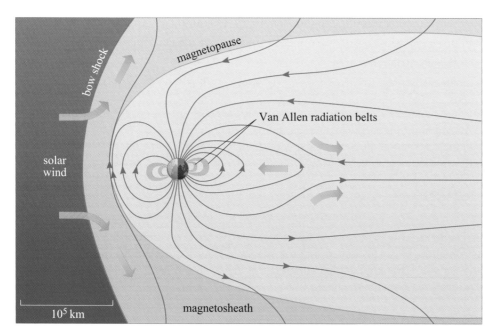

Figure 5.49 The Earth's magnetosphere. The main characteristics are defined by the Earth's magnetic field lines that are contained within the magnetopause, which is separated from the bow shock by the magnetosheath region (into which the solar wind particles can flow). The Van Allen radiation belts form toroids around the Earth inside the magnetosphere. Note the much larger extent of Earth's magnetosphere compared to those of Venus and Mercury.

The inner belt consists largely of protons (hydrogen ions) and electrons and originates from the solar wind and the ionosphere. The outer belt is populated from the solar wind. The high-energy fraction of particles in these belts is a hazard to space travellers. The charged particles also affect electronic instruments.

Because the magnetic field of the Sun (and hence the IMF and the solar wind) varies with time, the magnetosphere also changes and these changes can affect us. Visually, the most dramatic affect is the production of aurorae (aurora borealis in the Northern Hemisphere (Figure 5.3) and aurora australis in the Southern Hemisphere). The colours of the aurorae are due to emission by atoms and ions in the upper atmosphere:

- oxygen atoms produce green and red emissions;
- nitrogen, as N_2 or N_2^+, produces purple or red emissions.

■   If the atoms or ions are emitting light, what effect must the varying magnetosphere have had on them?

❏   They must have been excited to a higher energy level (Box 5.1).

The changes in the magnetic field of the magnetosphere cause particles from the tail of the magnetosphere to move along the magnetic field lines into the atmosphere where they collide with atoms, raising them to higher energy states or even ionizing them. The particles from the tail are channelled into the polar regions forming ovals around the magnetic poles. The display we view from the ground is just a part of the full auroral oval that can be imaged by satellites (Figure 5.50).

Large variations in the solar magnetic field and solar wind, such as occur during solar flares, lead to extraordinary variations in the magnetic field on the Earth's surface. These are called geomagnetic storms. These storms interrupt long-range communications by causing the ionosphere to vary in height and density.

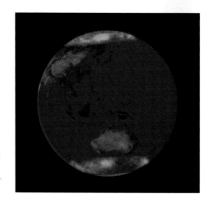

Figure 5.50 The POLAR satellite launched by NASA in 1996 is being used to study the Earth's magnetosphere. In September 2001 it captured a view of auroral ovals at both poles. (NASA)

■ Why does variation in the ionosphere affect long-range communications?

❏ The radio waves used are bounced off the ionosphere in order to cover long distances (Section 5.4).

During a geomagnetic storm some radio frequencies are absorbed rather than reflected and others follow unusual paths. Such storms can also interact with conductors such as power cables. A huge geomagnetic storm on 13 March 1989 left 6 million people in the Montreal region without electrical power for 9 hours.

QUESTION 5.10

Venus has a core containing molten material but only a very small or no magnetic dipole field. This is thought to be because of the slow rotation of the planet. Some astronomers have speculated that the planet has been slowed down by its atmosphere. This implies that rotation was faster in the past. If this is correct how would the early magnetosphere of Venus differ from the present one?

QUESTION 5.11

In the early 20th century, it was assumed that the detected CO_2 on Mars was a small fraction of the atmosphere, the bulk being nitrogen as on Earth. Why was it possible to take this point of view even though N_2 had not been detected?

QUESTION 5.12

It has been suggested that the transport to Mars of sufficient greenhouse gases would eventually make the planet able to support life. How would you expect the greenhouse gases to transform the atmosphere?

QUESTION 5.13

The temperature in the polar regions of the Earth can be very low. Would carbon dioxide freeze out if the temperature dropped to $-80\,°C$ (193 K)? (The partial pressure of carbon dioxide on Earth is 3.6×10^{-4} bar.)

QUESTION 5.14

How do (a) the ozone layer and (b) the magnetic field aid the continued existence of life on Earth?

5.8 Summary of Chapter 5

- Remote sensing of planetary atmospheres provides information about their composition. Infrared (vibrational) spectroscopy is applicable to all atmospheric gases except for those diatomic molecules that possess no electric dipole (i.e. homonuclear diatomic molecules).

- Analysis by direct access of planetary atmospheres has successfully employed two major techniques, gas chromatography and mass spectrometry. Gas chromatography is used to separate complex gaseous mixtures. Mass spectrometry provides the relative molecular mass of a substance, and can also give information about isotopic composition.

- The atmospheres of Venus and Mars are oxidized atmospheres whose major component is carbon dioxide, CO_2. The atmosphere of Earth is an oxidizing atmosphere. The main components are nitrogen, N_2, and oxygen, O_2. The volatile inventory for a planet includes estimates of volatile materials in surface water, polar ice-caps and rocks near the surface as well as those in the atmosphere. The volatile inventories of Venus, Earth and Mars have comparable fractions of CO_2. Venus is lacking in water when compared to Earth and Mars, and Mars is lacking in nitrogen when compared to Earth and Venus. Mercury has a very tenuous atmosphere that is mainly derived from the solar wind.

- The atmospheres of Venus, Earth and Mars show vertical variations of temperature, density and pressure. On this basis, the following regions of the atmosphere can be distinguished:

    The troposphere is the lowest region, in which surface heating causes convection to occur. The decrease of temperature with altitude is close to the adiabatic lapse rate.

    The stratosphere, unique to Earth, is the middle region heated by the absorption of solar UV radiation by ozone, O_3. The temperature increases with altitude in this region.

    The thermosphere is the highest region and is heated by direct absorption of solar UV radiation, mainly by O_2 (Earth) and CO_2 (Venus and Mars); the temperature increases with altitude.

    In the higher reaches of the thermosphere, the UV radiation causes a substantial fraction of the gases to be ionized, forming a plasma; this region is the ionosphere.

    The mesosphere lies below the thermosphere and is heated and cooled through emission and absorption of infrared radiation; the temperature decreases more slowly than the adiabatic lapse rate.

- The balance between incoming solar radiation and the thermal emission by a terrestrial planet determines the mean temperature of its atmosphere and surface.

- The greenhouse effect causes mean surface temperatures to be higher than effective temperatures. It arises from the absorption of infrared radiation emitted from the surface by the atmosphere.

- Cloud formation occurs on Venus mainly as H_2SO_4 droplets, on Earth as water droplets and ice particles (H_2O), and on Mars as ice (H_2O and CO_2) particles. Saturation vapour pressure diagrams allow us to predict the equilibrium conditions in which clouds will form.

- Differential heating of planetary surfaces by the Sun gives rise to convective cells in the troposphere. These are known as Hadley cells. Through the Coriolis effect, the rotation of the planet imposes a west-to-east motion on these cells. Venus rotates slowly, yet its single Hadley cell structure shows high east-to-west wind speeds at high altitudes (super-rotation). On Mars, the atmosphere further circulates by condensation flow (pole to pole) and by a thermal tide (flow from day-side to night-side).

- The interplanetary magnetic field (IMF) and associated solar wind interact with the terrestrial planets to form magnetospheres. For planets without a magnetic dipole field of their own, the IMF interacts with the ionosphere; for those with a magnetic dipole field, the IMF interacts with this field. In both cases the solar wind is deflected around the planet protecting the surface.

- The Earth, which has both a magnetic dipole field and an ionosphere, has a larger and more complex magnetosphere than the other planets, including two radiation belts encircling the Earth. Variations in the IMF and solar wind give rise to aurorae and can disrupt long-range communications and power transmission.

CHAPTER 6
THE GIANT PLANETS

6.1 Introduction

The previous chapter dealt with objects with definite surfaces – the terrestrial planets. We turn now to objects for which there is no discernible surface and where the greater part of the object (possibly all) is fluid (i.e. gas or liquid). These are the **giant planets**: Jupiter, Saturn, Uranus and Neptune (Figure 6.1).

We start by considering the overall structure of these planets. Much of the detailed evidence has come from instruments on board spacecraft, and it is hoped that even more information will be gathered by future missions. However, Earth-based instruments are by no means obsolete in this field and observations by space telescopes (in orbit around the Earth) have provided much valuable data. Ground-based observations were necessary as a starting point for data collection by the spacecraft. One advantage of Earth-based and space-telescope observations is that they can be used to study changes in a planet's appearance over a long time (in the case of Jupiter, hundreds of years), whereas fly-by and lander spacecraft observe for only a limited time.

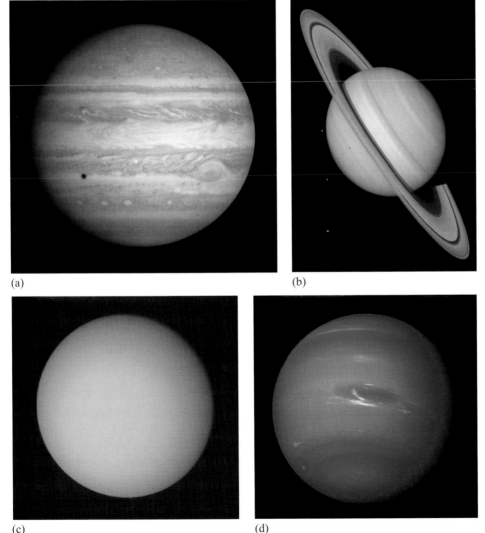

(a)

(b)

(c)

(d)

Figure 6.1
(a) Jupiter (mean radius 69 910 km),
(b) Saturn (mean radius 58 230 km),
(c) Uranus (mean radius 25 360 km) and
(d) Neptune (mean radius 24 620 km). (NASA)

We have a fairly accurate picture of the composition and structure of the outer layers or atmospheres of these planets, because we can detect and positively identify molecules in them. Our knowledge of the interiors, however, is less certain and is based on indirect measurements and modelling. As none of the four planets has an accessible surface (if they have any surfaces at all), we do not know where the base of the atmosphere is. The radii of the planets are therefore often defined as the distance from the centre of the planet to the 1 bar pressure level (1 bar being approximately the pressure of the Earth's atmosphere at sea-level). This level corresponds roughly to the base of the directly observable layers of the planets. In general, when we refer to the atmosphere of a giant planet we mean the layers of the planet from the 1 bar pressure level outwards. In some cases, for example the results of the Galileo probe measurements (see Box 6.1), we shall include layers slightly deeper than this.

In 1997 the Cassini spacecraft planned jointly by NASA, ESA (the European Space Agency) and ASI (the Italian space agency) was launched. This is due to reach Saturn in July 2004 and study the planet, its rings, magnetosphere and its satellites. As Cassini passed Jupiter in December 2000, observations of Jupiter were made by Galileo and Cassini simultaneously.

You will start this chapter by studying the evidence that is available for the nature of the interiors of these planets and how the data can be interpreted. You will then move on to study the atmospheres.

For an atmosphere, we can determine not only its composition and structure but also how it moves. Are the processes determining this motion the same as those on the terrestrial planets, or does the far greater depth of fluid on the giant planets mean that different processes occur? Can the way gas is transported around the atmosphere tell us anything of what goes on in the interior of the planet? We shall be looking at the observed wind motions and seeing if we can draw any conclusions from them.

Finally, you will be studying the magnetic fields of these planets and how these interact with, for example, the solar wind and satellites of the giant planets.

Because of the difficulties in collecting data from these bodies, you will often find that there is no one clear-cut explanation and that several plausible models will fit the data that we have. Consequently, there are some sections in which you will not be asked to learn an accepted explanation but to consider whether a particular model is ruled out by the information available. This is particularly true of Uranus and Neptune. Do not worry, therefore, if you leave this chapter feeling you are not quite sure, for example, what the internal structure of Uranus is. Nobody is sure!

6.2 The structures of the giant planets

6.2.1 Models of the structure

You saw in Chapter 2 that we are not even sure about the composition of the interior of the Earth. We have less observational data on the giant planets and so our knowledge of their interiors is much less certain. From the data we have and laboratory experiments on how the density of materials varies with temperature and pressure, possible models can be put forward. The current models of the planets predict a layer structure. Basically, Jupiter and Saturn would have a rocky core, a layer of metallic hydrogen surrounding this and a thick outer gaseous (hydrogen and helium) envelope, and Uranus and Neptune would have an inner icy–rocky core with

BOX 6.1 VOYAGER AND GALILEO

These space missions have provided a huge amount of data on the giant planets; you will find that many of the images in this chapter were obtained from them. Such missions require years of planning – involving scientific, technical and political considerations. Because of the distance of the giant planets from Earth, such spacecraft are launched years before they can provide the data they are planned to measure. The NASA Voyager missions comprised two spacecraft, Voyager 1 and Voyager 2, which between them studied all four giant planets and many of their satellites and are now heading out of the Solar System. The Voyager project was conceived in the 1960s when it was realized that, in the late 1970s, the four giant planets would be in such positions that a single spacecraft could encounter all of them, an opportunity that would not occur again for 175 years. The original aims of obtaining data on the four giant planets was magnificently achieved by 1989, indeed for Uranus and Neptune the Voyager-2 data remain the most comprehensive set we have. The decision was then made to continue collecting data in order to learn about the boundary of the Solar System. In 10 to 20 years from the time of writing (2003) the spacecraft are expected to leave the heliosphere – the region under the influence of the Sun's magnetic field – and head for interstellar space. We give below a timetable of launch dates and main encounters.

- **1972** First project director starts work.

- **1977** August 20, Voyager 2 launch.

- **1977** September 5, Voyager 1 launch.

- **1979** March 5, Voyager 1 closest approach to Jupiter.

- **1979** July 9, Voyager 2 closest approach to Jupiter.

- **1980** November 12, Voyager 1 closest approach to Saturn.

- **1981** August 26, Voyager 2 closest approach to Saturn.

- **1986** January 24, Voyager 2 closest approach to Uranus.

- **1989** August 25, Voyager 2 closest approach to Neptune.

- **1998** February 17, Voyager 1 outdistances Pioneer 10 to become the furthest human-made object in space.

- **2020** Predicted last date at which data can be received from the Voyagers.

The NASA Galileo mission was designed to study Jupiter and its satellites in greater detail. On its way to Jupiter, Galileo flew past Venus and the Earth using these planets to change its direction and send it towards Jupiter with sufficient speed. As the spacecraft approached Jupiter, it separated into two – the Galileo probe and the Galileo orbiter. The probe descended into the atmosphere of Jupiter giving the first direct measurements of the composition and wind speed of the atmosphere. The orbiter studied Jupiter remotely and went on to obtain data on Jupiter's magnetosphere and satellites. One of the final manoeuvres, made in 2002, was to put the spacecraft on an orbit which will result in Galileo entering the atmosphere of Jupiter, where the spacecraft will be destroyed.

- **1989** October 18, Galileo spacecraft launch.

- **1990** February 10, observations on Venus.

- **1990** December 8, fly-by of Earth.

- **1991** October 29, encounter with asteroid Gaspra.

- **1992** December 8, observations of Earth.

- **1993** August 28, encounter with asteroid Ida.

- **1994** July, observations of collision of comet Shoemaker–Levy with Jupiter.

- **1995** December 7, Galileo probe enters atmosphere of Jupiter. Galileo orbiter starts tour of Jovian system. (Note that 'Jovian' is the adjective for Jupiter.)

- **1997–1999** Fly-by observations of Europa.

- **1999** Fly-by observations of Callisto and Io.

- **2000** January, fly-by of Europa.

- **2000** February–May, fly-by of Io.

- **2000** May–September, fly-by of Ganymede.

- **2000** October–December, observations of Jupiter.

- **2001** October, fly-by of Io.

- **2002** January, final fly-by of Io.

- **2002** November 5, fly-by of the inner satellite Amalthea.

- **2003** September 21, end of Galileo mission – spacecraft collides with Jupiter.

an outer gaseous envelope (Figure 6.2). Current modelling does not however, assume that icy and rocky materials are completely differentiated from each other or from the gaseous component. The innermost layers are generally thought to consist of icy and rocky material. Some models distinguish two regions within this layer of differing density. At the centre of the planets, the temperatures and pressures are, according to the models, extremely high: 8000 K to 16 000 K and 13×10^6 bar to 50×10^6 bar. Although high, these figures are not unrealistic given that the corresponding figures for the Earth are 4300 K and 3.3×10^6 bar. The giant planets contain far more rocky material than any terrestrial planet but the icy–rocky layers occupy only a small fraction of their total volumes. Outside these layers the composition is predominantly hydrogen and helium, although rocky and particularly icy materials are still present. For much of the volume, the temperatures and pressures are still high but gradually decrease. Atmospheric temperatures are low.

Remember that the 'atmosphere' is above the 1 bar level and the 'interior' is below the 1 bar level.

The layer structure with rocky materials concentrated towards the centre supports the view that differentiation has occurred subsequent to the formation of the planets. The evidence is not sufficiently detailed to say whether the rocky materials themselves have differentiated as in the Earth to give a central core of iron. That such differentiation has occurred is probable but the models used to try to reproduce the observational data use only an average composition for the rocky materials, based on a mixture of materials such as iron, iron sulfide, and silicates, that on the available evidence, are present in the interiors of terrestrial planets.

Figure 6.2 shows sketches of the interiors of Jupiter, Saturn, Uranus and Neptune. These represent one interpretation of the data currently available. Estimated temperatures and pressures at the boundaries of the layers are given. The nature of the layer labelled 'helium plus metallic hydrogen' will be explained when we consider Jupiter and Saturn in more detail. Not all these boundaries are as sharp as they appear in Figure 6.2. The outermost layers are gaseous and the innermost are liquid but in descending through the interior, a point would be reached where the conditions are such that the material is in a fluid phase where gas and liquid are indistinguishable. It is therefore probable that these planets not only have no sharp solid–gas (i.e. solid-to-gas) boundary, but have no sharp liquid–gas (i.e. liquid-to-gas) boundary either. In both of these respects, the giant planets differ from the terrestrial planets.

■ From the temperatures marked on Figure 6.2, does the presence of icy materials imply that the layer is cold?

❏ No. All the temperatures shown are extremely high except for the tops of the outermost layers, i.e. the bases of the atmospheres.

The term 'icy material' simply denotes volatile compounds such as water, ammonia and methane and does not imply that the material is cold.

Note that the regions are shown by volume and that this does not represent the mass ratios of the different regions. The *cumulative* masses of the layers are labelled on Figure 6.2. For example, the inner core of Jupiter contains 3 to 3.5 Earth ($M_\oplus$) masses of material (i.e. about one-hundredth of the mass of Jupiter) but it occupies only one-thousandth of Jupiter's volume. Then the inner core of Jupiter *and* the next layer out together contain 5 Earth masses of material.

Before we go on to look at the interiors in more detail, let's consider the evidence on which such models are based.

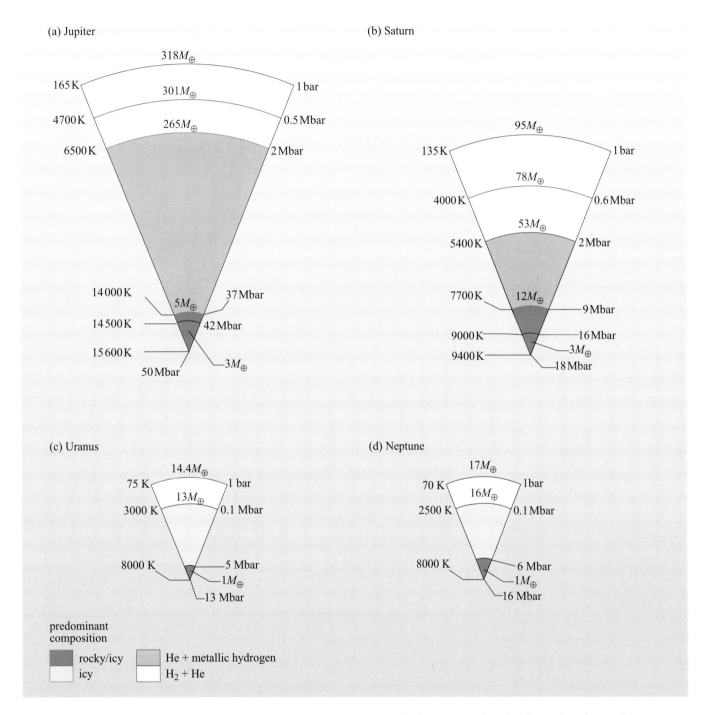

Figure 6.2 Cross-sections through (a) Jupiter, (b) Saturn, (c) Uranus and (d) Neptune, showing the various layers that are thought to exist in the interiors of these planets. Note that in each case, the 1 bar level is considered to be the *inner* boundary of the atmosphere. Although the outer two layers on Jupiter and Saturn are both labelled 'H$_2$ + He', there is a distinction between them as we will see shortly. There is also a distinction between the innermost rocky/icy layers. The cumulative masses contained within a layer are shown also (note $M_\oplus$ means Earth-mass).

6.2.2 Obtaining evidence

The first and incontrovertible piece of evidence is the average density of each planet as a whole. The giant planets are not quite spherical, being somewhat flattened at the poles. However, the volume can be calculated if the polar and equatorial radii are measured (as given in Appendix A, Table A1). From this we obtain the mean radius which is the value the planet's radius would have if it were spherical.

■ Saturn has a mean radius of 58 230 km. Assuming Saturn is a sphere, what is its volume?

❏ The volume of a sphere is $\frac{4}{3}\pi R^3$, where R is the radius. Thus the volume of Saturn, V_{Sat} is:

$$V_{Sat} = \frac{4}{3}\pi R_{Sat}^3$$
$$= \frac{4}{3}\pi \times (58\,230\,\text{km})^3$$
$$= 8.3 \times 10^{14}\,\text{km}^3$$
$$= 8.3 \times 10^{23}\,\text{m}^3$$

The other quantity needed to calculate the density is the planetary mass. The masses of all four giant planets have been obtained by studying the paths of their satellites and of spacecraft flying past. These smaller bodies are affected by the gravitational attraction between them and the planet, which of course depends on the planetary mass.

QUESTION 6.1

The orbital period P of a satellite of Saturn, Enceladus, is 1.370 days and its average distance a from Saturn is 238 000 km. Assuming that the mass of Saturn M_{Sat} is much greater than that of Enceladus and that the satellite's orbit is roughly circular, the mass of Saturn can be estimated using the formula:

$$M_{Sat} = 4\pi^2 \frac{a^3}{GP^2}$$

where $G = 6.67 \times 10^{-11}\,\text{N m}^2\,\text{kg}^{-2}$ and is the gravitational constant. Calculate the mean density of Saturn.

The generally accepted densities obtained from the volumes and masses are given in Appendix A, Table A1. The value obtained for Saturn in Question 6.1 agrees very well with the value in this table.

The passage of spacecraft and satellites tells us more than just the mass of the planet. Because the planets are not spherical, it turns out that from the gravitational field we can also learn something of the distribution of density within the planet. For example, as the giant planets are not spherical but are squashed along the polar axis the gravitational pull on a body at the pole differs from that on one at the equator. Furthermore, two objects that are the same distance from the centre of the gas giant planet feel a slightly different gravitational pull depending on whether they are over the pole or over the equator (Figure 6.3).

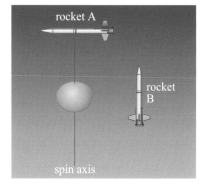

Figure 6.3 The gravitational pull on spacecraft A differs from that on spacecraft B although they are equidistant from the planet's centre.

How the gravitational pull differs depends not only on the degree of flattening but also on how material is distributed in the planet. More sophisticated analysis of how the gravitational field varies with latitude and longitude is possible and leads to the conclusion that there are denser and lighter materials, and that the denser materials lie near the centre, as in Figure 6.2.

Incidentally, this method is also the best way of measuring the internal variation of density for Venus for which there is no seismic data (Chapter 2) and Mars for which there is very little such data. The measurements we obtain do not give us a definitive model for the nature of the interior, but any model proposed must be able to reproduce the observed results. Models with different proportions of rocky/icy materials, hydrogen and helium in the different layers can be tested to see which fits the gravitational evidence most closely.

Another source of information on the interiors of the planets is measurements of the magnetic fields. Like the Earth, the four giant planets each have a magnetic field that behaves as though it originated in a magnetic dipole.

■ What properties does the Earth's core need to have to explain its magnetic dipole field?

❑ It has to be liquid, electrically conducting, and in motion.

Jupiter's magnetic field has been known and studied for many years, but Voyager 2 produced the first comprehensive measurements of the magnetic fields of Uranus and Neptune. The existence of the magnetic fields means that somewhere in the interior of each planet lies an electrically conducting liquid.

■ What is the electrically conducting liquid inside the Earth and where is it located?

❑ It is the outer core, which is liquid iron with about 4% nickel and some lighter elements.

Extensive mapping of the magnetic fields has allowed the region in which the magnetic field originates to be determined and this does not support a liquid-iron core as the main source of the magnetic field for the four giant planets. Figure 6.4 shows the main features of the magnetic fields. The bar magnets representing a dipole will approximately reproduce the observed fields.

For our purposes, the main conclusion from such measurements is that there must be an electrically conducting liquid layer in the metallic-hydrogen regions of Jupiter and Saturn, or in the icy-materials region of Uranus and Neptune.

More evidence on the nature of the interiors comes from the heat given out by the planets. The effective temperatures of the planets are very low: 120 K (Jupiter), 89 K (Saturn), 53 K (Uranus) and 54 K (Neptune). They are, however, except for Uranus, higher than would be expected based on their distance from the Sun, if the Sun were their only source of energy. The additional energy needed to produce the observed temperature is known as the **heat excess**. Jupiter, Saturn and Neptune must therefore have an internal source of energy, the origin of which will be discussed when we consider the individual planets. We shall also put forward one explanation for the apparent lack of an internal energy source in Uranus.

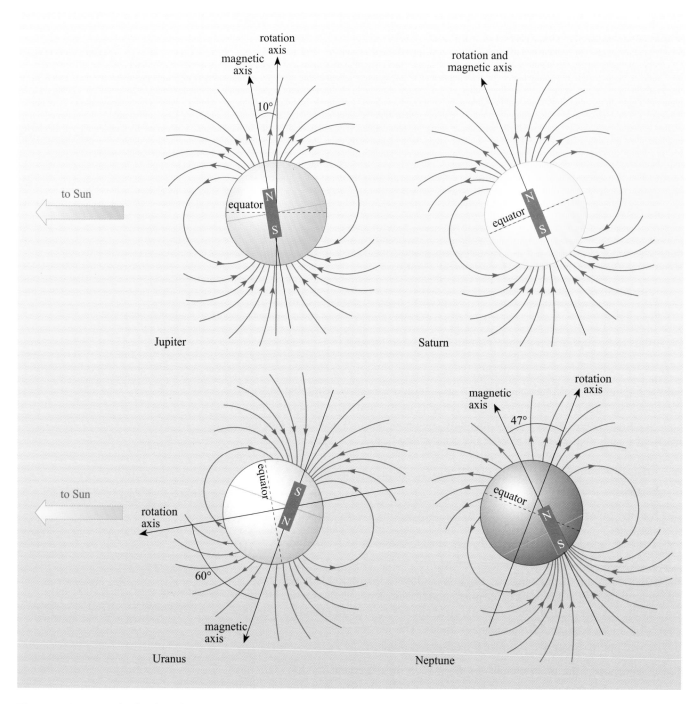

Figure 6.4 Magnetic dipoles of Jupiter, Saturn, Uranus and Neptune. The bar magnets indicate the position of the centre of the dipole and its polarity. The length is arbitrary. The lines represent observed magnetic field lines. The direction to the Sun as shown, is at one point in the orbit. However the directions of the rotation and magnetic axes are fixed with respect to the stars, not the Sun.

The amount of energy radiated by the planet can be used as a constraint in estimating the temperature of the interior. In the case of Jupiter, the core temperature was estimated by a recent model (1999) based on data from the Galileo orbiter to be between 15 600 K (Figure 6.2) and 16 000 K. Other models have suggested temperatures in excess of 20 000 K.

Finally, in the absence of evidence to the contrary, it is assumed that heavy elements are present in roughly solar ratios and that the planets are more likely to contain substances made from more abundant elements than less abundant ones. On these grounds, for example, ammonia (NH_3) is a more likely major constituent of icy materials than hydrogen fluoride (HF) because the solar abundance of nitrogen is greater than that of fluorine.

QUESTION 6.2

Outline the reasons why the element mercury (Hg) is not a suitable candidate for the material responsible for producing the magnetic dipole fields in the four giant planets. (The relative abundance by number of Hg is 12 on a scale that sets the abundance of H at 10^{12}. On this scale the relative abundances of some common elements are Si 3.5×10^7, N 1.1×10^8, Fe 3.2×10^7. Note that the density of Hg is about $1.36 \times 10^4\,\mathrm{kg\,m^{-3}}$.)

6.3 Jupiter and Saturn

6.3.1 Interiors

The densities of the giant planets are very low compared to the terrestrial planets; indeed that of Saturn is lower than the density of water at everyday temperatures on the surface of the Earth. However, the mass of material in these planets is so large that the interior must be at a very high pressure; in the case of Jupiter, the pressure increases from 1 bar at the edge of the atmosphere to 50 million times this at the centre of the planet. With such high pressures, the only way to account for the low overall density is that the composition of these planets is mainly of the light elements, hydrogen and helium, in some form.

■ Saturn is less massive than Jupiter. Saturn is also less dense than Jupiter. Does it follow that the two planets differ in composition?

❏ Not necessarily. The greater mass of Jupiter means that the material will be subject to greater self-compression (Chapter 2) and therefore will be more dense. From density measurements it is probable that Jupiter and Saturn have similar chemical compositions.

QUESTION 6.3

By considering the effect of self-compression only, an approximate value of the pressure at the centre of a planet, P_c is given by

$$P_c \approx \frac{2\pi G}{3}\, \rho_m^2 R^2$$

where G is the gravitational constant ($6.67 \times 10^{-11}\,\mathrm{N\,m^2\,kg^{-2}}$), ρ_m is the mean density of the planet and R is the radius of the planet. Calculate the pressure at the centre of (a) Earth, (b) Saturn and (c) Jupiter using this formula.

Take relevant values from Table A1 in Appendix A.

A further inference from density measurements is that Jupiter and Saturn each contain about 5 to 10 Earth masses of icy and rocky material. Measurements of the gravitational fields of these planets are consistent with most of this being in the central core. Models of the planets that fit these measurements, and agree with our current knowledge of how the density of materials varies with temperature and pressure, show a core of dense material. In the absence of further information, we can reasonably assume that this core is composed of common rocky and icy planetary materials, such as silicates, iron, iron compounds, water, ammonia and methane. The materials are probably further differentiated within the core to give metallic iron at the centre but current modelling methods use an average rocky material whose composition is based on solar abundances and the composition of typical rocky bodies. Observational data is not sufficiently precise to distinguish details of the core structure. The outer and inner core layers are similar but the outer layer has a higher proportion of helium.

Surrounding each core is a very thick layer taking up the greater part of the planet and labelled helium plus metallic hydrogen in Figure 6.5. The helium is believed to be present as a fluid, similar to that used on Earth in very low temperature studies and in superconducting magnets, but at much greater temperatures and pressures.

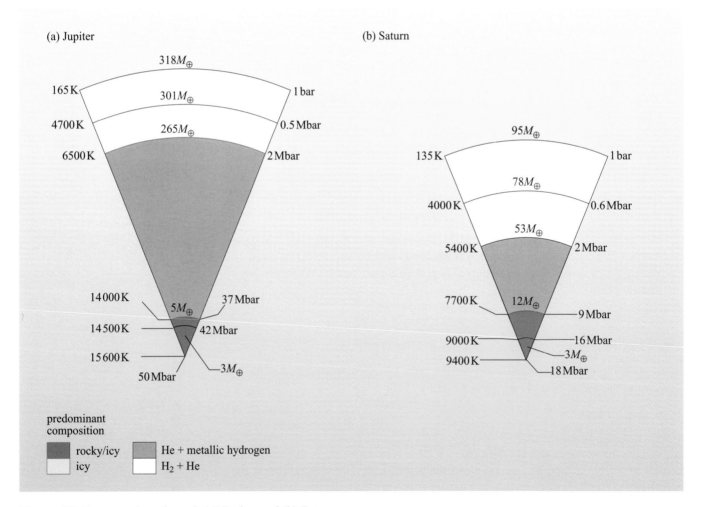

Figure 6.5 Cross-sections through (a) Jupiter and (b) Saturn.

The hydrogen is generally accepted to occur in a most unusual form – **metallic hydrogen**. On Earth, hydrogen is a colourless gas that is very difficult to liquefy and is certainly not metallic. To see how hydrogen can occur in a form described as metallic, read Box 6.2.

BOX 6.2 METALLIC BONDING

A simple picture of a metal such as aluminium, familiar to us on Earth, is of an orderly set of positively charged ions surrounded by electrons, which are shared. Imagine some aluminium ions arranged as in Figure 6.6.

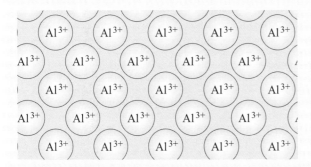

Figure 6.6 Aluminium ions (Al^{3+}) arranged as though in a crystal of aluminium. The electrons 'lost' from the atoms wander freely through the solid.

Now, the most common ion formed by aluminium is Al^{3+}, in which the aluminium atom has lost three electrons. Suppose the lost electrons are allowed to wander freely about the crystal. You may recall that two atoms can form a chemical bond by sharing electrons. In our picture of aluminium, the freely wandering electrons are shared by all the ions in the crystal, and thus serve to bond together all the ions. This sharing of electrons by a whole crystal constitutes **metallic bonding**. The electrons are responsible for many of the characteristic metallic properties of elements such as aluminium. In particular, they can move through the solid and, because they are electrons, they carry negative electrical charge with them. A moving electrical charge is an electric current, and so metallic bonding leads to electrical conductivity.

■ Aluminium is a solid at everyday temperatures and pressures, but there is one common liquid metal found at the Earth's surface. What is it?

❏ Mercury (Hg). You may have a mercury thermometer or barometer at home.

In liquid mercury, the situation is similar to that shown in Figure 6.6 except that because mercury (Hg) is a liquid, the ions are not so ordered (Figure 6.7).

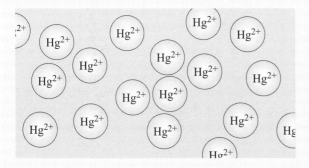

Figure 6.7 Mercury ions (Hg^{2+}) arranged as though in liquid mercury. The 'lost' electrons are free to travel through the liquid, but the arrangement of the ions is less regular than in solid aluminium.

At the very high pressures in the interior of Jupiter and Saturn the hydrogen nuclei are squashed together so that, instead of just two hydrogen nuclei sharing electrons to form H_2 (which is the normal form of the element hydrogen on Earth), large numbers of hydrogen nuclei share their electrons, and metallic bonding results as in liquid mercury. Metallic hydrogen is liquid at the temperatures and pressures inside these two planets.

Detailed measurements of Jupiter and Saturn show that the magnetic source is outside the core in the layer labelled 'metallic hydrogen' in Figure 6.5. The region where the magnetism originates is a spherical shell in this layer, and the magnetic dipole field produced by such a shell is centred in the planetary core. This provides an explanation for the magnetic field. The electrically conducting fluid is liquid metallic hydrogen. The heat of the interior will cause the liquid to be in motion through convection currents.

As we move further out from the core, temperatures and pressures are reached where hydrogen is more stable as H_2 molecules than as a metal. This is the case for the outer two layers. Recent (2000) experiments on hydrogen under extremely high pressure suggest that this transition is gradual and thus the boundary between metallic and molecular hydrogen in Jupiter is not sharp. As we rise through the outer layer, the hydrogen changes imperceptibly from liquid to gas. Although these layers are predominantly hydrogen and helium, they do contain some icy materials and a very small amount of rocky materials. The researchers who proposed this model distinguished the two H_2 + He layers (Figure 6.5) by the different fractions of rocky materials contained in these layers.

For Jupiter, the model that we have described accounts for the measurements of gravity, magnetic field and internal energy including the heat excess. Jupiter is large enough for the internal energy to be residual accretional heat (Chapter 2) which will have led to extremely high temperatures as Jupiter formed, and heat due to past differentiation of rocky and icy materials. Energy from these sources will have been comparatively slowly lost to space due to the low surface area of Jupiter relative to its volume.

■ What is the ratio of surface area to volume for a sphere of radius r?

☐ The surface area is $4\pi r^2$. The volume of a sphere is $\frac{4}{3}\pi r^3$. Therefore the required ratio is

$$\frac{\text{surface area}}{\text{volume}} = \frac{4\pi r^2}{\frac{4}{3}\pi r^3}$$

$$= \frac{3}{r}$$

So the surface-to-volume ratio gets smaller as r gets larger.

For Saturn, we have to introduce a further refinement to account for the heat excess. We might expect that Saturn should have had a lower initial temperature than Jupiter, and with a larger surface-to-volume ratio also, would have lost energy more rapidly and so cooled down long ago. However, a heat excess remains.

■   What is the main heat source for the interior of the Earth?

❑   Radiogenic heating.

The bulk of Saturn, however, is hydrogen (metallic *and* molecular) and helium, and the low density precludes sufficient abundances of radioactive isotopes for this to be a major form of heating.

QUESTION 6.4

Jupiter generates a heat flux of about 7 watts per square metre of surface at the 1 bar level. The decay of ^{40}K and other radiogenic isotopes produces 4.8×10^{-12} W kg^{-1} of heat in the Earth today. If Jupiter has a core of which $4M_\oplus$ (i.e. 4 Earth-masses) is rocky materials, what would be the rate of radiogenic heat generation in the core of Jupiter, assuming that the core is of similar composition to the Earth? Compare this with the observed heat flux.

The heat excess in Saturn may come from the separation of hydrogen and helium in the metallic hydrogen layer. In Jupiter, this layer is hotter and well stirred by convection. At higher temperatures, helium is soluble in hydrogen. As the temperature falls, helium becomes less soluble and the two liquids tend to separate out. If the layers are well-stirred, however, an emulsion is formed in which small droplets of helium are dispersed throughout the hydrogen. This situation is rather like olive oil and water. If you stir the two liquids vigorously you can get them mixed (as for example in salad dressing). If the stirring becomes less vigorous, larger droplets of one liquid form in the other and gradually sink to the bottom to form a separate layer. Convection stirs the layers in Jupiter so that the hydrogen and helium are still mixed, but the liquid helium in Saturn's metallic hydrogen layer is forming larger droplets which, because they are denser than the hydrogen, fall towards the centre of the planet and will eventually form a separate layer.

■   How does this separation cause the interior of Saturn to heat up?

❑   This is an example of heat from differentiation. The helium droplets release gravitational energy as they fall, and this energy is converted into heat.

QUESTION 6.5

Measurements of the gravitational field of Saturn indicate that the hydrogen/helium ratio in the molecular hydrogen layer is roughly solar and that helium is not depleted. Does this fit in with the attribution of the heat excess of Saturn to differentiation of helium from metallic hydrogen?

6.3.2 Atmospheres – structure and composition

Studying the atmosphere

The atmosphere of these planets is, as we said in the introduction, formally defined as the layer outside the 1 bar level. This coincides roughly with the layer above the outermost cloud layer and hence the layer we can view. Some of the studies we are going to discuss include measurements at slightly deeper levels but all are concerned only with a very small fraction of the radius of the planets.

The atmospheres of the giant planets were studied at radio, infrared (IR), visible and ultraviolet (UV) wavelengths by instruments on the Voyager probes. More detailed measurements of Jupiter were obtained by the Galileo Orbiter. The impact of the comet Shoemaker–Levy 9 with Jupiter's atmosphere (1994) resulted in ejection of material from Jupiter which was studied by Earth-based instruments. Initially, it was thought that the water detected in this way was from Jupiter but it proved difficult to establish how much material came from Jupiter and how much from the comet. The Galileo probe used a mass spectrometer and other instruments to sample the atmosphere from the 0.5 bar level to the 21 bar level.

As the Galileo probe sampled the atmosphere directly, you might consider the results from this probe to be definitive. However, the probe only entered Jupiter's atmosphere at one point and there is no guarantee that the atmospheric composition at this point is the same as the average composition. Indeed it has been suggested that the probe entered the atmosphere at a site which was particularly dry, i.e. lacking in water. Earth-based and space telescopes have also provided valuable information. All these sources of evidence have to be considered when putting together a picture of the nature of the atmosphere.

Because of the low temperatures of the atmospheres, most elements are present as molecules rather than atoms, and most of the molecules are identified from their vibrational spectra.

■ Vibrational spectroscopy was discussed in Chapter 5. What wavelength of radiation is usually associated with vibrational spectra?

❑ Infrared (IR).

It has been possible to detect molecules with relatively low abundances, such as ethyne (acetylene), C_2H_2, and phosphine, PH_3, through their absorption of IR radiation. These molecules act approximately as black-body emitters from layers at pressures of about 1 bar, and because the atmospheric temperatures are so low, the peak of the black-body radiation curve lies in the IR (see Box 5.5).

■ What do you suppose is the source of the IR radiation absorbed?

❑ The planets themselves.

The more abundant ammonia and methane were first detected through their absorption of visible sunlight reflected by the planetary atmosphere and received by telescopes on Earth. Methane and ammonia are colourless gases, and their electronic spectra lie in the UV. How then can we observe them by their absorption of visible light?

In vibrational spectra in the laboratory, the lines are usually produced by each molecule going from the lowest vibrational energy level to the next highest. It is possible for the molecules to go from the lowest level to higher levels, as in Figure 6.8, but this happens less often.

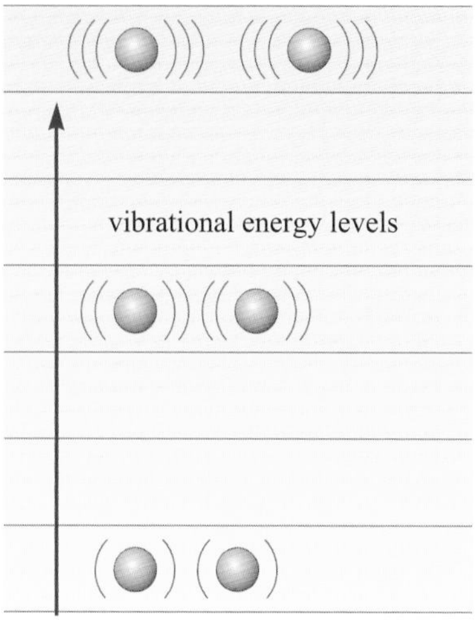

If there is a very large number of molecules in the path of the radiation, as is the case for planetary atmospheres, then there are enough molecules jumping several vibrational energy levels for the corresponding spectral lines to be observed. The absorption of red light by methane in the atmosphere of Uranus due to jumps of this sort is responsible for the blue–green tinge of the planet.

So far, we have not discussed the detection of the most abundant chemical species in the atmospheres, hydrogen (as H_2) and helium.

■ Can you spot a difficulty in detecting these elements through their vibrational spectra?

❑ Helium occurs as atoms, and so has no bonds to vibrate and no vibrational spectrum. Molecular hydrogen has vibrational energy levels but, as you saw in Chapter 5, homonuclear diatomic molecules, such as H_2, do not have a strong vibrational spectrum.

Figure 6.8 A molecular vibrational transition to a high vibrational energy level.

Because hydrogen is so abundant, it is actually possible to detect weak lines originating from higher energy vibrational levels. These were first observed telescopically in the 1960s. Hydrogen can also be detected in the UV through its electronic spectrum. The detection of helium is a major problem. Its spectrum is in a part of the UV not normally covered by common instruments, and in which it is difficult to work; helium was not detected on Jupiter until the Pioneer 10 encounter in 1973. The Galileo probe mass spectrometer measured abundances of noble gases including helium. Helium abundance has also been measured indirectly by Voyager and Galileo. These methods depend on assuming that the major component of the atmosphere not accounted for, must be helium atoms.

One method used by Voyager was to look at the shape of the lines in the vibrational spectrum of, say, methane (CH_4). One factor affecting the width of the lines is the pressure of the atmosphere. If we have direct measurements of the abundance of the major constituents of the atmosphere other than helium, then we can calculate the effect these would have on the line shape. Assuming the rest of the atmosphere is helium, we can then calculate what pressure of helium is necessary to give the observed line shape.

■ Why do we assume the rest of the atmosphere is helium rather than some other atom or molecule with no suitable spectrum?

❑ In the absence of evidence to the contrary, we assume roughly solar abundances, and helium is the second most abundant element in the Sun.

The second method involved sending radio waves through the atmospheres from a spacecraft and detecting them on Earth. Just as a ray of light travelling from air to water bends as it enters the water, so the radio waves are bent (or refracted) by the planetary atmosphere (Figure 6.9).

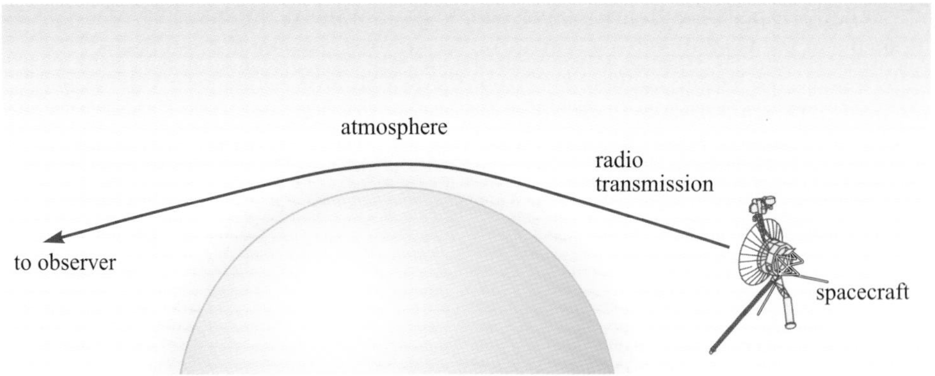

Figure 6.9 The path of the radio waves through a planetary atmosphere. The depth of atmosphere penetrated by the radio waves is exaggerated to show the deflection.

The amount by which the beam is deflected depends on the refractive index of the atmosphere which in turn depends on the average molecular mass of the gas mixture. Again, we assume that the undetected component is helium, and calculate how much helium is needed to give the average molecular mass measured. The identification of the missing constituent as helium was strengthened by the close agreement of the abundance of helium calculated by the two different methods.

The Galileo probe obtained an estimate of helium abundance from the refractive index of the atmosphere, again assuming this was a mixture of hydrogen and helium, as well as direct determination via mass spectrometry.

The first two indirect methods can also be used to estimate the temperature as both the line width and the deflection depend on temperature as well as pressure or average molecular mass. The deflection of radio waves was used by Voyager. The Galileo probe measured density and pressure. The temperature could then be obtained from these measurements using an established equation relating pressure, density and temperature. For example, let us see how the temperature can be estimated using a simple *ideal gas* model.

The ratio of the number of hydrogen molecules, H_2, to helium atoms, He, in the atmosphere of Jupiter is 0.864 : 0.136. The relative molecular mass of H_2 is 2 and of He is 4 and these gases make up most of the atmosphere so that we can calculate the average molecular mass ignoring other contributions.

Since 0.864 and 0.136 add up to 1, the average molecular mass is

$$(2 \times 0.864) + (4 \times 0.136) = 2.272$$

The mass of a hydrogen atom is 1.67×10^{-27} kg. So the average mass of a molecule in the atmosphere of Jupiter is $2.272 \times (1.67 \times 10^{-27})$ kg $= 3.79 \times 10^{-27}$ kg.

For ideal gases, the pressure, temperature and volume of *one mole* are related by

$$P = \frac{(1.38 \times 10^{-23})\rho T}{m} \tag{6.1}$$

where P is the pressure in pascals, 1.38×10^{-23} is a constant which has the units $J\,K^{-1}$, ρ is the density in $kg\,m^{-3}$, T is the temperature in kelvin and m is the average mass of one gas molecule in kilograms. 1 bar corresponds to 10^5 pascal. The density of Jupiter's atmosphere at the 1 bar level was measured as $160\,kg\,m^{-3}$. We can thus calculate the temperature at this level using Equation 6.1.

Rearranging Equation 6.1, the temperature is given by:

$$T = \frac{Pm}{(1.38 \times 10^{-23})\rho}$$

$$= \frac{10^5 \times (2.272 \times 1.67 \times 10^{-27})}{(1.38 \times 10^{-23}) \times 160}$$

$$= 440 \, \text{K}$$

Thus this ideal-gas method obtains a temperature of 440 K. In fact, Jupiter's atmosphere is not an ideal gas and in practice a more sophisticated equation is needed, and the temperature obtained is lower.

QUESTION 6.6

Outline reasons why it might be difficult to detect neon, Ne, in the atmospheres of the giant planets by remote methods.

QUESTION 6.7

At the 0.42 bar level on Jupiter, the number density (i.e. the number of molecules per unit volume) is $2.4 \times 10^{25} \, \text{m}^{-3}$. Use this information to estimate the temperature at this level. (Note that the number density is equivalent to ρ/m.)

Atmospheric composition

The chemical compositions of the atmospheres of Jupiter and Saturn are given in Table 6.1. The *abundance* of each molecule is given as the fraction of the total number of molecules.

Table 6.1 Chemical composition of the atmospheres of Jupiter and Saturn.

Atom or molecule	Abundance[a]	
	Jupiter	Saturn
H_2	0.864	0.80 to 0.85
He	0.136	0.15 to 0.20
CH_4 (methane)	1.81×10^{-3}	4.4×10^{-3}
NH_3 (ammonia)	1.7×10^{-10} to 8.6×10^{-6}	6×10^{-4}
H_2O	$(1 \text{ to } 30) \times 10^{-6}$	$(2 \text{ to } 20) \times 10^{-9}$
H_2S	$< 1 \times 10^{-7}$	$< 2 \times 10^{-7}$
C_2H_6 (ethane)	$(2.6 \text{ to } 8.6) \times 10^{-8}$	2×10^{-7}
C_2H_2 (ethyne)	$< 2 \times 10^{-6}$	1.1×10^{-7}
C_2H_4 (ethene)	6×10^{-9}	–
PH_3 (phosphine)	1×10^{-7}	2×10^{-7}
CO (carbon monoxide)	1×10^{-9}	1×10^{-9}
GeH_4 (germane)	7×10^{-10}	4×10^{-10}

[a] A dash indicates that the atom or molecule has not been detected. The number of significant figures given reflects the uncertainty, so that a number quoted as 2 implies that the quantity lies between 1 and 3, whereas 2.0 implies that it lies between 1.9 and 2.1.

There are several things to note that are relevant to theories of planetary evolution and to our models of the interiors of Jupiter and Saturn.

The most obvious point is that both atmospheres are predominantly hydrogen and helium, with a roughly solar ratio of hydrogen to helium. The predicted proportion of helium, assuming solar abundances, is 0.15. Another important point to note is the types of molecule found. In the atmospheres of the terrestrial planets, carbon, as you saw in Chapter 5, is found mainly as carbon dioxide, CO_2. The carbon is joined to oxygen and is said to be in an oxidized form.

▨ In the molecules listed in Table 6.1, what is carbon joined to?

❏ Mostly hydrogen (e.g. CH_4, C_2H_6, C_2H_2). There is also a small amount of CO, in which the carbon is joined to oxygen but to only one atom of oxygen not two as in CO_2.

Carbon in these molecules is in a reduced form.

▨ Are the other elements detected present in reduced forms?

❏ Yes. Nitrogen is present as NH_3, phosphorus as PH_3, and germanium as GeH_4.

As we said in Chapter 5, the atmospheres of the four giant planets are reducing atmospheres. Iron oxide, for example, left in such an atmosphere would eventually be reduced to iron, with the oxygen combining with hydrogen to form water. This is in contrast to Earth's atmosphere where, as you know, iron is oxidized to an oxide – rust!

Atmospheric profiles

Figure 6.10 shows how the temperature and pressure vary with depth in the atmosphere and the layers just below this, for Jupiter and Saturn. The curves shown are averages for the whole planet, as there are some variations in temperature from pole to equator at any one pressure.

Both curves show a fall in temperature with increasing altitude at the lower depths and an increase in temperature with altitude in the outer atmosphere. The atmospheres can thus be divided into a troposphere (dark area in Figure 6.10) and, above this, a thermosphere (pale area). In the troposphere the decrease in temperature with altitude corresponds approximately to the adiabatic lapse rate ($1.963\,K\,km^{-1}$ and $0.714\,K\,km^{-1}$ for Jupiter and Saturn respectively) and convection is present. The topmost layer of cloud has been identified as ammonia, and particles have been detected at the level labelled NH_4HS, but the nature of the lower cloud layers marked in Figure 6.10 has not been definitely confirmed and is based on models of the atmosphere.

The usual way to model the lower cloud layers is as follows. It is assumed that the elements are present in their relative solar abundances. The temperatures and pressures are known so we now need to decide in what form the elements will be present – atoms or molecules, gas, liquid or solid. The atmospheres are very well mixed and have been present for a very long time, and so they would be expected to have reached chemical equilibrium in the layers beneath the clouds. **Chemical equilibrium** is explained in Box 6.3.

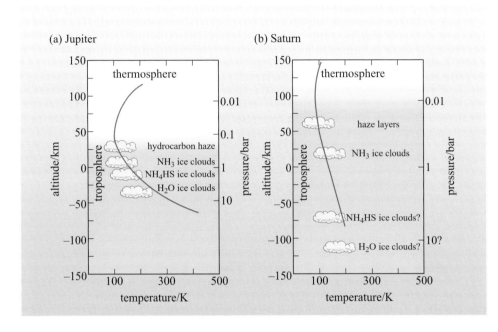

Figure 6.10 Vertical profiles of the atmospheres of Jupiter and Saturn. Zero altitude is taken to be where the pressure is 1 bar; the profiles extend below this in those cases where measurements are available. In each case, the troposphere is dark and the thermosphere is pale.

BOX 6.3 CHEMICAL EQUILIBRIUM

If we take a box containing a mixture of chemicals, say nitrogen, hydrogen and ammonia, and leave it for a very long time, ensuring that the temperature and pressure remain constant and that no chemicals or radiation leave or enter the box, then the chemicals will reach equilibrium. At equilibrium, chemical reactions will be taking place but the total rate of production of each compound equals the total rate of its destruction so that the amounts of the various compounds present will stay the same. For any chemical reaction occurring, the relative amounts of the compounds involved that are present at equilibrium, are given by the **equilibrium constant**, K, for that reaction. The value of K varies with the temperature, but does not depend on the total amount of chemicals present.

Let us take as an example of chemical equilibrium; the reaction between nitrogen, N_2, and hydrogen, H_2, to form ammonia, NH_3:

$$N_2 + 3H_2 = 2NH_3 \qquad (6.2)$$

We can put into our box any amount of nitrogen, hydrogen and/or ammonia. It does not matter whether we start with a mixture of nitrogen and hydrogen, or with ammonia, or with a mixture of hydrogen and ammonia, so long as both elements are present in sufficient amounts. After a very long time, we will have nitrogen, hydrogen and ammonia present in equilibrium amounts. If we measured the concentrations of the three compounds at equilibrium we could obtain the equilibrium constant, K. For the reaction in Equation 6.2, the equilibrium constant is given by

$$K = \frac{[NH_3]^2}{[N_2] \times [H_2]^3} \qquad (6.3)$$

where the square brackets [ ] denote concentrations and the exponents reflect the numbers in the chemical equation, e.g. $3H_2$ in Equation 6.2 translates to $[H_2]^3$ in Equation 6.3. The value of K is such that if the concentration of hydrogen is very much higher than that of nitrogen, as it is on Jupiter, then most of the nitrogen will be converted to ammonia. Thus we expect to find NH_3 rather than N_2 in Jupiter's atmosphere.

In a planetary atmosphere, there is not just one chemical reaction to consider but a very large number, and any one compound may be involved in more than one equilibrium reaction. Even in the simple case of a mixture of two elements, we have to think about all the compounds that might be formed.

For example, in the atmospheres of Jupiter and Saturn, several molecules containing just carbon and hydrogen are observed. We have to include equilibrium constants for the reaction between carbon, C, and hydrogen, H_2, to form methane, CH_4, ethane, C_2H_6, ethene, C_2H_4, and ethyne, C_2H_2. Given the equilibrium constants we can calculate the relative amounts of hydrogen, methane, ethane, ethene and ethyne. But in addition we have to consider the equilibria between carbon, hydrogen and other elements such as nitrogen and oxygen to form compounds such as carbon monoxide, CO, ammonia, NH_3, and water, H_2O. All these equilibria are linked so that a large number of equations have to be solved to obtain the abundances. Luckily this is just the sort of problem that computers are good at. From our knowledge of how chemicals react, we can choose, for any planetary atmosphere, a set of reactions involving the most likely molecules formed from the most abundant elements.

The calculations start where the temperature is relatively high (one very comprehensive study started at 2000 K). The equilibrium constants of all the reactions thought to be likely and the relative abundances of the elements, for example solar abundances, are fed into the computer and the relative abundances of the various possible molecules calculated. In the chosen region for starting the calculation, the temperature, as we have said, is high but the pressure is not very high so that all the icy materials will be gaseous.

We then imagine a parcel of this hot atmosphere rising into the cooler layers. As the temperature falls, the equilibrium constants change and so there is a slight reshuffling amongst the different molecules. A more dramatic change occurs when molecules condense out. Table 6.2 compares the predicted abundances of various molecules in the gas at the 300 K and 150 K levels in Jupiter's atmosphere. The observed abundances are also given for comparison.

Table 6.2 Predicted and observed abundances for selected molecules in Jupiter's atmosphere.

Molecule	Predicted at 150 K (1 bar level)	Observed at 1 bar level	Predicted at 300 K (4 bar level)	Observed at 4 bar level	Observed below the 4 bar level
H_2	0.89	0.864	0.89	0.864	0.864
He	0.11	0.136	0.11	0.136	0.136
CH_4	6×10^{-4}	1.8×10^{-3}	6×10^{-4}	1.8×10^{-3}	1.8×10^{-3}
NH_3	1.2×10^{-5}	$(2 \text{ to } 9) \times 10^{-6}$	1.5×10^{-4}	3×10^{-4}	7×10^{-4}
H_2O	8.8×10^{-13}	$<1 \times 10^{-6}$	10^{-3}	$<1 \times 10^{-6}$	5×10^{-5} (at 12 bar) 5×10^{-4} (at 19 bar)
GeH_4	~ 0	7×10^{-10}	$\ll 6.3 \times 10^{-9}$	7×10^{-10}	
H_2S	8.8×10^{-13}	$<1 \times 10^{-7}$	2.9×10^{-5}	$<1 \times 10^{-7}$	7×10^{-6} (at 8.7 bar) 7×10^{-5} (at $\approx$16 bar)

Ethane (C_2H_6) and ethyne (C_2H_2) are observed in the upper atmosphere (further out than the 1 bar level) but are not *predicted* to be present in observable quantities by this equilibrium model. The presence of these compounds can be explained by looking at the chemical reactions in the atmosphere triggered by the absorption of light.

The interaction of sunlight with methane leads to a series of reactions in which steady state abundances of ethane and ethyne are maintained. This is a similar situation to the one which maintains ozone in the Earth's stratosphere.

■ Which molecules are predicted to have a very marked drop in abundance on going from 300 K to 150 K?

❏ H_2O and H_2S.

■ Why do you suppose that these molecules are particularly affected?

❏ These are molecules believed to form clouds at layers between 300 K and 150 K (Figure 6.10: the H_2S is incorporated into the NH_4HS cloud).

The NH_3 abundance also drops, but not as dramatically because the NH_3 clouds lie at or near the 150 K level.

The water and ammonium hydrogen sulfide (NH_4HS) layers in Figure 6.10 are predicted from this type of calculation. Consider, for example, the NH_4HS layer. Ammonium hydrogen sulfide, when pure, is a white solid formed by the reaction of ammonia with hydrogen sulfide:

$$NH_3(g) + H_2S(g) = NH_4HS(s) \qquad (6.4)$$

Equation 6.4, like all chemical reactions, has an equilibrium constant K, given by:

$$K = \frac{[NH_4HS]}{[NH_3][H_2S]} \qquad (6.5)$$

and this tells us how much hydrogen sulfide must be present if the ammonia is to react and form appreciable amounts of ammonium hydrogen sulfide. The Voyager instruments failed to detect H_2S. The Galileo probe detected H_2S at lower altitudes (a relative abundance of 6×10^{-6} at a pressure of 8.7 bar and 6.6×10^{-5} at deeper levels where the pressure was 16 bar or greater). This could mean that the H_2S condenses out at a lower level than predicted, either as NH_4HS or another compound.

The Galileo probe was not able to positively identify the ammonium hydrogen sulfide and water cloud layers shown in Figure 6.10 although it made measurements down to the 21 bar level which lies below where they were predicted to be formed. The nephelometer – an instrument designed to detect the presence of particles (liquid drops or small crystals) in the atmosphere – found a tenuous cloud at the 0.5 bar level, another at 1.34 bar, a less tenuous but still thin cloud at 1.6 bar and an even more tenuous structure between the 2.5 bar and 3.6 bar levels. The most substantial of these at 1.6 bar could be the NH_4HS cloud but the chemical composition of the particles could not be determined. The density of all the measured clouds was much less than predicted by the model assuming solar relative abundances of the elements, but gave some agreement with a model in which the abundances of ammonia and particularly hydrogen sulfide were very much less than solar. It is possible, however, that the Galileo probe descended through a 'gap' in the clouds and that the results were atypical of the planet as a whole. Indeed, the Galileo Orbiter results indicated that lower cloud layers were patchy.

Figure 6.11 Jupiter's clouds showing true colour. (NASA)

One possible reason for the low observed abundance of H_2S lies in the colour of Jupiter's clouds (Figure 6.11). Ammonia, ammonium hydrogen sulfide and water ice are all colourless. Hydrogen sulfide can undergo chemical reactions that lead to the formation of compounds in which several sulfur atoms are joined together. Some examples of such compounds are given with their colours in Figure 6.12.

If these substances are formed it would explain both the colour of the clouds and the lack of H_2S. There are, however, other possible candidates for the origin of the colour, including phosphorus compounds and carbon compounds, and the spectral evidence is inconclusive.

QUESTION 6.8

Methane is a relatively abundant molecule in the atmospheres of Jupiter and Saturn. However, no methane cloud layers are predicted there. What might be the reason for this?

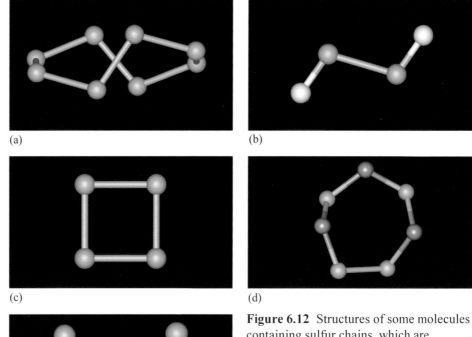

(a) (b) (c) (d) (e)

Figure 6.12 Structures of some molecules containing sulfur chains, which are coloured. Molecules such as these may be responsible for the colour of the clouds on Jupiter. (a) S_8, (b) H_2S_2, (c) S_4^{2+}, (d) $S_4N_3^+$, and (e) S_4N_4. Molecules (a–d) give a yellow colour, whereas molecule (e) gives an orange colour.

6.3.3 Winds and storms

When we look at the giant planets through a telescope, what we see is the outermost layer of clouds, but these clouds are not stationary features. They are circulating around the planets, sometimes with speeds in excess of those found for hurricanes on Earth, and the speed changes with latitude. Wind speeds have been measured directly by the Galileo probe and by tracking clouds using images from Voyager, the Galileo Orbiter and the Hubble Space Telescope. Before we look at the results, however, you need to be clear what we mean by a wind speed on these planets.

If we wanted to measure the **wind speed** on Earth by tracking clouds, we could stand on the surface and measure the speed at which the clouds passed us. The giant planets probably do not have surfaces, and certainly no surface has been detected. So think for a moment about the speed at which clouds in our atmosphere are observed to travel by someone standing on the Earth and the speed observed by someone in a passing spacecraft.

■ Would the speed of the cloud observed by the person on Earth be the same as that observed from a passing spacecraft?

❏ No. Standing on the surface we would be rotating with the surface so that we would measure the speed of the cloud relative to the surface. The spacecraft measurement would be with respect to a point outside the Earth and so would include the Earth's motion relative to the spacecraft due to its rotation as well as that of the cloud with respect to the surface.

From cloud images of Jupiter we obtain the speed relative to the spacecraft or telescope. So how are we going to allow for the rotation of the planet?

An ingenious method has been employed to measure the rotation of the planetary interior. Around the planets are charged particles, electrons and ions – the ionosphere which you will study further in Section 6.3.4. These particles are accelerated by the magnetic field of the planet and, as a consequence, emit radiation. Because the axis of the magnetic field is inclined to the rotation axis, the electrons and ions experience a magnetic field that varies with the planet's rotation. This leads to bursts of radiation whose interval gives the rotation period and thus the expected speed of rotation at the 'surface' (1 bar level). The period of rotation of all four giant planets has been measured in this way by monitoring bursts of radio-frequency radiation. Figure 6.13 shows the intensity of radio-frequency (481 kHz) emission from Uranus as a function of time for five consecutive rotations of the planet. Note that particular types of feature, for example the period of minimum intensity labelled 'bite-out', occur at the same time in each rotation. The rotation period was obtained by noting how often such a feature occurred.

As we do not have more direct measurements of the rotation periods, the values obtained using this method are used to determine wind speed.

The measured wind velocities from equator to pole are shown in Figure 6.14 for Jupiter and Saturn. The term **wind velocity** is used when we want to indicate the direction of motion as well as its magnitude (speed). Here we distinguish winds in the same direction as the rotation of the planet (positive velocity) from those in the opposite direction (negative velocity).

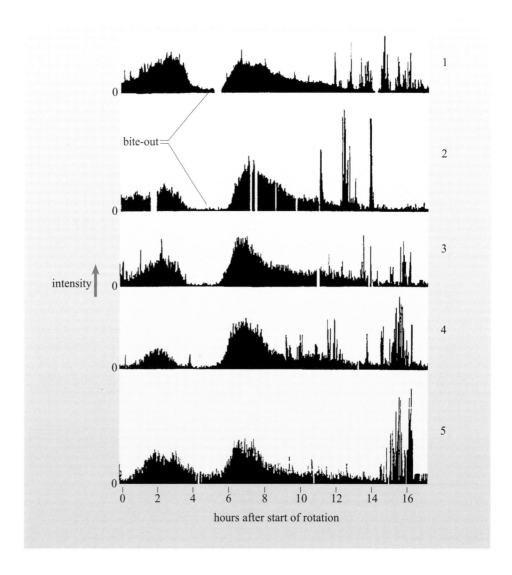

Figure 6.13 The 481 kHz frequency emission from Uranus shown for five consecutive rotations. It is seen that the rotation period is approximately 17 hours. (Warwick *et al.*, 1986)

The wind velocities shown were obtained by tracking cloud images. The direct measurements from the Galileo probe gave somewhat higher wind speeds for Jupiter ($160 \, \text{m s}^{-1}$ to $220 \, \text{m s}^{-1}$). Further analysis of Voyager images indicated that cloud features large enough to be studied from Earth interact with the prevailing wind and move more slowly than the jet stream constraining them, so that the Galileo speeds are probably more accurate. Galileo, however, only measured wind speeds at one latitude and so the data in Figure 6.14 remain the best for studying the variation of velocity with latitude.

Jupiter

The visible layer of Jupiter shows a pattern of alternate dark and light bands parallel to the equator, called **belts** and **zones**, respectively (Figure 6.15).

It has been shown that in the dark belts the atmosphere is rising and in the light zones it is sinking. These observations lead to a picture of a series of convection cells (Figure 6.16). Convection cells, as you may recall from Chapter 5, are a feature of atmospheric motion on the terrestrial planets.

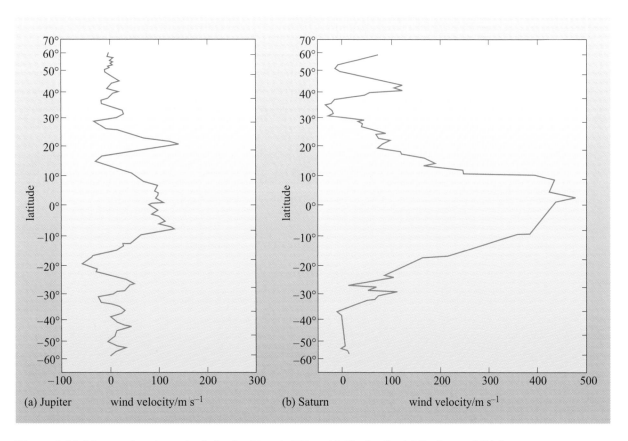

Figure 6.14 Measured east–west wind velocities at different latitudes for (a) Jupiter and (b) Saturn.

Figure 6.15 Close-up of Jupiter showing belts and zones. (NASA)

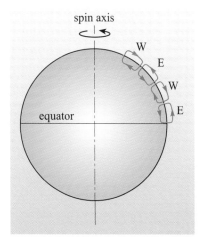

Figure 6.16 Schematic convection cells for Jupiter. Note that the high rotation speed of Jupiter leads to a succession of small cells. E means that there is flow to the east, and W means flow to the west.

As the atmosphere rises it expands and cools; ammonia condenses out and forms white clouds, which give the zones their light appearance. The atmosphere cools radiatively and begins to sink. The atmosphere at the top of the belts therefore contains less ammonia and as it sinks it gets hotter so that the remaining ammonia does not condense out and we can see the coloured material underneath. It has been suggested that the lack of water in the region studied by the Galileo probe was a consequence of this region being one of sinking atmosphere from which water as well as ammonia had condensed out as it was rising. Subsequent remote measurements suggest that the surrounding regions contained 100 times as much water as the region through which the probe travelled.

What is driving these convection cells and why are there so many?

■   On Earth, convection cells are formed when air is heated at the Earth's surface. What sources of heating are available on Jupiter?

❏   There is no solid surface as far as we know and certainly none in the layers where sunlight penetrates, but the cloud layers absorb solar radiation (and emit IR). The internal energy of the planet will contribute. (There is also some heat given out when gases condense to form clouds.)

For the atmospheres of the giant planets, the internal energy is the most significant source of heating. Convection would be the most efficient way of transporting this heat from the interior to the outer atmosphere. The vertical temperature profile of the troposphere shows that the rate of cooling across this region is close to the adiabatic lapse rate, so we can be fairly sure that convection cells occur here. It is possible that the cells extend deep into the interior, but we have no measurements of motion below the troposphere or of how far into the interior the troposphere extends. However the wind speeds measured by Galileo remained fairly constant with altitude and this is consistent with deep-seated convection cells powered by the internal energy of the planet.

The Coriolis effect (Section 5.6) plays a major role on the giant planets. Its role on Earth arises from the rapid rotation of the planet. Despite its greater size, Jupiter rotates more rapidly than Earth. The following question asks you to calculate the speed at which a windless atmosphere on Jupiter is moving with respect to an external non-rotating observer.

QUESTION 6.9

Jupiter has a radius (at the 1 bar level) of 71 490 km at the equator. Its period of rotation is 0.412 days. At what speed (with respect to a non-rotating observer in space) would a windless atmosphere at the equator be moving?

The equivalent figure for Earth is 465 m s^{-1}. The Coriolis effect is thus much greater for Jupiter than for the Earth. Whether it is sufficient to cause the banding is unknown.

One possible model considers Jupiter as a rapidly spinning, fluid sphere. Convection in the interior of such a sphere leads to a series of coaxial cylinders rotating at different speeds about the rotation axis. Each cylinder reaches the surface at a different latitude and forms a belt, as shown in Figure 6.17.

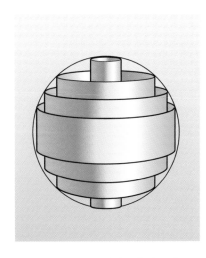

Figure 6.17 Coaxial cylinders in a fluid sphere, which may represent the belts on Jupiter.

The observation of coherent winds as rotating circles near the north pole of Jupiter has thrown doubts on this model as an explanation for the wind patterns at high latitudes. The rotation of a cylinder that would have to go through the core of the planet could not be maintained. It is still possible, however, that winds near the equator and at mid-latitudes arise from such coaxial cylinders.

The beautiful close-up images from Voyager (Figure 6.15) enable us to see that there are also regions of turbulent motion, particularly at boundaries of zones and belts.

The most famous region of turbulent motion is the Great Red Spot.

The Great Red Spot

The **Great Red Spot**, Figure 6.18, is a prominent feature of the Jovian atmosphere. It covers a vast area, being some 14 000 km from north to south and 26 000 km from east to west. (In either direction, it spans a distance much greater than the diameter of any terrestrial planet.) Because of its size and the contrast in colour between itself and the surrounding atmosphere, it can be observed by ground-based telescopes of only moderate power. The first record of the Great Red Spot dates from 1830, but the earliest sighting of a large spot on Jupiter is thought to be by the Italian–French astronomer Jean Cassini (1625–1712), Figure 6.19, in 1665.

Thus the Great Red Spot has been present for at least 170 years and possibly for more than 300 years. This would not be surprising if it were a surface feature, such as the mountains of Earth or Mars, but close-up time-lapse pictures show clearly that it is a region of circulating atmosphere, i.e. a giant storm. It is rotating anticlockwise about its centre and completes one rotation every six days. Its rotational energy is so great that it survives encounters with smaller storms.

Figure 6.18 The Great Red Spot on Jupiter. The image shows a region about 40 000 km across. (NASA)

Figure 6.19 Jean-Dominique Cassini was born in Italy in 1625. In his youth, he was intrigued by astrology. Later he discarded astrology as a folly but continued to study astronomical objects, accepting a position to set up the Paris Observatory. (Painting by Duragel, courtesy of the Observatoire de Paris)

Close to the Great Red Spot are three white vortices (see Figure 6.18). Between the visits of Voyager and the Galileo Orbiter these drifted eastwards relative to the spot and closer together. The Great Red Spot itself has moved westwards. The features surrounding the Great Red Spot appear to be in continual change. Figure 6.20 shows the way the Great Red Spot and its surroundings have changed from 1992 to 1999.

Both Voyager and Galileo detected convective thunderstorms near the Great Red Spot. Those viewed by Voyager occurred roughly once every ten days and lasted a few days each. Voyager could not determine the altitude of the storms, but the Galileo Orbiter could. Figure 6.21 shows a thunderstorm imaged by the Galileo spacecraft on 26 June 1996. The white cloud in the centre is 1000 km across and 25 km higher than the surrounding clouds. Its base extends 50 km below the surrounding clouds, placing it at the level of the predicted water cloud.

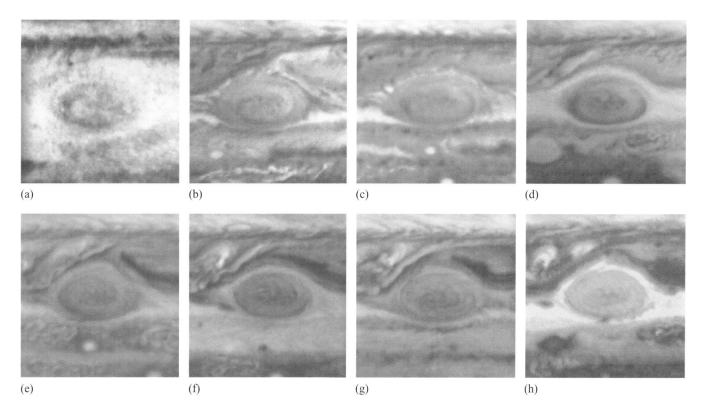

(a) (b) (c) (d)

(e) (f) (g) (h)

Figure 6.20 The Great Red Spot imaged by the Hubble Space Telescope over the period 1992–1999. (NASA)

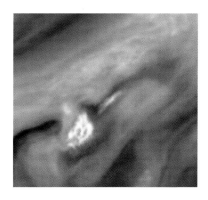

Figure 6.21 Thunderstorm on Jupiter imaged by the Galileo Orbiter. (NASA)

Computer modelling has shown that large, stable eddies can be produced regardless of the process maintaining the winds. Thus the presence of the Great Red Spot does not allow us to decide between competing models of Jupiter's atmosphere. More detailed analysis of how the Great Red Spot interacts with the surrounding atmosphere may enable us to narrow down the possible models.

There remain many unsolved problems. For example, why is there only one Great Red Spot, rather than one in each hemisphere? The colour of the spot is also puzzling. The temperature of the Spot as measured by IR techniques is the same as that of the surrounding white ovals, which are also storms. Yet on Jupiter as a whole, the coloured layers are deeper and at a higher temperature than the white clouds. Is the colour due to the same material as in the coloured cloud layers? If so, why does it occur at a different depth in the Great Red Spot?

We shall have to leave these questions unanswered and turn now to Saturn.

Saturn

The wind pattern on Saturn is in some ways similar to Jupiter. Saturn's atmosphere is banded, and contains streams moving at different velocities. Figure 6.22 is a close-up of Saturn's atmosphere showing flow patterns similar to those on Jupiter.

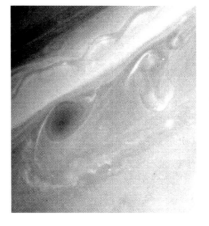

Figure 6.22 Close-up of Saturn's atmosphere imaged by Voyager 2. (NASA)

The pattern is notably different from that on Jupiter, however. A major difference is the stream of very high-speed wind near the equator. As you can see in Figure 6.14, the wind speed reaches $500\,\text{m s}^{-1}$ (which is about two-thirds of the speed of sound in that region of the atmosphere of Saturn). The highest wind speeds on Jupiter are about $220\,\text{m s}^{-1}$. None of the models put forward to explain the banding in the atmospheres of Jupiter and Saturn can account for this.

The other main difference is that, unlike on Jupiter, the changes in wind velocity on Saturn are not closely allied to zone boundaries. The contrast between belts and zones is much more muted. A true-colour image shows a fuzzy, yellowish planet with only very faint markings (Figure 6.1b). If we enhance the contrast, the banded structure becomes apparent as in Figure 6.23.

The different colour stripes do not correlate particularly well with different wind velocities.

Saturn's eddies are short-lived, especially compared to storms like Jupiter's Great Red Spot.

Eddies appear on Saturn, as for example in Figure 6.22. The largest found was 6000 km long, considerably smaller than the Great Red Spot.

Figure 6.23 Image of Saturn from the Very Large Telescope of the European Southern Observatory showing banded structure. (European Southern Observatory)

(a)

(b)

(c)

Figure 6.24 The Great White Spot on Saturn. These images (a–c) from the European Southern Observatory show the rapid expansion of the Great White Spot during October 1990. (European Southern Observatory)

Figure 6.25 Hubble Space Telescope image of Saturn showing a storm as a white feature near the equator. (NASA)

In 1990, a **Great White Spot** appeared on Saturn and over the course of about a month spread to encircle the entire equatorial region (Figure 6.24a–c). A similar feature had been observed earlier at roughly 30-year intervals (1933, 1960) but was not present when either Voyager spacecraft flew by (December 1980 and August 1981). (The 1933 sighting, incidentally, was first reported by an amateur astronomer, the Scottish comedian Will Hay.) The Great White Spot forms at the north boundary of the equatorial band, thus reinforcing the theory that such features form when two streams of different velocity meet. The atmosphere is rising in this region, and the white colour is due to ammonia crystallizing out. In 1994, a similar, but smaller, feature was recorded by the Hubble Space Telescope (Figure 6.25). There is as yet no accepted explanation for these storms.

QUESTION 6.10

Some of the wind velocities in Figure 6.14 are shown as negative and yet the entire atmosphere is rotating in the same direction. Explain this.

6.3.4 Magnetospheres

Jupiter

The strength of the magnetic dipole of Jupiter is about 20 000 times that of the Earth. Additionally, as Jupiter is further from the Sun, the strength of the interplanetary magnetic field is less and the density of charged particles from the solar wind is lower. These two factors result in Jupiter having the largest magnetosphere in the Solar System; when the solar wind is weak the magnetopause can be 100 Jovian radii from the centre of Jupiter and the magnetotail can extend beyond the orbit of Saturn. Electrons escaping from Jupiter's magnetosphere have even been detected at the Earth. Figure 6.26 shows the main features of the Jovian magnetosphere.

The Galilean satellites of Jupiter are so called because they were first observed by Galileo in 1610.

Broadly, the magnetosphere resembles a greatly magnified version of the Earth's magnetosphere, but is even less like a sphere, more closely resembling a disc. As for the Earth, the magnetosphere contains ionized matter (plasma) and there are toroidal belts resembling the Van Allen belts. The Galilean satellites – Io, Europa, Ganymede and Callisto – lie within the plasma belts and are continually bombarded by charged particles.

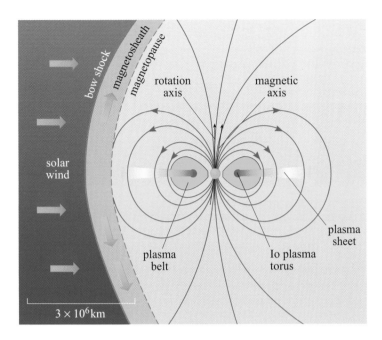

Figure 6.26 Magnetosphere of Jupiter.

A unique feature of the Jovian system is the Io plasma torus. On the approach to Jupiter, the Voyager spacecraft detected UV emission from sulfur and oxygen ions in a torus around Jupiter which enveloped the orbit of Io. As the spacecraft flew through the plasma torus it was able to identify the ions as O^+, O^{2+}, S^+, S^{2+}, S^{3+} and an ion which could be SO_2^+ or S_2^+.

■ The plasma science instrument which identified the ions did so on the basis of mass-to-charge ratio. Why could it not distinguish between SO_2^+ and S_2^+?

❑ These ions have the same charge and both have the same relative molecular mass of 64.

Io has a thin atmosphere composed mainly of SO_2 and other molecules deriving from its volcanic activity (Chapter 3). Charged particles from the plasma heat the atmosphere, and energetic atoms and molecules are thrown out to form a neutral cloud around Io. Further interaction of this cloud with magnetospheric plasma produces the ring of plasma which stretches around Io's orbit (i.e. the plasma torus). Io is in fact a substantial source of the plasma in Jupiter's magnetosphere. As the atoms are ionized, the electron and the newly-formed ion are separated, the ions being picked up by Jupiter's magnetic field and led to the ionosphere of Jupiter. This produces an electrical current of several million amps.

■ When high-energy particles enter the Earth's atmosphere, they cause atoms to be excited. What phenomenon does this give rise to?

❑ Aurorae. The excited atoms emit radiation when returning to the ground state. In the case of Earth, oxygen atoms and nitrogen molecules emit red, green and violet light (Chapter 5 Section 5.7).

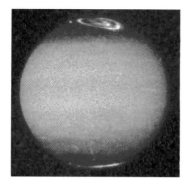 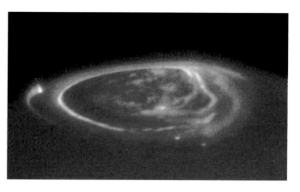

Figure 6.27 Auroral oval around the north pole of Jupiter imaged by the Hubble Space Telescope (left) and a close-up view (right). (NASA)

The Galileo spacecraft captured images of aurorae in Jupiter's atmosphere (Figure 6.27).

■ What is the most abundant molecule or atom in the atmosphere of Jupiter?

❑ Hydrogen, H_2.

Auroral emissions in the visible spectrum on Jupiter are mainly due to H and H_2, not O and N, and so will not be similar in colour to those seen on Earth.

Plasma particles can also excite atoms in the neutral cloud around Io. Red, green and violet light emitted by these excited atoms was imaged by the Galileo spacecraft (Figure 6.28).

■ Which atoms might emit the red and green light in this image?

❑ Red and green were emitted by O atoms in the Earth's atmosphere and O atoms from the dissociation of SO_2 will be present in the neutral cloud. So O atoms are probably the source of aurorae on Io.

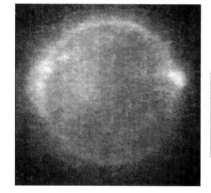

Figure 6.28 Io's aurora imaged by Galileo spacecraft. The blue light is thought to be due to volcanic plumes. (NASA)

The first evidence for Jupiter's magnetosphere came from observations of non-thermal (Chapter 5, Box 5.1) radio-frequency radiation in 1955. It was soon realized that this radio emission was due to energetic charged particles moving in a strong magnetic field and thus that Jupiter must have a significant magnetic field. Highly energetic electrons spiralling around the magnetic field lines give rise to radio emission which is beamed out from the region of the magnetic pole. Since the magnetic axis and rotation axis do not coincide, the variation of this emission can be used to determine the rotation speed of the planet as we noted in Section 6.3.3. Radio emission with a wavelength of order 10 m is modulated by Io.

Saturn

The magnetosphere of Saturn is similar to that of Jupiter but smaller in extent and without the perturbations caused by Io. Auroral ovals have been observed in the UV region by the Hubble Space Telescope (Figure 6.29).

Unlike those on Earth, Jupiter and Io, aurorae on Saturn have not been observed in visible light.

Whether the major satellite Titan has a role in populating the plasma of the magnetosphere is debatable as its orbit means that it is sometimes outside the magnetosphere.

Figure 6.29 Auroral ovals around both poles of Saturn imaged by the Hubble Space Telescope in the UV. (J. T. Trauger (Jet Propulsion Laboratory) and NASA)

QUESTION 6.11

Describe the interaction that produces the bow shock of the magnetosphere for Jupiter.

6.4 Uranus and Neptune

6.4.1 Composition and interior

These two planets are believed to be very similar to each other but different from Jupiter and Saturn. Uranus and Neptune are considerably smaller than Saturn so that their higher densities tell us that they have a higher ratio of heavier elements to hydrogen and helium.

■   If Uranus and Neptune had a similar elemental composition to Jupiter and Saturn, how would you expect their densities to compare?

❏   The densities of Uranus and Neptune would be much lower because the smaller masses of these planets would result in less self-compression.

Theories of planetary formation suggest that Uranus and Neptune began life with rock/ice kernels, but there is no direct evidence for differentiated rocky cores today. Measurements on the gravitational field are consistent with the density increasing towards the centre, but this could be due to the increasing self-compression of icy materials. It is however difficult to use the density to distinguish between icy materials and mixtures of rocky materials, hydrogen and helium. Most models indicate that complete separation of icy and rocky materials has not occurred. However our knowledge of other bodies in the Solar System would suggest that there is a core of mainly rocky materials. The boundary between the icy layer and the hydrogen–helium layer is probably not sharp, and the model illustrated in Figure 6.30 assumes icy materials are found in the outer layers.

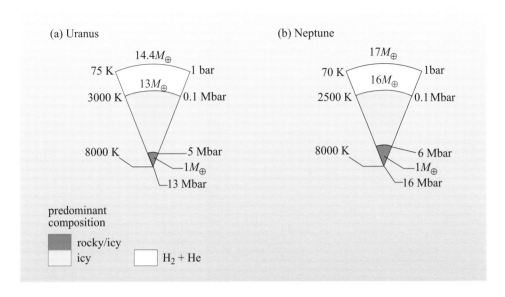

Figure 6.30 Cross-sections through (a) Uranus and (b) Neptune.

R_U = radius of Uranus,
R_N = radius of Neptune.

The magnetic fields of Uranus and Neptune originate in a magnetic dipole source that is some way from the centre of the planet and whose axis is considerably tipped with respect to the rotation axis (Figure 6.4). When it was found that the magnetic axis of Uranus was at 60° to its rotation axis, it was suggested that Voyager may have caught the planet as its magnetic field was changing its polarity. A change of magnetic field polarity is not unknown in planetary science: the magnetic field of the Earth is known to have changed polarity many times.

When Neptune's magnetic field was also found to be tilted, this time at 47° to the rotation axis, then it was felt there must be some other explanation as the coincidence of two planets simultaneously undergoing a change of polarity was too unlikely. Some calculations indicate that, if the electrically conducting shell is thin ($0.1R_U$ or $0.2R_N$) then the magnetic axis has to be tilted.

The position of the magnetic dipole source (roughly one-third of the way out from the centre for Uranus and just over halfway out for Neptune) puts it well outside any rocky core that may be there, in a layer believed to be mostly liquid icy materials.

A mixture dominated by water, ammonia and methane does not look a very promising candidate for an electrically conducting layer. However, experiments on water under high pressure and extrapolation to the temperatures and pressures found in Uranus and Neptune suggest that there will be much greater ionization of water in these layers than in the more familiar surroundings of the Earth's surface. In addition, ammonia dissolved in water forms NH_4^+ (ammonium) ions. The most popular view, therefore, is that the currents giving rise to the magnetic fields of Uranus and Neptune are carried by the ions NH_4^+, H_3O^+, and OH^-.

The final line of evidence that is available is the internal heat of the planets. Uranus and Neptune have almost identical effective temperatures, despite Neptune's greater distance from the Sun. These temperature measurements lead to the conclusion that Neptune emits more heat than it absorbs (as do Jupiter and Saturn) but that Uranus does not. The lack of a heat excess for Uranus is a problem. The two planets are in other respects very similar and so would be expected to have similar internal energy sources and indeed the presence of the magnetic field suggests a hot interior. It has been proposed that Uranus does have an internal energy source but that energy from the interior is only transferred slowly to the upper atmosphere from where it can be radiated to space. The slow rate of transport may be a consequence of the unusual inclination of Uranus' rotation axis (Figure 6.4). The source of the internal energy could be heat of differentiation from the continuing differentiation of rocky and icy materials but the evidence on the current extent of differentiation is, as you have seen, sparse.

QUESTION 6.12

Laboratory work on a mixture of icy materials at very high pressures (around 2×10^6 bar = 2 Mbar) has led to a suggestion that Uranus and Neptune do not have rocky cores at all but are composed entirely of icy materials plus hydrogen and helium. Does such a suggestion conflict with the observational evidence?

6.4.2 Atmospheres

Composition

The chemical compositions of the atmospheres of Uranus and Neptune are given in Table 6.3. The *abundance* of each molecule is given as the fraction of the total number of molecules.

Table 6.3 Chemical composition of the atmospheres of Uranus and Neptune.

Atom or molecule	Abundance[a]	
	Uranus	Neptune
H_2	0.83 ± 0.025	0.80 ± 0.025
He	0.15 ± 0.025	0.18 ± 0.025
CH_4 (methane)	0.02 ± 0.015	0.02 ± 0.015
NH_3 (ammonia)	–	–
H_2O	–	–
C_2H_6 (ethane)	–	3×10^{-5}
C_2H_2 (ethyne)	3×10^{-8} to 2×10^{-7}	3×10^{-7}

[a] A dash indicates that the atom or molecule has not been detected.

As with Jupiter and Saturn, the atmospheres are predominantly hydrogen, as H_2, and helium. The methane abundances above the cloud layers are 3×10^{-5} for Neptune and $<10^{-7}$ for Uranus. The abundances quoted in Table 6.3 relate to depths below the visible cloud layers.

■ What does this suggest about the composition of the outer cloud layers on these planets?

❑ The outermost cloud layers on Uranus and Neptune are composed of methane.

The temperatures on Jupiter and Saturn are too high for methane to condense at the atmospheric pressures of these planets, but at the lower atmospheric temperatures on Uranus and Neptune, methane will condense. Ammonia and water, the other major icy materials, have not been detected.

■ Does this necessarily mean that Uranus and Neptune are depleted in ammonia and water?

❏ No. It is likely that these materials have condensed out at lower depths than those investigated.

QUESTION 6.13

No oxygen-containing molecules have been detected in the atmosphere of Uranus, but this does not mean that there is no oxygen in Uranus. Why not?

Atmospheric profile

The atmospheric profiles of Uranus and Neptune are given in Figure 6.31.

As for Jupiter and Saturn, the observed atmosphere can be divided into two regions – a troposphere and a thermosphere. The decrease of temperature with altitude in the atmosphere of Neptune is about that of the adiabatic lapse rate, $0.853 \, \text{K km}^{-1}$. On Uranus, however, the rate of cooling is less than the adiabatic lapse rate, which is $0.676 \, \text{K km}^{-1}$.

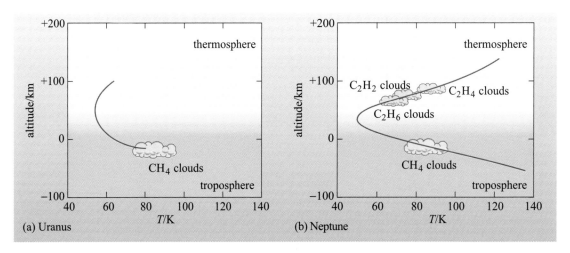

Figure 6.31 Atmospheric profiles of (a) Uranus and (b) Neptune. The C_2H_2, C_2H_4 and C_2H_6 clouds are tenuous and form a 'hydrocarbon haze'.

■ If the rate of cooling is less than the adiabatic lapse rate, what does this imply about heat transfer in the troposphere of Uranus?

❏ Energy is not transferred through convection.

This will reduce the efficiency of heat transfer from the interior of Uranus to the outer atmosphere and this may be the cause of the low effective temperature that we noted above. Models of Uranus suggest that convection does occur in deeper layers.

In Figure 6.31 only the methane clouds and a hydrocarbon haze that lies above these clouds have been positively identified. Models suggest that there may be a lower hydrogen sulfide (H_2S) cloud.

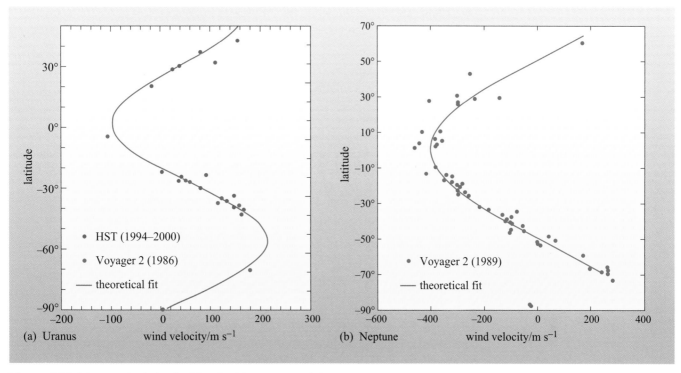

Figure 6.32 Measured wind velocities for (a) Uranus and (b) Neptune. The curves indicate an idealized picture based on the available data. (Note the different scales.) (Hammel *et al.*, 2001)

Winds and storms

Figure 6.32 shows wind velocities on Uranus and Neptune. As you can see from this figure, the data are much less complete and less certain than for Jupiter and Saturn (Figure 6.14).

Data on atmospheric motion on Uranus are sparse owing to the lack of discernible features (as for Venus, there is total coverage of clouds with uniform hue) so we consider Neptune first.

■ Why is it important to be able to distinguish features such as spots in the cloud layers?

❑ In remote studies, the velocities of the cloud layers are measured by tracking features across the planet as it rotates. If the cloud layers appear uniform then we cannot track features.

Neptune, like Jupiter and Saturn, has a banded structure, in this case patterned in blue and white (Figure 6.33). Wind velocities were obtained by tracking the white clouds seen in this figure.

Unfortunately, many of the features at the time of the single Voyager fly-by (Voyager 2, August 1989) were relatively short-lived, and wind velocities obtained from small-scale, short-lived features and the longer-lived, large-scale features do not always coincide. This may be due to the small-scale features moving relative to the wind. There is sufficient information to draw some conclusions, however.

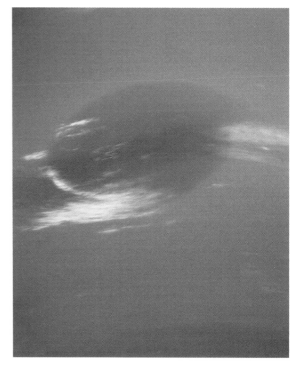

Figure 6.33 Neptune showing bands and white cloud features. (NASA)

■ Look carefully at the wind-velocity scales in Figures 6.14 and 6.32, and state one obvious difference between the winds on Neptune and those on Jupiter and Saturn.

❏ The wind velocities on Neptune, except near the poles, are negative. That is, the atmosphere is rotating more slowly than the interior.

On Jupiter and Saturn there are both positive and negative wind velocities, but the majority are positive. The equatorial wind velocity on Neptune, as well as being negative, is also very high, at least $-300\,\mathrm{m\,s^{-1}}$. Time-lapse images obtained from the Hubble Space Telescope and the NASA Infrared Telescope Facility in Hawaii have been examined and indicate an equatorial wind velocity of $-400\,\mathrm{m\,s^{-1}}$. It seems then that Neptune has an equatorial stream like Saturn, but in the opposite direction.

Voyager 2 observed a large dark spot, possibly a giant storm, when it passed Neptune. By 1996, Hubble Space Telescope observations showed that this had vanished and a new, smaller spot had emerged. These spots showed little visible evidence of jet streams confining them above and below, as happens with the Great Red Spot of Jupiter. Continued observations by space and ground-based telescopes have shown that large, short-lived storms are a feature of the atmosphere. Figure 6.34 for example, compares storm features in the atmosphere as imaged by the Hubble Space Telescope in 1996 and 1998.

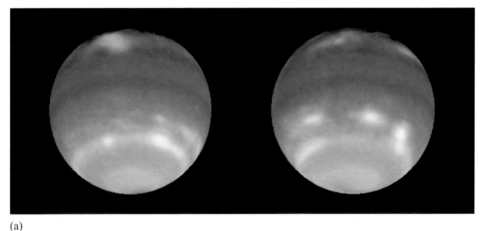

(a)

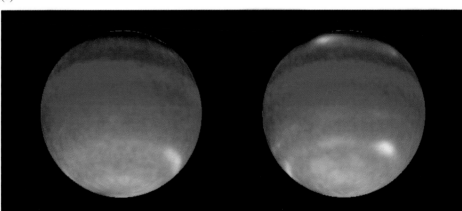

Figure 6.34 Views of Neptune in (a) 1996 and (b) 1998 showing how the white features associated with storms changed in this period. (NASA)

(b)

A distinguishing feature of Uranus when compared to Jupiter, Saturn and Neptune, is the orientation of its rotation axis.

■ Find the tilt of the rotation axis of Uranus from Appendix A, Table A1 (axial inclination).

❑ The planet is effectively tipped over by 97.9°, so that for large fractions of the Uranian year, a pole receives more solar radiation than the equator. Figure 6.35 shows the inclination to the Sun at the time of the Voyager 2 observations (January 1986).

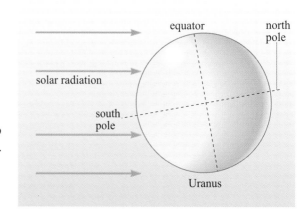

Figure 6.35 The tilt of Uranus, showing how one pole receives more solar radiation than the equator at certain times during the planet's year. Uranus was at this point in its orbit when Voyager 2 passed by.

The rotation axis of Uranus is always at the same angle to the ecliptic plane but, like the Earth and all other planets in the Solar System, the hemisphere facing the Sun varies as the planet orbits. With the rotation axis almost in the ecliptic plane, Uranus at some times in its orbit has a pole facing the Sun and at other times the equator faces the Sun. When Voyager 2 passed, the south pole was facing the Sun but after one-quarter of an orbit, the equator would be facing the Sun. When a pole is facing the Sun, rotation about the rotation axis does not move the other pole into the sunlight, but when the equator is facing the Sun all the planet receives some solar radiation. Averaged over the 84-year orbital period, Uranus receives more solar energy at the poles than at the equator. This could lead to a convection cell originating at the pole and producing negative wind velocities at mid-latitudes through the action of the Coriolis effect.

■ What other features of Uranus might cause the pattern of atmospheric circulation to differ from that on the other three giant planets?

❑ Uranus has a negligible heat excess. The internal heat source may be a significant factor in atmospheric circulation.

Figure 6.36 Uranus in false colour showing banded structure, imaged by Voyager 2. (NASA)

At the time of the Voyager 2 fly-by, Uranus appeared to be enveloped in a blue–green fog in which virtually no features are discernible, as in Figure 6.1c. In false colour (Figure 6.36) it is possible to make out a banded structure similar to those on the other three giant planets. Note that the banding runs parallel to the equator as on the other giant planets, despite the unusual tilt of the rotation axis. Such methods made it possible to find a few features, such as small spots, which were used to obtain wind speeds, but only for latitudes greater than 20° and only in the southern hemisphere. More recent images from the Hubble Space Telescope (Figure 6.37) have identified about 20 clouds.

These clouds have been tracked using the Hubble Space Telescope and the Keck 10 m telescope in Hawaii and wind velocities were obtained for the northern hemisphere as well as the latitudes studied by Voyager 2. Figure 6.32 showed the wind velocities obtained from all three sources. The measured velocities are almost all positive. Extrapolation as indicated in Figure 6.32 suggests negative velocities within 20° of the equator. As with the other giant planets, wind speeds can be very high, up to 200 m s^{-1}. The orange feature seen on the white band in Figure 6.37, for example, is moving at 140 m s^{-1} relative to the planet.

Figure 6.37 Enhanced colour image of Uranus in 1998 from the Hubble Space Telescope. (NASA)

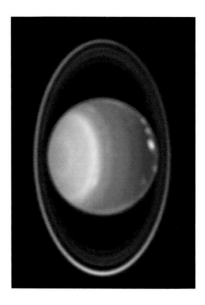

6.4.3 Magnetospheres

As well as discovering the unusual tilting and position of the magnetic dipoles of Uranus and Neptune, Voyager 2 also mapped their magnetospheres. These have the same main features as those of Earth, Jupiter and Saturn:

*   a bow shock where the solar wind and the interplanetary magnetic field (IMF) are resisted by the planetary magnetic field,

*   a magnetopause,

*   a plasma sheet,

*   radiation belts forming a torus around the magnetic equator,

*   an extended magnetotail.

The magnetic fields of the planets are aligned by the magnetic dipole (Figure 6.4) but the planets are rotating rapidly about their rotation axes. Since the angle between the rotation and magnetic axes is large, the planetary magnetic field is constantly changing with respect to the IMF and the field lines in the magnetotail are wound into a corkscrew. Figure 6.38 shows the magnetospheres as they were at the Voyager approach. Note the positions of the magnetic poles and the radiation belts relative to the solar wind.

Aurorae have been detected for Neptune but do not form neat ovals round the poles.

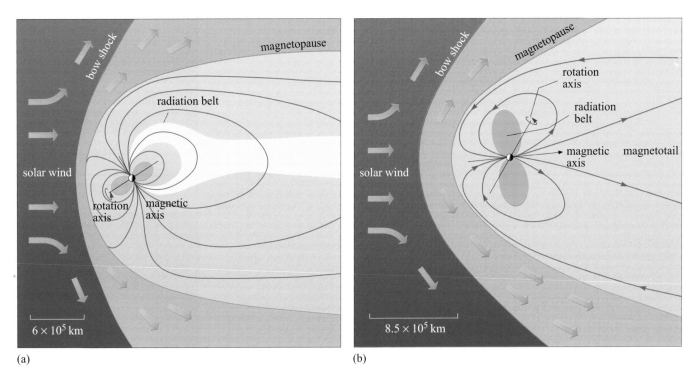

Figure 6.38 Magnetospheres of (a) Uranus and (b) Neptune at the time of Voyager 2's approach.

QUESTION 6.14

Until the Voyager encounter, there was no direct evidence for the magnetic field of Uranus. If the planet did not have a magnetic field, what would its magnetosphere be like?

QUESTION 6.15

The average density of Jupiter is very similar to that of Rhea, the icy satellite of Saturn. Rhea is thought to be composed of one-half rocky and one-half icy materials. How is the same average density compatible with very different chemical compositions of the two bodies?

QUESTION 6.16

Nitrogen molecules in the atmosphere of Titan were detected by the UV spectrometer on Voyager 1. Why would it be difficult to detect the nitrogen molecules using ground-based instruments on Earth?

QUESTION 6.17

List some observations on heat sources and heat transport on Jupiter that must be taken into account when developing theories to explain the pattern of winds.

QUESTION 6.18

The Earth is protected from harmful energetic particles by its magnetosphere. Is the same true for Jupiter?

QUESTION 6.19

Before the Voyager 2 mission reached Uranus and Neptune, it was suggested that Uranus had a hot, molten, rocky core, which gave rise to a magnetic field, but that the core in Neptune was solid and so Neptune would have no magnetic field. How far does this model fit the current observational data? (Consider magnetic field measurements and other measurements.)

QUESTION 6.20

Given the composition of the atmospheres of Uranus and Neptune, which atoms might give rise to auroral emissions on these planets?

6.5 Summary of Chapter 6

- Data on the interiors of the giant planets can be obtained from measurements of density, gravitational field, magnetic field, emitted heat and atmospheric composition.

- Jupiter and Saturn probably do not have a definite liquid or solid surface. Current models of Jupiter and Saturn distinguish five layers. The two innermost layers constitute a core of rocky and icy materials. This core is surrounded by layers that are mostly hydrogen and helium, which account for most of the planets' mass. The layer adjacent to the core in Jupiter and Saturn is predicted to contain hydrogen in a metallic state. The deep interiors of both Jupiter and Saturn are very hot (over $15\,000$ K in the case of Jupiter).

- Uranus and Neptune may not have a definite liquid or solid surface. They may have rocky cores but current models suggest that rocky and icy materials are not completely differentiated. Surrounding the core is a mantle of mainly icy materials and around this is a layer of mainly hydrogen and helium. Overall, these two planets are less dominated by hydrogen and helium than Jupiter and Saturn and the layers are probably less differentiated in composition.

- Jupiter, Saturn and Neptune emit more energy than they receive from the Sun (heat excess). The heat excess of Jupiter is thought to be due to the continuing escape of original accretional heat and heat of differentiation. Saturn's heat excess is thought to have an additional contribution from helium droplets separating out from metallic hydrogen and sinking. Neptune and Uranus are both thought to have internal energy arising from continuing heat of differentiation. The cause of the heat excess for Neptune is still debatable. The cause of the lack of heat excess for Uranus may be associated with its unusual spin-axis inclination.

- The magnetic fields of Jupiter and Saturn are believed to originate in the shell of liquid metallic hydrogen. The magnetic fields of Uranus and Neptune are thought to originate in a shell of liquid icy material containing the ions H_3O^+, OH^- and NH_4^+.

- The atmospheres of the giant planets have hydrogen, H_2, and helium as their major components. Other molecules detected are reduced forms of the heavier elements, for example CH_4 and NH_3.

- Most of the molecules in the atmospheres are detected by IR or UV spectroscopy. The Galileo probe used mass spectrometry to obtain the relative abundances of molecules in the region of atmosphere it entered.

- The outermost cloud layer can be identified as ammonia on Jupiter and Saturn. Clouds of methane have been observed in the atmospheres of Neptune and Uranus.

- Models assuming chemical equilibrium can predict the composition of the lower cloud layers, but these compositions have not been positively identified by observation. The Galileo probe detected a very tenuous cloud which could be part of the ammonium sulfide cloud layer.

- The variation of temperature with depth on the giant planets divides the atmospheres into two regions. In the lower part (the troposphere) the temperature decreases the further out from the centre we go. The decrease is close to the adiabatic lapse rate, except for Uranus where the rate of decrease is slower. In the upper layers of the atmosphere (the thermosphere) the temperature increases with distance from the centre.

- Wind velocities on the giant planets are measured, remotely, by tracking the movement of cloud features. These measurements will give a velocity that includes the rotation speed of the planetary interior and so this has to be subtracted. The rotation speed of the interior can be measured from radio bursts.

- On Jupiter and Saturn, there is evidence for a series of deep convection cells giving rise to the observed pattern of wind velocities. On Jupiter, major changes in wind velocity correlate with the boundaries between different coloured bands. There is no such correlation on Saturn.

- Jupiter and Saturn have positive wind velocities at the equator. Equatorial wind velocities can be very high; up to $500\,\mathrm{m\,s^{-1}}$ on Saturn. The Galileo probe measured wind velocities on Jupiter directly. These were higher than the values obtained by Voyager but were only for one latitude. Neptune has a large negative equatorial wind velocity, and extrapolation suggests that Uranus has a negative equatorial wind velocity too.

- Jupiter, Saturn, Uranus and Neptune all have large magnetospheres produced when the solar wind and IMF interact with the planetary magnetic fields. The main features of the magnetospheres are similar to those of the Earth's magnetosphere. Io contributes to Jupiter's magnetosphere. The large angles between the magnetic and rotation axes of Uranus and Neptune cause the magnetic field lines to vary substantially with time.

CHAPTER 7
MINOR BODIES OF THE SOLAR SYSTEM

7.1 Introduction

In this chapter, you will be taking a closer look at the minor bodies of the Solar System. Although most asteroids and comets may seem relatively tiny compared to the planets, they have an important role to play in shaping the appearance of planetary surfaces. You considered this when looking at impact cratering processes in Chapter 4. Furthermore, the study of fragments of asteroids that land on the Earth as meteorites, can give us crucial information on the elemental abundances of the material that formed the solar nebula from which the planets were made. This was discussed in Chapter 2, and will be considered again in Chapter 9. Similarly, the study of comets, both remotely using telescopes and via the dust particles that they release, gives us information about the processes involved in the formation of the Solar System. For these reasons, the minor bodies of the Solar System can be of major importance.

Before we look at the minor bodies themselves, we need to consider the orbits of bodies in the Solar System in more detail than we have up until now. This is important because, due to the gravitational influence of the planets, the orbits of minor bodies can change significantly, and rapidly enough, to transport the minor body from one region of the Solar System to another – for example, from the asteroid belt to the Earth. Understanding orbits is also the key to understanding the motion of the planets, their satellites, tidal heating process, and even the structure of the ring systems around the giant planets.

7.2 Orbits and Kepler's laws

The German astronomer Johannes Kepler (1571–1630) worked at a time when it was generally believed that the orbits of celestial bodies were based on circles. Complex schemes using many circles were devised to try and explain the apparent path of the planets across the heavens. However, in trying to reconcile actual observations of the movements of Mars with his notions of how he thought the Solar System should be constructed, Kepler realized that the apparent motion of Mars could be described simply by an ellipse. From this starting point, Kepler went on to formulate three laws of planetary motion, which remain fundamental to understanding the functioning of the Solar System. They apply not only to the movements of the planets around the Sun, but to all bodies orbiting under the influence of gravity. Kepler was working at a time when the absolute distances of the planets from the Sun and from the Earth were unknown, so he worked entirely in relative terms – astronomical units (you met these in Chapter 1). **Kepler's first law** states that:

Planets move in elliptical orbits, with the Sun at one focus.

An **ellipse** is shown in Figure 7.1a. The 'length' and 'width' of the ellipse are defined by the major axis and the minor axis respectively. We usually refer however to the **semimajor axis** a and the **semiminor axis** b (Figure 7.1a) which are simply

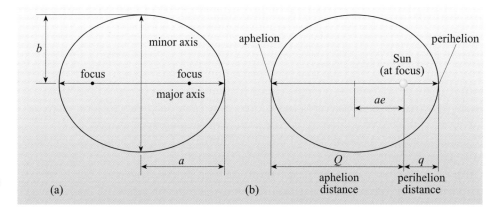

Figure 7.1 (a) An ellipse defined by its major axis and its minor axis. (b) An elliptical planetary orbit. The Sun is located at one of the foci.

half the lengths of the major and minor axes. Two *foci* lie along the major axis (foci is the plural of focus). The positions of the foci depend on the degree of flattening (or elongation) of the ellipse. The flatter the ellipse, the further the two foci are apart. The distance of a focus from the centre is given by ae where e is called the **eccentricity** of the ellipse. The eccentricity e is a dimensionless number (i.e. it has no units) whose value can lie between 0 and 1. For a circle $e = 0$, whereas for an extremely flattened (or elongated) ellipse the value of e approaches 1.

Figure 7.1b shows the ellipse again, now representing the orbit of a planetary body around the Sun. We see that the Sun is located not at the centre, but at a focus (Kepler's first law). The point on the orbit that is closest to the Sun is known as the **perihelion**, and the point that is furthest from the Sun is the **aphelion**. The semimajor axis represents the orbiting body's mean distance from the Sun. Given an orbit defined by a semimajor axis a and eccentricity e, it is useful to be able to calculate the distances from the Sun of the perihelion and aphelion points, that is the **perihelion distance** q, and the **aphelion distance** Q, respectively. These quantities are related to a and e by the following expressions:

$$q = a(1 - e) \tag{7.1}$$

$$Q = a(1 + e) \tag{7.2}$$

As e has no units associated with it, these equations will work regardless of what units q, Q and a are in (as long as they are all in the *same* units as each other). If the value of the perihelion distance is held constant, and the eccentricity is increased, then the semimajor axis (and thus the aphelion distance) must also increase. This is shown in Figure 7.2 where a circle and several ellipses with the same perihelion distance are plotted.

QUESTION 7.1

Pluto has a semimajor axis of 39.48 AU and an eccentricity of 0.249. What are Pluto's perihelion and aphelion distances?

QUESTION 7.2

Neptune has a semimajor axis of 30.07 AU and an eccentricity of 0.009. What are Neptune's perihelion and aphelion distances?

QUESTION 7.3

In general, we consider Pluto to be the outermost planet in the Solar System. Considering your answers for Questions 7.1 and 7.2, is this always the case?

Kepler's second law is less obvious than his first. It states that:

A line connecting the Sun to a planet would sweep out equal areas of space in equal times.

Figure 7.3 shows an orbital ellipse with two such equal areas highlighted. Study this figure, and think about what implications the second law has for the speed of a planet along different parts of its orbit.

▪ Referring to Figure 7.3, how does the speed of a planet between A and B compare with its speed between X and Y?

❑ If the shaded areas in Figure 7.3 are swept out in the same time, the planet must move more slowly from A to B than from X to Y.

As a planet on an elliptical orbit approaches the Sun, it speeds up. As it recedes, it slows down. The speed at any particular position depends on the distance of the planet from the Sun at that time. Similarly, bodies with larger semimajor axes move more slowly. Furthermore, a planet with a large semimajor axis has further to travel to complete one orbit. It might be expected then, that the time to complete one orbit (i.e. the **orbital period**) would be related to the semimajor axis. This relationship has been formulated as **Kepler's third law**, which states that:

The square of a planet's orbital period is proportional to the cube of its semimajor axis.

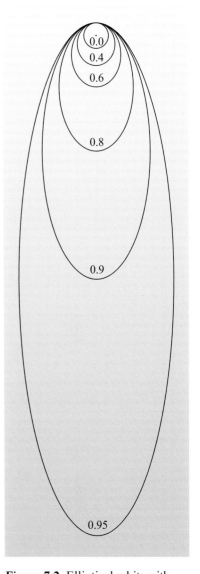

Figure 7.2 Elliptical orbits with different eccentricities (the eccentricity value is shown). If the perihelion distance is kept constant, the semimajor axis increases as the eccentricity increases.

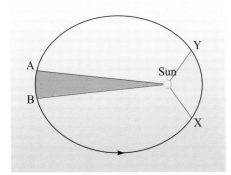

Figure 7.3 Kepler's second law states that a planet sweeps out equal areas in equal times. Two equal areas are shown. The planet's speed along its orbit is not uniform, but varies with its distance from the Sun.

If we were stating Kepler's third law mathematically, we would write

$$P^2 = ka^3 \qquad (7.3)$$

where P is the orbital period and a is the semimajor axis. The constant of proportionality, k, depends on the mass of the body that is being orbited. Thus k will be the same for all planets orbiting the Sun, but will have a different value for a satellite orbiting a planet. (Indeed you used a more explicit form of this equation, given in Question 6.1 in Chapter 6, to calculate the mass of Saturn from the orbital period of its satellite Enceladus.) When considering a planet orbiting the Sun however, the numerical value of k, can be simplified by carefully choosing the units for P and a. It is most convenient to work in years, and astronomical units. As an example, consider the Earth, for which $P = 1$ year (yr), and $a = 1$ AU. Thus k, which is simply P^2/a^3, equals $(1 \text{ yr})^2/(1 \text{ AU})^3 = 1 \text{ yr}^2 \text{ AU}^{-3}$. So in doing a calculation, k can be essentially ignored as its numerical value is 1.

QUESTION 7.4

Mercury has a semimajor axis of 0.39 AU. What is Mercury's orbital period?

Kepler's laws apply to the motion of a small body around a large body. The small body can thus be described as being in a **Keplerian orbit**. If however other massive bodies gravitationally influence the motion (such as Jupiter influencing the orbit of Mars) then subtle modifications (called *perturbations*) to the orbit are continuously made. However, at any one moment, the Keplerian orbit is an excellent approximation to the actual motion of the body.

So far, we have defined an orbit by just two parameters, the semimajor axis, a, and the eccentricity, e. However, we also need to be able to describe the orientation of the orbit with respect to the rest of the Solar System. You will remember from Chapter 1 that the Earth and the other planets orbit the Sun in approximately the same plane. We referred to this as the **ecliptic plane**. In fact, the ecliptic plane is *defined* by the plane of the Earth's orbit. Thus the Earth orbits *exactly* in the ecliptic plane, and the other planets orbit approximately in the ecliptic plane. However, other planetary bodies, for example asteroids or comets, might have orbits that are not confined to the ecliptic plane. This is described by a quantity called the orbital **inclination**, i. This is the angle between the plane of an orbit and the ecliptic plane, and is shown in Figure 7.4. The inclination angle is defined to be between the values 0° and 180°. Orbits that have a general motion around the Sun in an anticlockwise direction, as viewed from above the Sun's north pole, are called **prograde**, whereas orbits that have a general motion around the Sun in a clockwise direction, are called **retrograde** (Figure 7.5).

> If you have studied some astronomy, you might recognize that the ecliptic plane is defined by the apparent motion of the Sun across the sky throughout the year.

Prograde orbits have values of inclination between 0° and 90°.

Retrograde orbits have values of inclination between 90° and 180°.

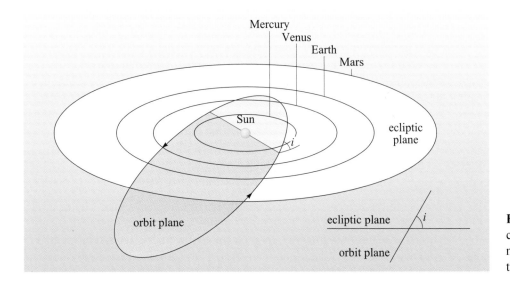

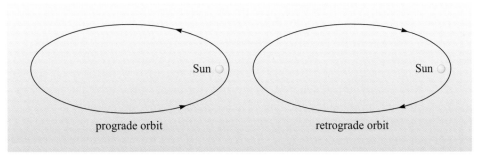

Figure 7.4 An orbit need not be confined to the ecliptic plane but may be inclined at an angle called the orbital inclination, *i*.

Figure 7.5 Prograde and retrograde orbits, as would be viewed from above the north pole of the Sun. The arrows indicate the direction of motion of a planetary body on these orbits.

- ■ If two bodies in space collide, what type of orbit does each of the bodies need to have (prograde or retrograde) to obtain the maximum impact speed?

- ❏ Just like vehicles on a road, the maximum relative impact speed occurs when there is a head-on collision. This can only be the case when one body has a prograde orbit, and the other has a retrograde orbit.

7.2.1 Tidal heating

Now we have looked at orbits more closely than before, we can consider in more detail, how the mechanism of tidal heating operates. You will remember from Chapters 2 and 3 that tidal heating played a crucial role in some instances. In particular, Jupiter's satellite Io is the most volcanically active body in the Solar System, and the major heat source that drives this comes from tidal heating. You considered previously a satellite orbiting a planet. For example, we can think of the Moon orbiting the Earth. The mutual gravitational pull acts to distort both the Moon, and the rotating Earth, creating tidal bulges that try to align themselves towards and away from the neighbouring body. On Earth, this gives rise to the ocean tides, and even manages to distort the shape of the Earth by around 1 m as it rotates. This continuous kneading of Earth gives rise to frictional heating (i.e. tidal heating).

However we saw previously that virtually all of the satellites of the planets have *synchronous rotation*, i.e. their rotation period equals their orbital period, such that the same face points to the planet at all times. In this case, the satellite is essentially not rotating with respect to the planet. Thus the tidal bulge produced by the gravitational pull of the planet does not move, and so does not produce a frictional heating effect in the satellite. Indeed, the satellites of the planets have reached their current state of synchronous rotation because the action of tidally-induced frictional heating slows the rotation of the satellite down, due to rotational energy being converted to heat. The slowing continues until the satellite reaches synchronous rotation. At this point, we might expect tidal heating to stop. However it is clear that bodies like Io are undergoing significant tidal heating *now*. Why is this?

The answer lies in the effect of orbital eccentricity, and Kepler's laws. The example in Figure 7.6 shows a synchronously rotating satellite at ten places on its elliptical orbit, separated at equal time intervals (and because of Kepler's second law, equal areas are swept out in these equal time intervals). The red radial lines drawn from the planet to the satellite represent the direction of the gravitational force, and thus the line along which the tidal bulge of a satellite will act. The blue arrows show the direction of a face of the satellite. When the satellite is at its closest point to the planet, the face points along the gravitational force direction. We know that the planet will then rotate 360° in each complete orbit, and so during each of the ten equal time intervals, it will rotate just 36° (i.e. each successive blue arrow is rotated by 36°). However, you can see that, as a result, for most of the orbit the face does not then precisely point along the line of gravitational force. In fact only at two points in the orbit does it do this. As the satellite orbits, the face moves 'forward' of the red line, and then moves 'behind' the red line. If you were watching the satellite while standing on the planet, you would see the face turn slightly one way for half the orbit, and then turn back the other way for the second half. This is called **libration**. As the satellite *librates*, the gravitational bulge is 'dragged' back and forth, causing frictional heating.

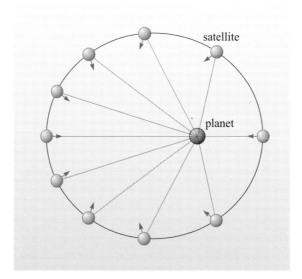

Figure 7.6 The figure represents a synchronously rotating satellite at ten places on its elliptical orbit, separated at equal time intervals. The planet rotates 360° in each complete orbit, and so during each of the ten equal time intervals it rotates 36° (i.e. each successive blue arrow is rotated by 36°).

Thus bodies that are in synchronous rotation can undergo significant tidal heating *only* if they are in orbits that have eccentricities that are not zero. This is the case for many of the satellites of the giant planets, and this mechanism gives rise to the tidal heating within satellites such as Io.

QUESTION 7.5

In the following cases, decide whether the satellite might undergo tidal heating.

(a) A newly formed satellite that is in a circular orbit, and has a rotation period that is much less than its orbital period.

(b) An ancient satellite that has acquired synchronous rotation, and has a circular orbit.

(c) An ancient satellite that has acquired synchronous rotation, and has a significant orbital eccentricity.

7.3 Asteroids

Asteroids are by far the most abundant named objects in the Solar System. Over one hundred thousand asteroids have been detected, with over thirty thousand having well determined orbits, most of these occupying the **asteroid belt** between about 2 and 4 AU from the Sun (between the orbits of Mars and Jupiter, Figure 7.7). The total mass of all the bodies in the current asteroid belt is only about one-thousandth of an Earth mass, although originally, a few Earth masses of material would have been available in the solar nebula in the region. In the 19th and early 20th centuries, astronomers thought that the asteroid belt represented fragments of a single planet which had somehow disintegrated catastrophically. However the asteroids are now thought to represent fragments of *many* small planetary bodies that never managed to accrete into one single body. This is due to the strong gravitational influence of the newly formed Jupiter 'stirring up' the asteroid population, causing collisions which would repeatedly break up the bodies and so impede the formation of one single large object.

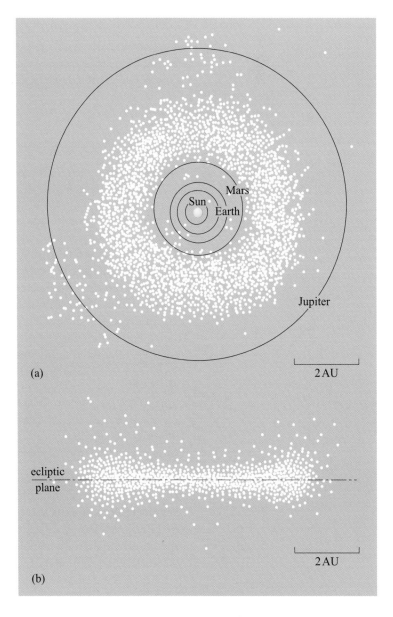

(a)

(b)

ecliptic plane

2 AU

2 AU

Jupiter

Sun
Mars
Earth

Figure 7.7 (a) A representation of the asteroid belt. It is seen that the asteroid belt is actually a diffuse cloud, or swarm of orbiting bodies. (b) A cross-section through the belt, shown on the same scale. Each individual asteroid shown moves in an orbit inclined to the ecliptic plane, so that sometimes it is above it, and sometimes below. You can imagine that collisions between asteroids will be quite common.

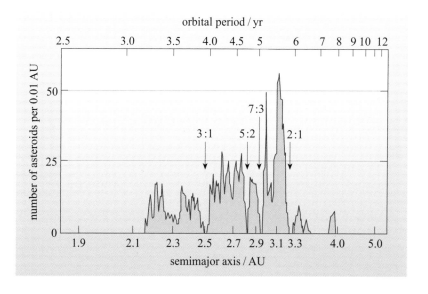

Figure 7.8 Variations in the abundance of asteroids within the asteroid belt, showing the Kirkwood Gaps at values of semimajor axis which correspond to orbital resonances with Jupiter.

Jupiter continues to exert a strong influence on the asteroid belt. When we plot the number of asteroids with a given semimajor axis interval against semimajor axis, a striking pattern emerges, as shown in Figure 7.8. There are spaces, or gaps, where there are very few asteroids with a particular value of semimajor axis. These gaps, known as **Kirkwood Gaps** after their discoverer, are not random. They occur when the orbital period associated with a given value of semimajor axis, is a simple fraction of the orbital period of Jupiter. This is known as **orbital resonance**. For example, if an asteroid had an orbital period half that of Jupiter, then it would be said to be in a 2 : 1 resonance (we say, 'two-to-one resonance'), i.e. it makes two orbits around the Sun for every one orbit Jupiter makes. Sometimes you can confuse yourself depending on whether you think in terms of the asteroid's period being half that of Jupiter, or Jupiter's period being double that of the asteroid. It is of course the same thing, but it can cause people to be confused as to whether to write 2 : 1 or 1 : 2. One way to avoid confusion is to consider the number of times the bodies make complete orbits of the Sun. Then use an often followed convention where: the number on the left side signifies the number of orbits that the body *closer to the Sun* would make in a given period of time, and the number on the right signifies the number of orbits the body *further from the Sun* would make in the same time. Thus the number on the left will be larger than the number on the right.

We can readily calculate the value of the semimajor axis associated with an object in the 2 : 1 resonance using Kepler's third law. We know that Jupiter has a semimajor axis of 5.20 AU (from Appendix A, Table A1) and thus an orbital period of 11.86 years (using $P^2 = ka^3$ from Section 7.2). So an object in the 2 : 1 resonance has an orbital period one-half of this, i.e. 5.93 years, which corresponds to a semimajor axis of 3.28 AU.

QUESTION 7.6

At what semimajor axis value would you expect to find a gap in the asteroid belt semimajor axis distribution corresponding to a 3 : 1 resonance with Jupiter?

The effect on a small body that is orbiting the Sun, and is also in orbital resonance with a large body, can have two outcomes. For the small body in the resonance, there will be times when it is being accelerated forward in its orbit by the gravitational pull of the larger body, and other times when it is being decelerated. The cumulative effect of these forces is to distort the orbit of the smaller object, until it no longer has a resonant period, and its former orbit remains unoccupied. This is the process at work to produce the Kirkwood Gaps, where many of the resonances are cleared of objects. However, another possible outcome of some resonances, is that the small object gets locked into its orbit, and the gravitational influence of the larger body essentially holds the smaller object in its orbit for long periods of time. These *stable resonances* can be very important, and we will return to this point in Section 7.4.

QUESTION 7.7

If you travelled to the distance from the Sun equal to the semimajor axis associated with a Kirkwood Gap, might you find any asteroids there? (*Hint*: think about the effect of orbital eccentricity.)

The changing of a body's orbital elements (or orbital parameters) is called **orbital evolution** and is the key to understanding the distribution of various minor bodies throughout the Solar System today. Orbital evolution means that over time (this could mean thousands or millions of years) a minor body could change its orbit significantly within the Solar System. A good example where orbital evolution is critical, is in the *Near Earth Asteroid* population.

Near Earth Asteroids (NEAs), are bodies that have orbits which come near (or indeed cross) the orbit of the Earth. You might have already noticed a few of these objects in Figure 7.7. There are almost 2000 NEAs currently known. Some objects in the NEA group can come very close to the Earth, and could collide with the Earth at some time in the future. This subset, of which around 400 are currently known, are called **Potentially Hazardous Asteroids** (PHAs). Figure 7.9 shows the orbits of known PHAs in relation to Earth's orbit. Looking at the PHA orbits, it is perhaps not surprising that the Earth occasionally suffers an asteroid impact!

It is thought that some members of the Near Earth Asteroid population might in fact be more related to comets, and so you may see reference to Near Earth Objects (NEOs) instead of Near Earth Asteroids (NEAs).

■ Today, well over 4 billion years after the origin of the Solar System, there are still numerous asteroids that could collide with the Earth. The lifetimes of these asteroids must be short relative to the age of the Solar System, because they are rapidly removed by collisions with the terrestrial planets. What does this imply?

❑ The supply of Earth-crossing asteroids must somehow be replenished.

Figure 7.9 The orbits of known Potentially Hazardous Asteroids (PHAs). The orbit of Earth is also shown.

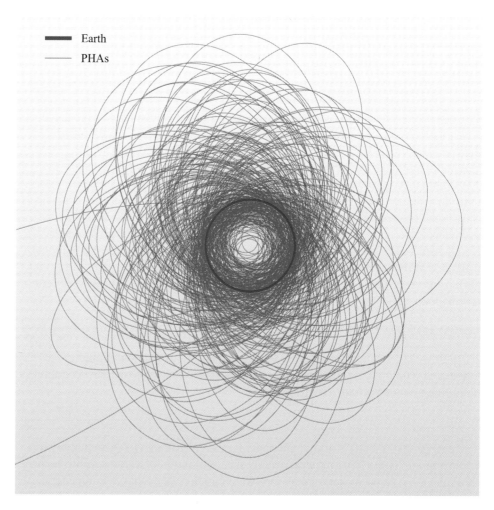

The very fact that we see NEAs today means that the NEA population is being continually replenished, and this happens because of the orbital evolution of objects in the inner asteroid belt. The long-term gravitational effects of Jupiter (and even Mars) give rise to a slow 'conveyor belt', which delivers bodies to the inner Solar System (although you should also appreciate that it can be a two-way process – bodies that are already in the inner Solar System can evolve outwards again). Some of the objects that make it into the inner Solar System might eventually hit one of the terrestrial planets.

7.3.1 Asteroid sizes

The largest main belt asteroid, discovered in 1801, is (1) Ceres (pronounced 'series') which has a diameter of 913 km. The next biggest is (2) Pallas, with a diameter of 523 km. (Note that the asteroids are numbered, and so the full name is, for example, (1) Ceres, although often, you will see only the name being used.) As we go smaller and smaller, the asteroids become more numerous. So while there is only 1 asteroid larger than, say, 600 km (i.e. Ceres), there are 7 larger than 300 km, 81 larger than 150 km, and so on. Note that for each reduction in size the number rises steeply. This behaviour is described by a size distribution. This concept will sound familiar to you after considering impact crater size–frequency distributions in Chapter 4 (Box 4.1). It is exactly the same concept, except we are now thinking in terms of asteroid diameter rather than crater diameter.

Figure 7.10 shows the cumulative size distribution of known asteroids in the asteroid belt. We see that there are many more small asteroids than large ones. The data 'flattens out' at small sizes (10 km or smaller) but this partly due to *observational selection*; we simply have not yet discovered all the small asteroids. The gradient of the dashed line in Figure 7.10 is significant when considering where most of the material in the asteroid belt is concentrated. In other words, we could ask, is most of the material (i.e. the mass) to be found in the few largest asteroids, or is most of it distributed amongst the numerous small bodies? It turns out that, if all the data followed the same slope as the dashed line shown in Figure 7.10, the total mass of objects contained in each logarithmic diameter step would be approximately the same. For example, the total mass of all asteroids with diameters between 1 and 10 km would be the same as those with diameters between 10 and 100 km. If however the slope of the data was shallower than the dashed line (i.e. more towards the horizontal), this would indicate that the largest bodies accounted for most of the mass contained in the asteroid belt. Conversely, if the data were steeper than the dashed line, most of the mass would be contained in the smaller bodies. The data in Figure 7.10 lies close to the dashed line in the middle region of the plot, but if we were to take all the data together, a best fit straight line would be somewhat shallower than the dashed line. Thus most of the mass in the asteroid belt is concentrated in the few largest asteroids.

As many of the impact craters seen on planetary bodies are caused by the impact of asteroids, it follows that the impact crater size distribution must broadly reflect the asteroid size distribution in some way. So if we expect a large asteroid to make a large crater, and a small asteroid to make a smaller crater, then because there are far more small asteroids, we would expect to see far more small impact craters on planetary surfaces. Indeed, this is what you found in Chapter 4, with the crater size–frequency distribution.

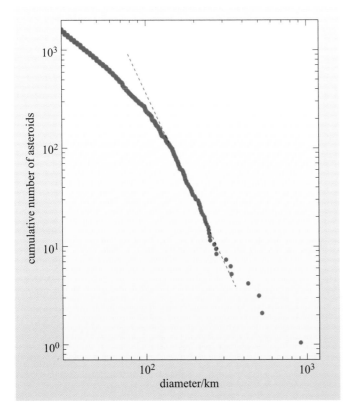

Figure 7.10 The cumulative size distribution of the known bodies in the asteroid belt, plotted logarithmically. The graph tells us the number of asteroids that have diameters greater than a given value.

7.3.2 Asteroid types

Not all asteroids are the same. The composition will depend on how and where an asteroid was formed, and what thermal, physical and chemical processing has happened to it since. Different types of asteroid are sorted into **taxonomic classes**, and the basis for deciding what class a body belongs to, comes from observational astronomy.

One useful parameter that we would like to know is how reflective the asteroid's surface is. In other words we would like to determine the *albedo* (a concept you came across in Chapter 5), and more precisely how the albedo changes at different wavelengths of light. However, much of the time, we cannot easily determine absolute values of the albedo, as we do not have an accurate knowledge of how big the asteroid actually is. In other words, you cannot always tell if you are observing a reflective small object, or a less reflective but larger object. What we can do however is to determine the *relative* efficiency with which the asteroid reflects sunlight, as a function of the wavelength of the light. This is called a **reflectance spectrum**. For example, a body that simply reflected all the sunlight equally would have a *neutral* reflectance spectrum, whereas a body that reflected light more efficiently at longer wavelengths would have a more *red* appearance. It is the precise nature (particularly the slope) of the reflectance spectrum that identifies the taxonomic class. Figure 7.11 shows typical reflectance spectra associated with three taxonomic classes. Because we do not know the absolute reflectance values, we plot the different asteroids simply over the top of each other, forcing the relative reflectance value to be 1.0 at the wavelength of about 0.55 μm, i.e. a representative value for visible light. For example, a relative reflectance value of 2.0 at 1.0 μm means that the body reflects light with double the efficiency at 1 μm as it does at visible wavelengths (where the relative reflectance value is 1.0). This would be true for an object with an albedo of, for example, 0.04 at visible wavelengths and 0.08 at 1 μm. It would be equally true for an object with an albedo of 0.3 at visible wavelengths and 0.6 at 1 μm.

The taxonomic classes themselves are due to compositional differences. There are many classes, and sub-divisions, but we need only mention a few here. C-type asteroids (carbonaceous types) are rather dark (i.e. non-reflective – see the typical albedo values for the different taxonomic classes in Table 7.1). They have neutral reflectance spectra, and contain carbon-rich rocky material. S-type asteroids are generally a stony (or stony–metallic) mix and are more reflective and somewhat more red. E-types are often highly reflective and appear to be predominantly composed of the mineral enstatite (magnesium silicate $MgSiO_3$).

Table 7.1 The typical albedo values of selected taxonomic classes of asteroids.

Taxonomic Class	Albedo
E	0.25 to 0.60
S	0.10 to 0.22
C	0.03 to 0.07
M	0.10 to 0.18
P	0.02 to 0.06
D	0.02 to 0.05

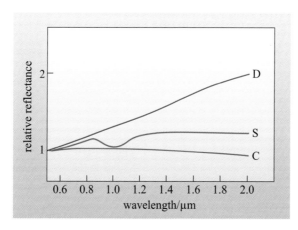

Figure 7.11 The reflectance spectra of a few taxonomic types of asteroids. If a body reflected all wavelengths equally (i.e. it was neutral) the spectra would have a value of 1.0 throughout; C-types approximate this behaviour. D-types however reflect longer wavelengths better, and so would appear red. S-types have a distinctive 'S-shape' feature in their spectra.

D-types (dark type) are extremely dark and red. M-types (metallic type) are thought to be made of mostly iron and nickel, with P-types (pseudo-M type) also thought to have a major metallic component in the composition. It is thought that C and D-types are probably quite primitive (least processed) bodies, whereas E-types, S-types and M-types are likely to be fragments from a larger body, which underwent differentiation (as discussed in Chapter 2) so producing a metallic core, and a rocky mantle. Such fragments might collide with Earth and be collected as meteorites. Figure 7.12 shows that different classes of asteroid generally occupy different regions in the asteroid belt. This is a consequence of the fact that the region of formation (distance from the Sun) affected the composition of the asteroid.

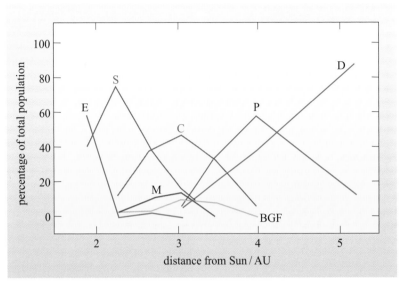

Figure 7.12 Distribution of some of the major classes of asteroid within the asteroid belt as a function of distance from the Sun. B, G and F-types are sub-classes of C-types.

7.3.3 Asteroids up close

Until relatively recently, what we knew about asteroids was based on ground-based observations. But now a handful of spacecraft missions have come very close to asteroids allowing us to learn about these minor bodies in much greater detail than before. Table 7.2 details some close fly-bys of asteroids by spacecraft. The Galileo spacecraft, while en route to Jupiter, flew by (951) Gaspra (Figure 7.13), obtaining the first ever high-resolution image of an asteroid. A rather irregularly shaped body peppered with impact craters was seen. Galileo's second, much larger, asteroid target, (243) Ida, showed a similar scenario (Figure 7.14) with many impact craters and irregular features.

Table 7.2 Parameters of the asteroids that have had spacecraft fly-bys. The sizes indicated refer to the major and minor axes of the body. Porosity describes the relative volume of voids within the object (so a completely solid body has a porosity of 0%). Remember that a is the semimajor axis of the orbit and e is the eccentricity of the orbit.

Asteroid	Spacecraft	Encounter date	Asteroid size/km	Taxonomic class	Density /kg m^{-3}	Porosity	a /AU	e
(951) Gaspra	Galileo	29 Oct 1991	19×12	S-type	$2500 \pm 1000?$	30%?	2.21	0.17
(243) Ida	Galileo	28 Aug 1993	58×23	S-type	2600 ± 500	30%	2.86	0.05
(253) Mathilde	NEAR	27 Jun 1997	59×47	C-type	1300 ± 200	80%	2.65	0.27
(433) Eros	NEAR	14 Feb 2000[a]	33×13	S-type	2700 ± 30	25%	1.46	0.22
(9969) Braille	Deep Space 1	29 July 1999	2.2×1	?	?	?	2.34	0.43
(5535) Annefrank	Stardust	2 Nov 2002	8×4	?	?	?	2.21	0.06

[a]NEAR went into orbit around Eros on this date. It remained there for a year and then landed on the surface of Eros on 12 February 2001.

Figure 7.13 (above) Image of main belt S-type asteroid (951) Gaspra (19 km × 12 km) taken by the Galileo spacecraft. (NASA)

Figure 7.14 (above) Image of main belt S-type asteroid (243) Ida (58 km × 23 km), taken by the Galileo spacecraft. The rather jagged shadow line is a consequence of the asteroid being illuminated only from one direction (from the Sun). This causes a great contrast between the sunlit areas and the shadow areas, which look as black as the background space. (NASA)

QUESTION 7.8

Look at Figure 7.14, the image of the 58 km × 23 km asteroid, (243) Ida. Remembering the types of crater that you met in Chapter 4, how would you describe the craters on Ida?

Surprisingly, Ida was also found to have a much smaller satellite asteroid (named Dactyl) orbiting around it (Figure 7.15). Binary asteroids that orbit each other now appear to be more common than was previously thought, with several examples being discovered recently by ground-based telescopic studies.

Gaspra and Ida are S-type asteroids, but a C-type asteroid, (253) Mathilde (Figure 7.16), was encountered by the NEAR (or NEAR Shoemaker) spacecraft, on its way to its main target (433) Eros (an S-type Near Earth Asteroid). On reaching Eros (Figure 7.17), NEAR went into orbit about the asteroid (a major technical achievement) and spent a full year taking scientific data.

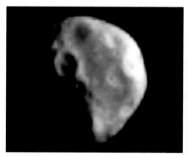

Figure 7.16 Image of main belt C-type asteroid (253) Mathilde (59 km × 47 km), taken by the NEAR spacecraft. (NASA)

Figure 7.15 Image of Ida's satellite asteroid, Dactyl (1.6 km × 1.2 km) taken by the Galileo spacecraft. (NASA)

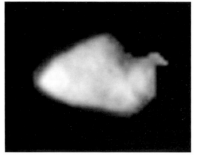

Figure 7.18 Image of the main belt asteroid (5535) Annefrank (8 km × 4 km) obtained by the Stardust spacecraft. (NASA)

Figure 7.17 Image of near Earth S-type asteroid (433) Eros (33 km × 13 km) taken by the NEAR spacecraft. (NASA)

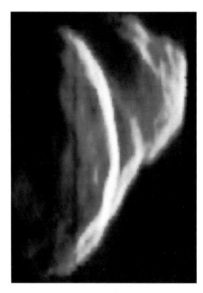

Figure 7.19 The radar image of Near Earth Asteroid (4179) Toutatis (5 km × 2 km). Toutatis is extremely elongated. (Calvin J. Hamilton)

The asteroid (9969) Braille was encountered by the technology-proving spacecraft, Deep Space 1. Indeed the fly-by was the closest yet undertaken, being just 15 km from the asteroid. However imaging of the asteroid was not very successful. Asteroid (5535) Annefrank however, was encountered by the Stardust mission in November 2002, returning the image shown in Figure 7.18.

A close fly-by of a spacecraft is not the only way that detailed information on the shape of an asteroid can be determined. By using some of the world's most powerful radio transmitters in conjunction with some of the world's largest radio telescope dishes (for example the huge Arecibo radio telescope in Puerto Rico), radar techniques can be used to image the asteroid. This technique involves sending radio wave pulses towards an asteroid, and then receiving a reflection, or echo, back on Earth. By complex processing of the returned signals, an image representation of the asteroid can be constructed. Such an image, of the asteroid (4179) Toutatis, is shown in Figure 7.19. Repeated observations of an asteroid as it rotates, allows a full 'shape model' to be derived, an example of which is shown in Figure 7.20.

We have seen that the asteroids imaged so far, are non-spherical. A good description of them might be 'potato-shaped'. This is not unexpected. Observations from the ground often show the brightness of asteroids increasing and then decreasing regularly. This behaviour is illustrated in Figure 7.21, which is an example of an asteroid's **lightcurve**. As the light we see from the asteroid is simply reflected light from the Sun, the amount of light we receive is related to its albedo, and the cross-sectional area of the region of the asteroid that is illuminated. So if a 'potato-shaped' body is spinning, then you will see a changing cross-sectional area, and hence a changing brightness. You can convince yourself of this by simply looking at an irregularly shaped body (e.g. a potato) and turning it while considering how the cross-sectional area changes. The *period* of the lightcurve tells us how long the asteroid takes to spin, and the *amplitude* of the lightcurve (the difference between the maximum and minimum brightness) depends on how elongated the body is. Thus the lightcurve tells us the spin rate, and the ratio of the longest side to the shortest side.

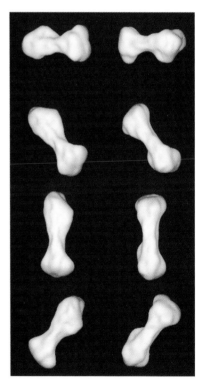

Figure 7.20 The 'shape model' of main belt asteroid (216) Kleopatra (220 km × 95 km) derived from radar observations. Kleopatra bears a remarkable resemblance to the kind of bone favoured by dogs! (NASA)

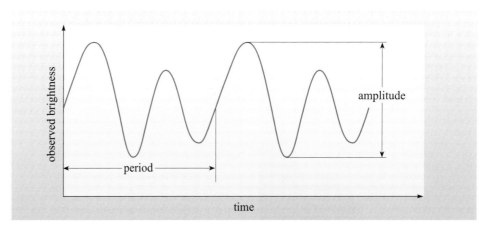

Figure 7.21 A schematic of an asteroid lightcurve, showing the amplitude (difference between the maximum and minimum brightness) and the period (time taken for one full axial revolution of the body). Note that a lightcurve produced by the irregular shape of an object is double peaked (rather than a sine wave), and that a spherical object would give an amplitude of zero (i.e. we would say it had a 'constant lightcurve').

Some asteroids have lightcurves that display very little variation, indicating that the asteroid is spherical (or near spherical). Some of the largest asteroids are thought to have undergone differentiation in a similar way to the terrestrial planets (as you met in Chapter 2). During this process the body maintains a spherical shape due to compression under its own gravity. However, subsequently, when the asteroids had cooled so that they were solid throughout, impacts could fragment and break parts off the parent asteroid, creating much more irregular shapes. This said, most asteroids that are larger than a few hundred kilometres in diameter, tend to be approximately spherical due to gravitational compression.

The motion of binary asteroids with respect to each other, or indeed the motion of a spacecraft around an asteroid, can be used to derive masses for the asteroids, and thus (if the asteroid size is known) indicate the density of the asteroid. S-type asteroids Ida and Eros were found to have densities of around $2600 \, \text{kg m}^{-3}$ and $2700 \, \text{kg m}^{-3}$ respectively. Remembering that S-types are predominantly rocky, one might have anticipated a density somewhat higher than this. For example, typical stony meteorites (which are thought to come from fragmented S-type asteroids) have densities of around $3400 \, \text{kg m}^{-3}$. The lower densities of Ida and Eros suggest that the bodies might be slightly *porous*, i.e., the asteroids are not solid rock, but have a structure with voids in it (similar to what you would find for a pile of rocks). Even more surprising, the density of the C-type Mathilde was found to be around $1300 \, \text{kg m}^{-3}$ which appears very low indeed. Additionally, ground-based observations of another C-type asteroid, Eugenia, have indicated a similar density of $1200 \, \text{kg m}^{-3}$. In fact these results suggest that C-type asteroids might be up to 80% porous. Even recent observations of some M-type binary asteroids have indicated a much lower density than might have been expected, again suggesting high porosity. These results show that some (perhaps most) asteroids are probably not solid lumps of rock and metal, but are more like a **rubble pile** of fragments of all sizes, bound together by their own gravity.

After the NEAR spacecraft had been orbiting Eros for a year, the decision was made to manoeuvre NEAR such that it landed on the surface of Eros (even though it was not designed to do this!). This allowed some images to be obtained during the descent, of unprecedented resolution. Figure 7.22 shows three images taken just before 'touch down'. Boulders and pebbles are clearly seen, with many of the boulders being partly buried by regolith (i.e. fine particle 'soil'). To the lower left of Figure 7.22c, there are fewer boulders and a smoother dusty area. These type of regions have been nicknamed *ponds* and are thought to be areas where fine regolith has gathered, covering larger boulders beneath the surface. Figure 7.23 also shows views of the surface of Eros. 'Ponds' are also seen in Figures 7.23a and c, with a more 'rugged' appearance (i.e. more boulders, and less regolith covering) being seen in Figure 7.23d.

The NEAR data show that impacts that produce boulders and other smaller particles play a large part in determining the nature of the asteroid surface. Asteroids can no longer be thought of as lumps of bare rock, but are often collections of smaller fragments, or at least can suffer significant *fracture* due to impacts. Future spacecraft missions will further investigate the nature of asteroids. For example, the Japanese MUSES-C mission has been designed to go to a Near Earth Asteroid, and attempt to bring some small samples of the surface back to Earth, allowing detailed chemical composition analyses to be done in the laboratory.

QUESTION 7.9

Imagine an asteroid of diameter 1 km, of unknown taxonomic class, is about to hit the Moon. Will it make any difference to the impact crater produced, whether the asteroid was S-type, or C-type?

(a)

(b)

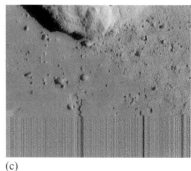

(c)

Figure 7.22 The last images from the descent sequence of NEAR. Part (a) shows a region 54 m across taken at a range of 1150 m, (b) shows a region 12 m across taken at a range of 250 m, and (c) shows a region 6 m across taken at a range of 120 m; this is the final image obtained before the loss of signal (the lines at the bottom of the image indicate when signal was lost). The spacecraft probably landed about 7 m to the left of the edge of image (c). (NASA)

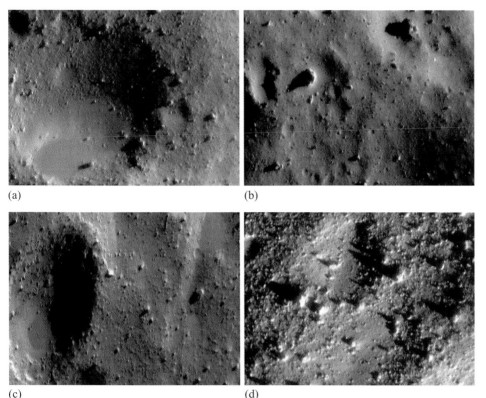

(a)

(b)

(c)

(d)

Figure 7.23 Four images of the surface of Eros, where regolith appears to have collected in depressions on the surface. A 'pond' is particularly evident in the lower left region of Figure (a). Figures (a) and (b) show regions about 550 m across. Figures (c) and (d) show regions about 230 m across. (NASA)

7.4 The Kuiper Belt

In Chapter 1 we considered the layout of the Solar System, and saw that a large population of planetesimals lies beyond Neptune – the **Kuiper Belt**. These objects were first discovered in 1992 – but not by chance. A small group of astronomers had been specifically searching for these objects for a number of years. Before considering the properties of Kuiper Belt objects in detail, let us take a step back and consider why astronomers suspected the existence of the Kuiper Belt.

Consider the layout of the outer Solar System. The gas giant Jupiter contains about 318 Earth masses of material. Saturn contains 95 Earth masses, and then Uranus and Neptune contain about 14 and 17 Earth masses respectively. As we go further from the Sun, the amount of material that accreted into the gas giants appears to decrease. However, Uranus and Neptune still contain a huge amount of material from the solar nebula. But what happens beyond Neptune? Do we expect the significant mass contained in the planetary bodies to just stop at Neptune? We do of course have Pluto, but Pluto accounts for just 0.002 Earth masses. So perhaps we might expect a lot of material from the solar nebula to be contained in bodies beyond Neptune.

It is this sort of reasoning that led astronomers to hypothesize that the Solar System most likely continues beyond Neptune and Pluto, if not with major planets (gas giants) then in a multitude of smaller bodies. And so the existence of the Kuiper Belt (named after the astronomer Gerard Kuiper who did some work on this concept in the 1950s) was suspected for many years. But technology took a long time to catch up. Bodies like Pluto (and smaller) so far from the Sun, appear extremely faint, and they move very slowly with respect to the fixed star background. Even with large telescopes they are difficult to identify. A hint of things to come, was seen in 1977 when an unusual asteroid, named (2060) Chiron (pronounced 'kye-ron'), was discovered in an orbit far beyond the asteroid belt. Chiron was observed to have a semimajor axis of 13.6 AU and an eccentricity of 0.38 indicating a perihelion distance of 8.4 AU and an aphelion distance of 18.8 AU. This meant Chiron occupied the space between Jupiter and Uranus, crossing the orbit of Saturn. Another object on an even more surprising orbit, (5145) Pholus, was also discovered (a = 20.3 AU, e = 0.57) and this object crossed the orbits of Saturn, Uranus and Neptune (Figure 7.24). This new class of objects was called **Centaurs**, for which a loose definition would be objects with orbits that cross those of the giant planets. But objects in these sorts of orbits must be quite short lived, as they would suffer strong gravitational perturbations, or even close approaches and impacts with the giant planets. So if we see them today, it means they must be being replenished (a bit like the case of Near Earth Asteroids). But from where do they come? The most likely answer is that Centaurs are actually objects that started in the Kuiper Belt and have undergone orbital evolution, slowly making their way into the inner Solar System.

You can see that the expectation that the Solar System did not just abruptly stop at Neptune and Pluto, and the observations of Centaurs, meant that astronomers felt reasonably sure that other bodies must occupy space beyond Neptune. It was also becoming apparent that a belt of comet-like bodies beyond Neptune could be the source of short-period comets (this will be discussed further in the next section). It was astronomers Dave Jewitt and Jane Luu who found the first object, designated 1992 QB$_1$, while using a relatively newly introduced CCD camera, so confirming the existence of the Kuiper Belt (see Figure 7.25).

Somewhat before Gerard Kuiper published his ideas, a British astronomer, Kenneth Edgeworth, published some short communications outlining the same ideas, but this only really came to light after the term 'Kuiper Belt' was generally accepted. However, you may see references to the 'Edgeworth–Kuiper Belt'.

You can appreciate how difficult it is to observe Kuiper Belt objects when you realize that a typical Kuiper Belt object is 100 000 000 times fainter than a typical star in a constellation we see with the naked eye.

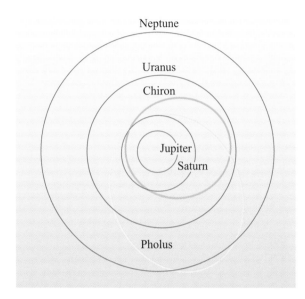

Figure 7.24 The orbits of Centaur objects (2060) Chiron and (5145) Pholus. For comparison, the orbits of the giant planets are also shown.

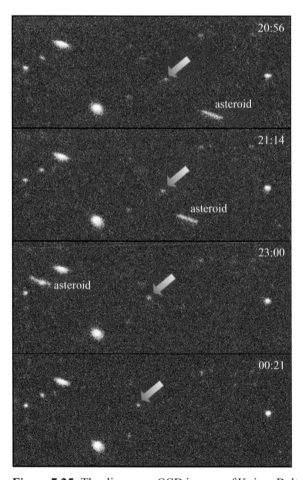

Figure 7.25 The discovery CCD images of Kuiper Belt object 1992 QB$_1$. The four frames are taken over almost 4 hours, and in this time the object has moved (mostly due to the Earth's motion rather than the Kuiper Belt object's motion) so proving that it is not a star. Further observations confirmed the orbit, and that the object was orbiting the Sun at a distance of 44 AU (semimajor axis). (Jewitt and Luu, 1992)

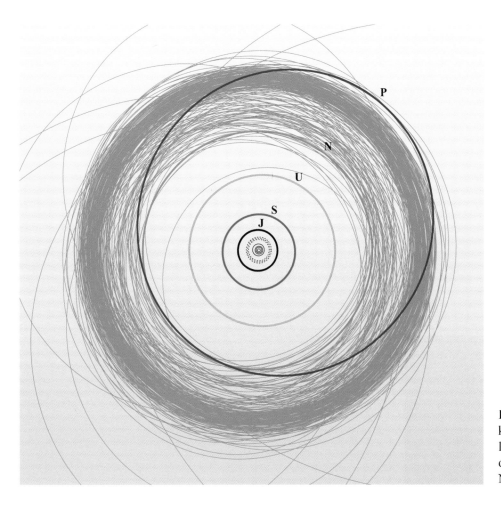

Figure 7.26 The orbits of the known Kuiper Belt objects. The labels J, S, U, N and P denote the orbits of Jupiter, Saturn, Uranus, Neptune and Pluto, respectively.

Two more Kuiper Belt objects were discovered in 1993, and four more in 1994; since then the discovery rate has grown greatly as more astronomers have made the effort to look for these bodies. Several hundred objects have now been identified (the orbits are shown in Figure 7.26), many of which have quite well determined orbits. It is now clear that, based on their orbits, there are 3 distinct classes of Kuiper Belt objects. This is shown by plotting the orbital eccentricity against the semimajor axis, as has been done in Figure 7.27. Firstly, there are many objects that have a semimajor axis of (or close to) 39.4 AU. The significance of this becomes clear, when we note that the semimajor axis of Neptune is 30.0 AU, and so if we compare orbital periods (using Kepler's third law) we find that this type of object is in a 3 : 2 orbital resonance with Neptune. Thus Neptune orbits 3 times around the Sun for every 2 orbits of these particular Kuiper Belt objects.

Orbital resonances were discussed in Section 7.3, and it was noted that some resonances can result in objects being quickly removed from a particular semimajor axis (like the Kirkwood Gaps) and some can be locked in to a resonance for a long time. The 3 : 2 resonance with Neptune is the latter type (i.e. an example of a *stable resonance*), and so objects stay in these orbits for a long time.

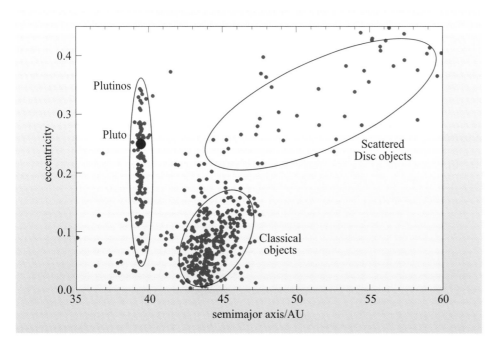

Figure 7.27 The distribution of eccentricity versus semimajor axis for the known Kuiper Belt objects. The dynamical groups (Plutinos, Classical objects and Scattered Disc objects) are shown. Note the position of Pluto.

Referring to Figure 7.27 you will notice that Pluto sits right in the middle of all the 3 : 2 resonance Kuiper Belt objects. This is no coincidence. Pluto is indistinguishable from the other 3 : 2 resonance Kuiper Belt objects in terms of its orbit. This leads us to ask the question, is Pluto really worthy of the designation 'planet'? In fact, Pluto is just another member of the Kuiper Belt. Its fame as the ninth planet has only arisen because it is larger than other objects so far discovered, and thus considerably brighter and easier to detect. Because of this, Pluto was first identified in 1930, i.e. 62 years before the *next* Kuiper Belt object was found (1992 QB$_1$). Today we have identified Kuiper Belt objects with diameters over 1000 km (compared with Pluto's diameter of 2370 km) and we may even find larger bodies over the next few decades.

Some astronomers and planetary scientists have called for Pluto to be officially demoted from the status of a planet, and to join the ranks of bona fide Kuiper Belt objects. The arguments for doing so are not easily ignored, and certainly it fits into a tidier concept of the Solar System, where we have the small rocky terrestrial planets near the Sun and large gas giants in the outer Solar System. However, we have lived with the idea of a nine-planet Solar System for over 70 years, and old habits die hard. Furthermore, it might seem somewhat disrespectful of the many years of effort that astronomer Clyde Tombaugh expended before discovering Pluto in 1930 (although he also felt that Pluto was not quite the 'Planet X' that he had been looking for, and he spent a further 10 years looking for a more massive planet). And so, it has generally been agreed that Pluto will retain its status as the ninth planet.

Returning to the consideration of the 3 orbital classes of Kuiper Belt object, we have seen that there are the Pluto-like 3 : 2 resonance objects. This group is often called the *Plutinos* (i.e. 'mini-Plutos'). The next group are objects that have somewhat larger semimajor axes (generally greater than 42 AU) and small eccentricities. These objects are known as the *Classical* objects, and as the first object discovered, 1992 QB$_1$, was one of this class, members of this group are sometimes referred to as '*Cubewanos*' (i.e 'QB_1-os'). Finally there is a group of objects with large eccentricities. These are known as *Scattered Disc* objects, and their large eccentricities mean that their aphelia (plural of aphelion) can be at large distances (up to 100 AU or more). Some Scattered Disc objects have perihelia (plural of perihelion) that come within the orbit of Uranus, and so begin to look rather like Centaurs, and so the distinction between Centaurs and Scattered Disc Kuiper Belt objects is rather vague.

QUESTION 7.10

Of the three dynamical classes in the Kuiper Belt (Plutinos, Classical and Scattered Disc), which group of objects are most likely to have close approaches with (or impact) the giant planets?

The total mass of the Kuiper Belt is probably in the region of about 0.1 Earth masses. This can be compared to the asteroid belt which is thought to contain currently only about 0.001 Earth masses of material.

7.4.1 Observational properties of Kuiper Belt objects

Because these objects are so faint, it is hard to deduce a great amount of information from observations. However we can at least obtain a broad picture of the physical nature of the bodies. We can obtain reflectance spectra, as for asteroids, and thus learn something of the surfaces of the bodies. The first outer Solar System bodies to be investigated in this way were the Centaurs, Chiron and Pholus. They could not have been more different. Chiron had a neutral spectrum (not unlike a C-type asteroid) whereas Pholus had the steepest (that is, reddest) spectrum of any Solar System object ever observed. Figure 7.28 shows the reflectance spectra, in comparison with those of the main asteroid types that you met previously.

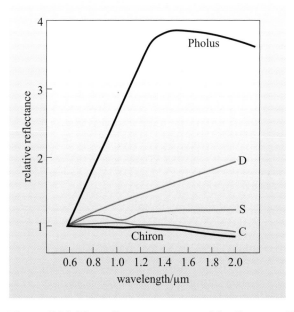

Figure 7.28 The reflectance spectra of the Centaurs Chiron and Pholus, with some asteroid types shown for comparison. Pholus shows the reddest spectrum of any Solar System body.

Observations of Kuiper Belt objects (and indeed other Centaurs), show that their spectra have great diversity, with slopes essentially lying anywhere between those of Chiron and Pholus (i.e. Chiron and Pholus by chance happen to represent the most extreme examples of spectra). This behaviour probably suggests that Centaurs and Kuiper Belt objects are indeed the same types of body – the Centaurs being objects from the Kuiper Belt slowly moving inwards due to orbital evolution. But how can we get such diversity within the Kuiper Belt group? Surely if all the objects formed together from much the same material, they should look the same?

Figure 7.29 An artist's impression of the proposed New Horizons mission. A spacecraft would take about 10 years to reach Pluto, and then could continue, hopefully performing a fly-by of a Kuiper Belt object. (Copyright Johns Hopkins University)

The answer might lie in the way the surfaces of these bodies age, and are subsequently altered by impacts. It is thought that the surface material in Kuiper Belt objects will chemically change over time due to the exposure to the Sun's radiation and also galactic cosmic rays. The material might slowly form complex organic molecules at the surface which turns the appearance very dark and red (e.g. giving a Pholus-type reflectance spectrum). This dark layer might only extend to less than a metre from the surface. With more time, the surface might become more neutral again due to the organic molecules eventually breaking down. In addition, impacts to the surface will excavate 'fresh' sub-surface material which could coat the surface. And so the object returns to its initial neutral colour. If this scenario is correct, then the diversity in reflectance spectra shows that the objects are at various stages of the ageing and impact re-surfacing process – a cycle that may occur many times during the lifetime of an object. On the other hand, it is also possible that the colour diversity is unrelated to impacts, but a consequence of differing original composition. Or perhaps some other as yet undetermined factors are involved.

What is clear is that *some* impacts *must* occur in the Kuiper Belt, and this would probably give rise to fragmentation of the objects (rather like in the asteroid main belt). Recent lightcurve observations of object 2000 WR$_{106}$ (named Varuna) suggest a reasonably elongated object (i.e. another ubiquitous 'potato-shaped' object!) with a rotation period of about 6 hours. Consideration of the expected material strengths of the object has led to a conclusion that Varuna is most likely a fragmented rubble pile which is distorted into an elongated shape due to its relatively rapid rotation. Furthermore, recent observations have identified binary objects, somewhat like the Pluto–Charon system (or some asteroid binaries).

Our understanding of the nature and evolution of Kuiper Belt objects is far from complete. Although ground-based observations will continue to give us information about the objects, a spacecraft mission to a Kuiper Belt object would clearly be highly desirable. NASA's proposed New Horizons mission (Figure 7.29) may offer such an opportunity. While the main aim of a mission like this would be to perform a close fly-by of Pluto, the spacecraft would continue into the Kuiper Belt and perhaps perform a fly-by of a Kuiper Belt object. We will have to wait to see what surprises a future mission of this type might bring us.

7.5 Comets

Our discussion of Kuiper Belt objects leads us naturally to comets, because as you will see, the two types of object may have much in common.

Comets have been observed for as long as the stars and the planets. However, their unexpected appearance in the sky, and the fact that they are generally only observable for a relatively short duration (weeks or months) set them apart from other heavenly bodies. An apparition of a comet was once generally thought to be a portent of terrible events to come. It is certainly easy for us to understand how spectacular a comet might have appeared to ancient civilizations, when we see images such as the ones shown in Figure 7.30. The bright *head* of the comet, called the **coma**, leads into a wispy *tail* which can extend significantly across the sky (for example several times the angular width of the Moon). So what exactly gives rise to this spectacular object in the sky?

As we said in Chapter 1, the comet itself is a relatively small body, rich in ices and small solid particles (i.e. dust). As the ices are heated when the comet comes near to the Sun, the ices turn directly to gas (i.e. the ices *sublimate*) and escape from the surface into space, carrying the solid dust particles with them. This process is seen in action, in Figure 7.31. Figure 7.31a shows one of the most famous comets, Halley's Comet, as seen from Earth in 1986. Figure 7.31b shows the cometary body itself, usually referred to as the **cometary nucleus**, as imaged by a spacecraft that passed within 600 km of the body. The image shows *jets* of gas and dust leaving the nucleus, which initially might typically be a few km or a few tens of km across, to form the tenuous coma and tail structure which can be hundreds of thousands (or even millions) of km in size. Another example of a cometary nucleus is seen in Figure 7.32.

If comets are ejecting significant amounts of gas and dust, then it is clear that the cometary emission process causes the comet to lose mass (i.e. decay). For example, Halley's Comet (which is quite an active comet) loses over 10^{11} kg of mass during each perihelion passage, which would be equivalent to about a metre

Figure 7.30 Examples of comets as observed from Earth. (a) Comet Hale–Bopp seen in 1997 and (b) Comet Hyakutake in 1996, were both naked-eye comets, i.e. one could see them in the sky without the aid of a telescope. These images however were taken using a telescope. Note that in (a) the tail appears in two parts. The bluish part is called the *ion tail*, and is seen due to the gaseous ions from the comet emitting light, while travelling directly away from the Sun under the influence of the solar wind. The other part is called the dust tail, and is due to light reflected and scattered from small dust particles that were ejected from the comet and are now also orbiting the Sun. (David Malin Images)

(a)

(b)

(a)

(b)

Figure 7.31 Halley's Comet. (a) The image shows what was observed from Earth in 1986, whereas, (b) shows the nucleus (15 km × 7 km), imaged by the European Space Agency's Giotto spacecraft in 1986. ((a) NASA; (b) European Space Agency)

of material from the surface. After a few thousand more perihelion passages, the comet will have decayed considerably. Indeed comets may decay far faster than this, as they are occasionally seen to split and fragment. The ejection of nucleus fragments (metres or tens of metres across) is probably quite commonplace, and almost every cometary nucleus *may* undergo some sort of splitting event at least once during its lifetime. These effects mean that once a comet enters the inner Solar System, its lifetime is relatively short compared to the age of the Solar System – a comet might only last a few thousand or tens of thousands of years in the inner Solar System.

The ices in a comet are mainly water ice (H_2O). However other ices are present also, such as frozen ammonia (NH_3), carbon dioxide (CO_2), carbon monoxide (CO) and others. The ices are mixed with the dust particles, which are rocky particulates (mainly silicate minerals) forming what is often referred to as 'dirty snow'. This is probably a good description, since the ices are certainly not as compact and homogeneous as an ice cube from a freezer. We would expect a more fluffy and randomized structure, analogous to snow with fine dust mixed with it. The nucleus is expected to be a very porous, low density body (perhaps analogous to an icy version of a C-type asteroid). Indeed the bulk density of a cometary nucleus might be as low as one-tenth that of compact ice. As ices at the surface sublimate, some dust is left behind on the surface, so building a dark, unreflective layer. This layer could shield the icy layers beneath from sunlight, and so a cometary nucleus will not be releasing gas from its entire surface. Regions on the surface (so called 'active regions') with exposed ices may account for most of (e.g. 90% or more) the gas and dust output of the nucleus (indeed the nucleus of Halley's Comet shown in Figure 7.31b appears to be significantly outgassing from only around 10% of its surface). Cracks and fissures in the surface may give rise to 'jets' or active spots. A visualization of what we might expect for the structure of a nucleus is shown in Figure 7.33.

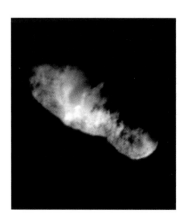

Figure 7.32 Comet Borelly (8 km × 3 km). The image was obtained in 2001, by NASA's Deep Space 1 spacecraft. It shows a nucleus similar to that seen for Halley's Comet. (NASA)

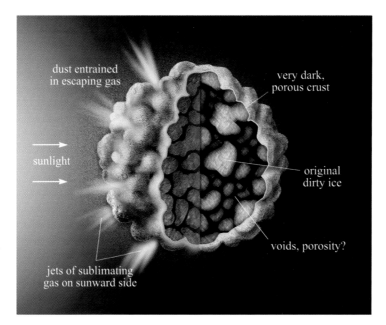

Figure 7.33 The expected structure of a cometary nucleus. The 'dirty snowball' of the nucleus is almost certainly very porous and low density, with a structure probably more akin to a rubble pile rather than a homogeneous body. (Copyright Don Davis)

QUESTION 7.11

If a 'typical' comet nucleus were to split apart while in the inner Solar System (i.e. when it is undergoing significant heating from the Sun), why might we expect to see an outburst of cometary activity associated with the event.

While we now understand something of a comet's outgassing behaviour due to the 'snowy' ices present, we have not discussed *why* comets are predominantly icy. In fact, they are icy due to their formation process. Comets are planetesimal bodies, formed by accretion from the material in the solar nebula. They are the leftovers from the planetary formation process.

◼ Remembering the division between the inner Solar System 'rocky' terrestrial bodies, and the outer Solar System icy satellites, where in the Solar System might comets have originally formed?

❑ Comets must have formed in the outer Solar System (i.e. in the regions associated with the giant planets) to have maintained their icy content. Thus comets must be the *outer* Solar System leftovers from the planetary formation process.

Indeed, icy planetesimals will have formed in the region associated with the giant planets (and further out, i.e. in the Kuiper Belt), where the temperatures in the solar nebula allowed ices to exist. While the huge majority of these cometary planetesimals will have accreted into planetary embryos and eventually formed the giant planets, some will have survived unscathed. However the survivors in the Jupiter to Neptune region will have undergone rapid orbital evolution due to the large gravitational influences of the newly forming giant planets. This process would have effectively scattered the comets. Some would have been thrown into the inner Solar System where many will have collided with the newly formed terrestrial planets. Some will have been thrown out of the Solar System altogether,

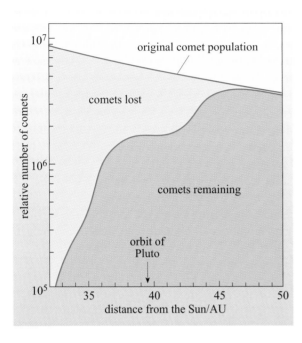

Figure 7.34 The relative number of cometary bodies in the outer Solar System.

leaving a huge number orbiting the Sun at great distances, so forming a huge cloud of comets around the Solar System. This cloud is the Oort cloud, which you met in Chapter 1. So while the Oort cloud was produced from comets formed in the giant-planets region of the Solar System, the comets that had formed near, and beyond Neptune, were far less gravitationally influenced by the giant planets (particularly Jupiter), and were not expelled like the Oort cloud comets. Some of these cometary bodies still exist today in the region beyond Neptune, i.e. they make up the Kuiper Belt.

Figure 7.34 indicates the relative number of comets that are thought to have originally existed in the outer Solar System (beyond Neptune), and what are thought to remain today, as a function of the distance from the Sun. The 'missing comets' have either been ejected or accreted by the giant planets. Beyond the orbit of Pluto the population is thought to be more representative of the original population, albeit with many of the 1 to 10 km sized comets accreting to form the 100 to 1000 km sized bodies, which we referred to in the previous section as Kuiper Belt objects.

It is tempting then to consider that Kuiper Belt objects (or indeed Pluto) must be just like very big comets. However, with large bodies (a few hundred kilometres or more in diameter) gravitational compression would play a significant role, increasing the density to nearer that of compact ice. Furthermore, radioactive elements within the large body will have supplied a heat source significant enough to fuel some chemical alteration, or melting of the inner material. In the end then, the larger bodies will be considerably more geologically processed than a small cometary nucleus. However, they will still retain much of the original cometary (icy) material.

If then, objects in the Kuiper Belt are somewhat cometary in nature, might we expect them to behave like comets and produce gaseous emission? The answer is yes, *but* we must consider carefully the distance of the body from the Sun. As the main component of the cometary ices is water ice, then a comet will produce its major outgassing when it is close enough to the Sun in order to heat the surface to a temperature sufficient to sublimate water ice. This happens at a distance of less than 3 AU from the Sun (and indeed most of the gas production from a comet occurs when it is closest to the Sun, at perihelion). It would seem then that a Kuiper Belt object, or indeed a Centaur, is not close enough to the Sun to produce water-driven sublimation. However, the other components in the cometary ices (e.g. NH_3, CO_2, CO) are more volatile than water ice, and sublimate at much lower temperatures. Hence some cometary outgassing could occur at much greater distances from the Sun than we might first expect. The first evidence of this came from the first Centaur object we considered, Chiron. When discovered in 1977, it was designated an asteroid. However, some years later, some images of Chiron showed that it appeared to have a faint 'cometary coma' associated with it, even when at a distance of 15 AU from the Sun. Whether Kuiper Belt objects, at 30 AU and more, can undergo routine cometary outgassing is not yet known. However, it is likely that some cometary outbursts might occur, related to impacts. If an impact of a much smaller body onto the Kuiper Belt object excavated some subsurface ices, then some of those ices would presumably be sublimated (due to the input of energy from the impact) producing some transient tenuous atmosphere. This scenario might well account for some of the diversity in colours (and reflectance spectra) seen in the Kuiper Belt object population. If a transient atmosphere recondensed on the surface, it might produce a thin *frost*. Although the frost layer would be thin, it could be enough to change the appearance (i.e. reflectance) of the body. Hence some of the diversity we see in the Kuiper Belt object population *might* be due to sporadic impact events.

It is the fact that cometary material represents relatively unaltered accreted material from the solar nebula that makes comets so scientifically interesting. Because they formed in the colder regions of the Solar System, they have not lost their volatile components (i.e. the ices), and as most of the comets are relatively small, they will not have been greatly affected by gravitational compression. Comets do indeed offer an excellent opportunity to sample pristine relatively unprocessed material reflecting the composition of the solar nebula. Space missions that will land on comets and sample the cometary material directly (for example the European Space Agency's Rosetta mission) will return valuable data on the precise composition of the cometary material, and thus give us an insight about the solar nebula. These data will also complement the studies of meteorites (which you will be looking at in detail in Chapter 9).

7.5.1 Comet orbits

In terms of the orbits, there are two distinct groups of comets. The first group are the **long-period comets**. These comets come from the Oort cloud and thus can have semimajor axes of hundreds or even thousands of AU. As the Oort cloud surrounds the Solar System in a spherical distribution, the comets can enter the Solar System from any direction, and so the orbital inclinations can take any value. Thus we see as many retrograde long-period comets as prograde ones. Furthermore, the large values of semimajor axis mean that the comets have very long orbital periods (hence the name) and so it is unlikely that we will have observed a long-period comet before, at its previous perihelion passage. Thus long-period comets appear unannounced – we cannot predict when and from what direction they will come.

The second group of comets have orbits which are confined to the planetary system, i.e. we might consider this group as having semimajor axes less than that of Neptune (which corresponds to an orbital period of less than 200 years). These **short-period comets** generally have orbits with relatively low inclinations (i.e. they are mostly prograde orbits), although they can often have quite large eccentricities (for example, a value of $e = 0.6$ might be typical). Most of the short-period comets are thought to come not from the Oort cloud but from the Kuiper Belt. Some bodies in the Kuiper Belt will slowly evolve inwards, to Centaur orbits (as discussed in Section 7.3), from where they might evolve into short-period comet orbits under the influence of Jupiter (mainly) and Saturn. Thus orbital evolution provides a mechanism for replenishment of the short-period comet population. It is evident that some mechanism of replenishment *must* exist, because short-period comets will be destroyed, either by collision with the planets or by sublimating themselves to nothing, over relatively short timescales (relative to the age of the Solar System). Figure 7.35 shows inclination as a function of orbital period, for the known comets with periods less than 100 years. The majority of these comets have prograde orbits, reflecting the fact that the objects in the Kuiper Belt also have prograde orbits. The few comets that are retrograde are likely to have been long-period comets that were gravitationally captured into short-period orbits by a close approach to Jupiter.

The exact future position of a comet in its orbit at a given time (and indeed the comet's orbit itself) is often very hard to predict accurately. As comets undergo outgassing as they approach their perihelion, the emission of gas and dust applies a small force to the nucleus (like a small rocket). These are referred to as *non-gravitational forces,* and are enough to alter the comet's orbit very slightly. Because we cannot know exactly the orientation of the cometary emission on the surface, we cannot easily predict whether the non-gravitational forces might act to slightly speed the comet up in its orbit or slow it down. Thus there is an inherent uncertainty in predicting *precisely* the behaviour of an active comet's orbit in the future. This may affect our prediction of the exact time of the next perihelion passage, to the level of days or even weeks.

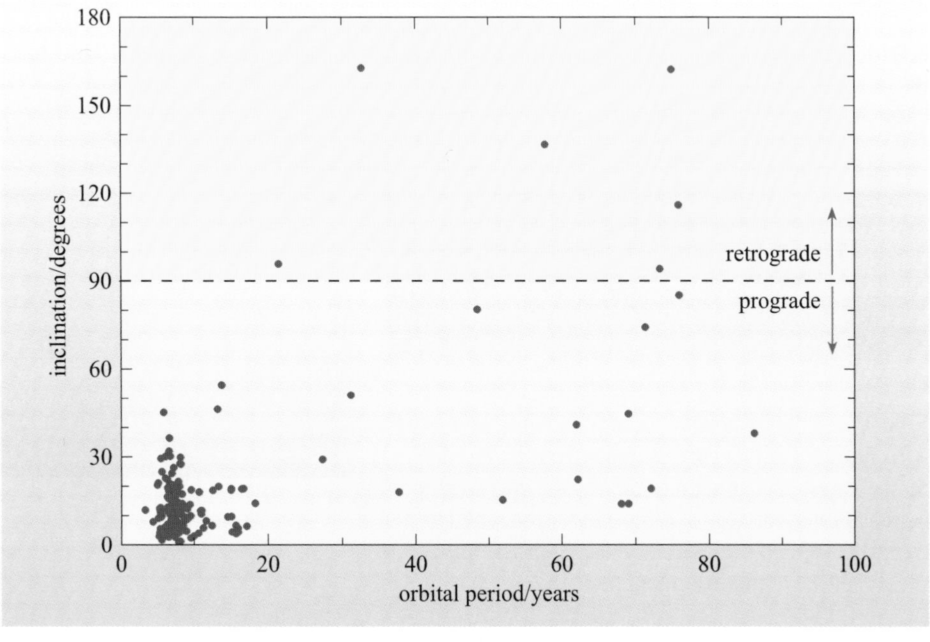

You may find that comets with periods between 20 years and 200 years are sometimes referred to as *intermediate-*period comets.

Figure 7.35 The distribution of inclination for short-period comets. Most have prograde orbits, i.e. inclination less than 90°.

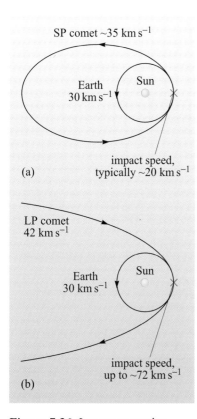

Figure 7.36 Impact scenarios between two different comet orbits and the Earth. (a) A low-inclination short-period comet orbit produces a relatively low impact speed, whereas (b) this possible long-period comet orbit produces a much higher impact speed.

As was discussed in Section 7.1, the potential impact speed of a minor body colliding with a planet crucially depends on its orbit. Thus, the possible impact scenarios involving long- and short-period comets can, on average, be somewhat different. As the majority of short-period comets have prograde orbits (like the planets), then impact speeds are somewhat limited. Figure 7.36a shows what might be regarded as a typical impact scenario between Earth and a low-inclination short-period comet. Earth orbits the Sun at around 30 km s^{-1}. If the impact is near the comet's perihelion, then the comet will be travelling somewhat faster than this, perhaps 35 to 40 km s^{-1} depending on the exact orbit. The collision is akin to cars travelling the same way on a road, coming together side by side, i.e. the impact speed is considerably less than the speeds of the individual cars. In this case the comet impact speed might be in the region of 15 to 20 km s^{-1}. However, if we consider an impact of a long-period comet, the comet orbit could well be retrograde (remember half the orbits are) and so a possible impact scenario can occur as shown in Figure 7.36b. The long-period comet can be travelling up to 42 km s^{-1} if its perihelion is near Earth's orbit, and thus a head-on collision with Earth (travelling at 30 km s^{-1}) can occur resulting in an impact velocity of up to 72 km s^{-1}.

One comet impact, which was well observed by scientists on Earth, happened in 1994. A comet, named Shoemaker–Levy 9, had been captured by the gravitational pull of Jupiter after a very close approach to the planet in 1992. As the comet effectively orbited Jupiter, it underwent tidal stresses. The tidal stresses were sufficient to split up the cometary nucleus into many fragments (the fact that this could happen gives an indication of the relatively low material strength of the cometary material). The fragments then collided with Jupiter in 1994. Figure 7.37a shows the Shoemaker–Levy 9 fragments before the impact, and the Figure 7.37b shows the 'scars' in Jupiter's atmosphere after several of the fragment impacts. We should hope that impacts like these do not happen in the near future on Earth!

QUESTION 7.12

It is clear that the inner Solar System is visited by comets from the outer Solar System (i.e. the Kuiper Belt or the Oort cloud), and so these comets could impact the Earth and other terrestrial planets. Why might the presence of Jupiter result in the number of comets impacting the Earth being lower than otherwise might be expected?

(a)

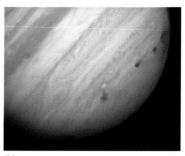

(b)

Figure 7.37 The impact of comet Shoemaker–Levy 9 with Jupiter. (a) The comet, imaged by the Hubble Space Telescope, is seen to have split into fragments which spread out along the comet's orbit. (b) The 'scars' in Jupiter's atmosphere left after the impact of several of the comet fragments. (NASA)

7.6 Interplanetary dust

You will now be familiar with the term 'dust' that we use to describe small particles of solid material. The term **interplanetary dust** literally means dust that is found between the planets. The Solar System is actually a rather dusty place, and in this section we'll look at some of the sources of the dust, and how it is distributed.

We have seen in the previous section, that as a consequence of cometary emission, the dust particles locked within the cometary nucleus can be ejected into space. The dust particles thus become free of the parent body, and become members of what is often called the **interplanetary dust cloud** (a term describing the collection of dust particles in the inner Solar System).

Comets are undoubtedly one of the most important sources of interplanetary dust, but cometary emission is not the only source of dust. Impacts on the surfaces of minor bodies and other planetary bodies (or indeed complete fragmentation of bodies due to impacts), will eject small fragments and dust particles into space. At this point, it might be useful to define the size regime that we are discussing when using terms such as 'dust' or 'fragments'. Although the division is vague, we usually describe particles that are less than about 1 mm in diameter as 'dust'. Larger particles are often called **meteoroids**, although objects metres, or tens of metres across might be better described as (for example) 'asteroidal fragments' (or indeed small asteroids).

Figure 7.38 An interplanetary dust particle, thought to be from a comet. This sample was collected in the stratosphere by a high-altitude aircraft, and is shown in this electron microscope image. The dust particle is very porous. When in the comet, the voids probably would have had ices in them, but these will have sublimated when heated by the Sun. (NASA)

■ If impact events are a source of dust, where in the inner Solar System might we expect impacts to occur relatively frequently, so producing a relatively large amount of dust?

❑ We would expect most impacts to occur in a region that has a high concentration of minor bodies, e.g. the asteroid belt.

The asteroid belt is indeed a major source of interplanetary dust, possibly as important as the production of dust from comets. However other sources of dust also contribute. All impact events will generate dust. For example, impacts on the satellites of a giant planet will produce dust particles that may predominantly stay within the giant planet's system (think of the rings of Saturn, which consist of solid fragments that themselves must be generating impact related dust). Impacts on the Moon must eject lunar dust particles, some of which might subsequently collide with the Earth. The volcanoes on Io eject solid material into the Jupiter system. Indeed, experiments designed to be sensitive to dust impacts have detected these tiny (about 0.01 μm across) volcanic dust particles streaming away from Jupiter, highly influenced by Jupiter's magnetic field. In Section 7.4, we considered impacts within the Kuiper Belt possibly ejecting subsurface icy material. It follows that dust particles, meteoroids and even larger fragments must occupy the Kuiper Belt.

The composition and structure of the dust particles depend on the source of the dust. For example, impact fragments from the surface of a rocky asteroid will be solid 'chunks' of the asteroidal material, i.e. rock. Cometary dust however has a more complicated structure. The dust in comets originally came from agglomerations of the tiny silicate dust particles present in the solar nebula. Larger particles can be a collection of smaller dust particles stuck together, often with ices in the voids. Figure 7.38 shows an interplanetary dust particle thought to come from a comet. The structure is very porous, the tiny (submicron) particles acting as the 'building blocks' of the larger particle are clearly seen.

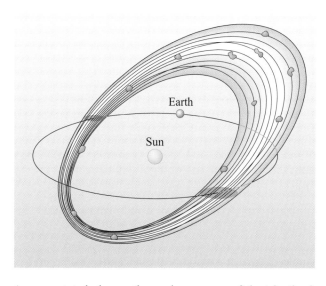

Figure 7.39 A meteoroid stream produced by the ejection of dust particles from a comet. The orbits of the dust particles are reasonably similar to the parent comet, producing a tube of dust in space, which is thickest at the aphelion region.

As was stated above, the main sources of dust in the inner Solar System are comets and the asteroid belt. Let us consider the emission of cometary dust in more detail. When dust particles leave the surface of the comet, they are then on their own. They orbit the Sun on their own orbital paths, (i.e. on Keplerian orbits) and have little or no interaction with other dust particles and the parent nucleus (the gravitational influence of the nucleus being very small). However, the orbits that the dust particles take are, perhaps not surprisingly, very similar to the parent comet's orbit. This means that around the comet's orbit, there is a multitude of dust particles whose orbits form a sort of tube. This tube of dust particles is called a **meteoroid stream**, and one such stream is represented in Figure 7.39. The slight differences in the orbits of the various dust particles (i.e. differences in the semimajor axes) mean that the dust particles have a range of orbital periods. Thus quite quickly (a matter of a few tens of orbits) the dust particles become uniformly distributed around the meteoroid stream.

All active comets will produce meteoroid streams. However in most cases, observing the stream directly can be virtually impossible. The exception is when the orbit of the Earth happens to pass *through* a meteoroid stream. Once a year the Earth passes through the stream, as the stream passes through the ecliptic plane. In fact, for a few streams, the Earth could pass through the stream twice per year, 6 months apart (this is the scenario shown in Figure 7.39). When the Earth passes through a stream, the meteoroids in the stream will collide with the upper atmosphere of the Earth. As a particle hits the atmosphere, friction between the meteoroid and air heats both air and the meteoroid (which usually completely vaporizes due to the heating). As the meteoroid passes into the atmosphere (at an altitude of around 100 km) the atoms of the air close to the path of the meteoroid are excited (heated) and emit light, producing a short-lived 'streak' of light. This streak of light is called a **meteor**, or a 'shooting star', and when many meteors are observed associated with the same meteoroid stream, the display is called a **meteor shower**. The typical size of a meteoroid that gives rise to a meteor visible with the naked eye is about the size of a grain of sand. There are at least 50 meteor showers throughout the year that have been identified, although only a few produce sufficient meteors to be particularly noteworthy. Table 7.3 lists the major showers seen in the Northern Hemisphere, and the likely hourly rate of meteors that an observer would see (observing in ideal conditions). Significant activity can occur for a number of days around the date of maximum rate. Also listed are the parent bodies associated with the shower.

Table 7.3 Major annual meteor showers. The name of the shower derives from the constellation where the 'radiant' lies. The radiant is the position in the sky that the meteors appear from.

Date of maximum rate	Name of shower	Hourly meteor rate	Parent body
Jan 3	Quadrantids (Bootids)	130	Unknown
Aug 12	Perseids	80	Swift–Tuttle
Oct 21	Orionids	25	Halley
Nov 17	Leonids	25[a]	Tempel–Tuttle
Dec 13	Geminids	90	(3200) Phaethon[b]

[a]This rate is usually what is observed, but every 33 years or so, this shower can display much higher rates (see text).

[b]When discovered, (3200) Phaethon was assumed to be an asteroid as no cometary coma was observed. However it is likely that there has been some activity in the past.

Occasionally, the Earth passes through a particularly well populated part of a meteoroid stream, and the rate of meteors seen in a shower can increase significantly. For several years, between 1997 and 2002, this was the case for the Leonid meteor shower in November, where a peak hourly rate of up to several thousand was seen. An all-sky image of the Leonids shower in 1998 is shown in Figure 7.40. If you have never seen a meteor, then we would highly recommend trying to view one of the showers listed in Table 7.3, on or within a few days of the dates given. However, on any clear night you will see meteors. These meteors are not associated with an identified shower, but are just the random background of dust particles entering the Earth's atmosphere (often called **sporadic meteors**). On average you will typically see a meteor every 15 minutes or so.

QUESTION 7.13

If you observed sporadic meteors throughout the night, you would notice that the rate increases significantly *after* midnight. Why might this be the case?

As you found with comet impacts in Section 7.5.1, meteoroids can collide with Earth at high speeds – depending on the orbits involved, speeds of between around 11 km s^{-1} and 70 km s^{-1} are possible. Imagine if a particle travelling at these speeds hit a spacecraft orbiting the Earth. The large kinetic energy involved means that even a relatively small meteoroid (for example 1 mm or so in size) could do a considerable amount of damage. Thankfully the risk of a major meteoroid impact is low, although evidence of small impacts is clearly seen on surfaces that have spent time in space and have then been retrieved. Figure 7.41 shows the result of a particle travelling at high speed hitting a solar array.

QUESTION 7.14

The kinetic energy of a body of mass m, travelling at a velocity, v, is given by $\frac{1}{2}mv^2$. Calculate and compare, the impact energies (i.e. the kinetic energies) of a small car travelling at 50 mph, and a meteoroid of mass 1 g travelling at 20 km s^{-1}. (Assume the car has a mass 1000 kg, and note that 50 mph is approximately 20 m s^{-1}.)

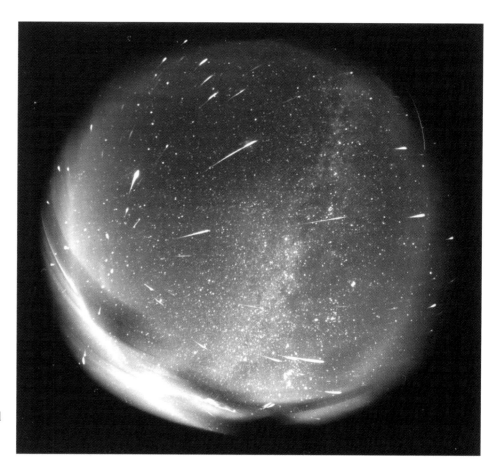

Figure 7.40 An all-sky image taken during the 1998 Leonid meteor shower, when the peak rate of meteors reached several hundred per hour. (Juraj Toth, Modra Observatory)

We have discussed various dust sources, such as dust in meteoroid streams, and dust in the asteroid belt, but dust particles are not confined to these well-defined regions. This is because of several mechanisms that can cause the orbits of dust particles to change significantly, so transporting them to different regions of the Solar System. The first of these mechanisms is the orbital evolution that you have met for asteroids and other minor bodies. Meteoroids undergo exactly the same effects, and so for example, asteroid fragments could be slowly transported from the asteroid belt to Near Earth Asteroid type orbits, in exactly the same way as we described for the Near Earth Asteroid population in Section 7.2.

The second important mechanism is that of **radiation pressure**. This effect arises because photons of light have momentum. When you shine light on an object, the photons can transfer their momentum to the object, in a way analogous to throwing heavy balls at an object in order to try and make it move. The force exerted by light in this manner is actually very small, but is most effective for very small objects. This is because the amount of light intercepted by the particle is directly related to its cross-sectional area, whereas the effect on the particle (the acceleration resulting from the radiation force) is inversely proportional to the particle's mass. Thus the radiation pressure has a maximum effect when the ratio of cross-sectional area to particle mass is maximized – and this ratio increases as the particle size decreases. However, the light ceases to effectively interact with particles that are very much smaller than the wavelength of light (typically 0.5 μm). Thus the effect of radiation pressure is a maximum for particles in the approximate diameter range 0.1 μm to 1μm. For these small dust particles, the force exerted by solar photons (a force which acts in a direction radially outwards from the Sun) can be larger than the

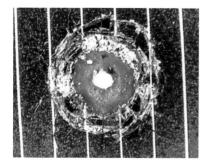

Figure 7.41 An impact crater found on the solar arrays of an Earth-orbiting satellite. The vertical lines are about 1 mm apart. The crater could have been made by a particle just 100 μm across.

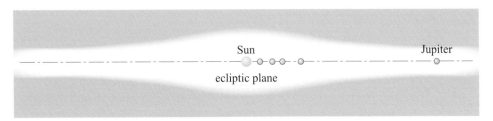

Figure 7.42 A diagram indicating the interplanetary dust cloud. The concentration of dust increases towards the Sun, as particles slowly spiral in towards the Sun due to the Poynting–Robertson effect.

force of gravity. The result is that these small particles, as soon as they are produced in an impact event, or emitted from a comet, can be pushed out of the Solar System, a bit like being blown away in a wind. (This is why the comet dust tail appears to 'flow' away from the Sun, see Figure 7.30a.)

A third important effect for particles larger than about 1 μm (diameter) is called the **Poynting–Robertson effect**. This is also related to radiation pressure, but has almost opposite results. As photons travelling radially outwards from the Sun hit a dust particle, one might expect the force to also act radially outwards. However, as the dust particle has some transverse motion (i.e. perpendicular to the motion of the photon), the resulting impact of the photon appears to come slightly from the side. This acts to slow down the dust particle slightly in its orbit (like a braking effect) which in turn causes the semimajor axis of the orbit to decrease. Thus over many orbits, the dust particle gradually spirals into the inner Solar System. Many of these particles reach the near-Sun region, where they are vaporized due to the intense heat. The Poynting–Robertson effect thus supplies a mechanism whereby dust in the asteroid belt slowly spirals in towards the orbit of Earth and the other terrestrial planets. In fact many of the meteors that are not associated with specific cometary meteoroid streams (i.e. the sporadic meteors), are caused not by cometary dust, but by small dust particles that originally came from asteroids in the asteroid belt. This mechanism also affects larger objects such as centimetre-sized fragments, although it takes longer for these larger objects to reach the orbit of the Earth. A 10 μm sized particle would take only about 10 thousand years or so to spiral in from the asteroid belt, whereas a 1 cm object would take over 10 million years.

These radiation pressure and orbital evolution effects ensure that dust particles slowly move away from the region of their source. This results in the Solar System being filled by a dust cloud (i.e. the interplanetary dust complex mentioned above), an impression of which is given in Figure 7.42. This dust cloud is even sometimes visible with the naked eye. About an hour after sunset (or an hour before sunrise) if the sky is clear and you are observing in a good dark environment, you might just be able to see a faint glow extending up from the horizon nearest the Sun. This is referred to as the **zodiacal light** (Figure 7.43), and is due to the dust particles in the interplanetary dust cloud reflecting and scattering sunlight, just like particles in a cometary coma.

Figure 7.43 A photograph, taken after sunset in 1997, of the zodiacal light (which extends up from the centre horizon) which is due to sunlight being scattered from the dust particles in the interplanetary dust cloud. Also seen in the photograph is comet Hale–Bopp. (J. C. Casado)

7.7 Summary of Chapter 7

- The motion of the planets under the influence of the Sun's gravity is described by Kepler's laws. (i) Planets move in elliptical orbits, with the Sun at one focus, (ii) A line connecting the Sun to a planet would sweep out equal areas in equal times, (iii) The square of a planet's orbital period is proportional to the cube of its semimajor axis.

- Kepler's laws also apply to a satellite orbiting a planet (or any body orbiting another large body under the influence of gravity).

- The size and shape of an orbit are described by the parameters, semimajor axis (a), eccentricity (e). The orientation of the orbit plane with respect to the ecliptic plane is described by the inclination (i).

- Asteroids are rocky and metallic minor bodies, most of which reside in the asteroid belt between Mars and Jupiter. Asteroids that have orbits that can come close to Earth are called *Near Earth Asteroids*.

- Asteroid diameters have a *size distribution*, quite similar to the size distributions of impact craters (Chapter 4). Many planetary impact craters were produced by the impact of an asteroid.

- Asteroids are given different taxonomic classes, which are related to their composition and determined by astronomical observations.

- A large belt of objects, called the Kuiper Belt, lies beyond the orbit of Neptune. Kuiper Belt objects are thought to be quite icy in nature. The planet Pluto represents the largest known member of the Kuiper Belt.

- Comets are icy bodies formed in the outer Solar System; leftovers from the planetary accretion process. They thus have very old, relatively unprocessed material within them.

- Comets eject gases (due to sublimation of the ices) and small rocky dust particles that give rise to meteoroid streams and interplanetary dust.

- Minor bodies in the Solar System undergo orbital evolution, such that over a long time (thousands and millions of years) the orbits are changed significantly.

- Orbital evolution offers a mechanism by which objects in the Kuiper Belt can slowly evolve into the inner Solar System, so renewing the short-period comet population. Similarly, Near Earth Asteroids come from the asteroid belt by the action of orbital evolution.

CHAPTER 8
THE ORIGIN OF THE SOLAR SYSTEM

8.1 Introduction

In the beginning there was a formless void of emptiness known as chaos; from this darkness emerged a black bird known as Nyx (the goddess of night). Eventually the bird laid a golden egg, out of which was born Eros, the god of love. The shell of the egg broke into pieces, one of which rose into the air and became the sky (which Eros called Uranus) and the other became the Earth (called Gaia).

This is one version of the Greek creation myth. It considers that we started with 'nothing' and evolved fairly rapidly towards the environment which we experience today. In fact, this is a feature of nearly all creation myths – the Sun, the Earth, its inhabitants, and by inference, the planetary system around us, all formed soon after a divine event had acted to add purpose to the pre-existing nothingness, or chaos.

The details of the traditional scientific view are somewhat different. The Universe was created about 13 Ga ago, in the *Big Bang* (the exact age is unclear although somewhere between 10 and 20 Ga is the current consensus). Clumps of material then formed into galaxies, and galaxies spawned stars. From that time until the present day, the cycle of stellar birth and death has continued remorselessly. Our own Solar System formed 4.56 Ga ago from materials that had been cycled in and out of stars several times (see Box 8.1).

There have been (and still are) many different theories of how our Solar System formed. A fundamental difference in approach, considers whether the processes that formed the Sun and the planets took place simultaneously in a single integrated event, or whether the planetary system was added to a pre-existing Sun, some time after the Sun's formation. These two approaches are referred to as *monistic* (single event) and *dualistic* (two separate events). An example of a dualistic theory of Solar System formation, would be the theory that another star passed close to the Sun, causing matter to be pulled from the Sun into a single filament, which then broke up along its length to form individual planets. (Indeed this theory was proposed in 1917.)

BOX 8.1 CYCLING OF STELLAR MATERIAL

The fact that Earth and all its constituent components (including life forms) are made largely of materials that have been produced in stars (which, necessarily existed before the Sun) may be somewhat surprising! In brief, except for hydrogen and helium, all the elements are made in stars. Now, we know that the Sun is about 4.56 Ga old, and it is predicted that it will last for a further 5 Ga or so. In other words, stars that are about the same mass as the Sun last for about 10 Ga. Smaller stars last longer, and larger stars (where most of the heavy elements are made) have shorter lifetimes. However, the relationship between size and lifetime is not linear. A star that has 3 times more mass than the Sun will have a lifetime of 0.5 Ga. One that is 25 times more massive will last for less than 10 Ma. It should thus be apparent, that in the billions of years before our Solar System formed, many generations of stars would have come and gone. The debris from these previous generations of stars is what the Solar System is made of.

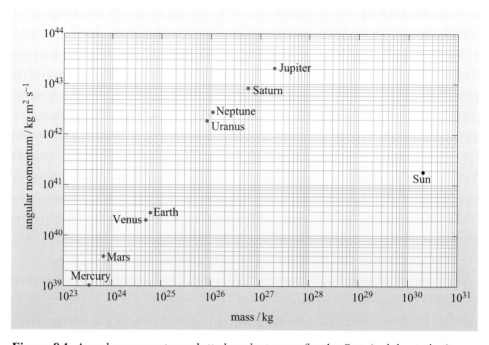

Figure 8.1 Angular momentum plotted against mass for the Sun (axial rotation) and the planets (orbital motion). The contribution to orbital angular momentum from Pluto and the satellites of the planets, and the contribution from axial rotation of the planets and satellites, are too small to show on this plot. Note that both axes are logarithmic. (The nearly straight-line relationship for the planets, between angular momentum and mass, is because angular momentum is directly proportional to mass, but (as a consequence of Kepler's third law, and the dependence of the planet's orbital velocity on the orbital radius) it depends only weakly on orbital radius (in fact on its square root). If you very carefully draw a line at 45° passing through the data point for Mercury, you will find that the data points for the other planets fall progressively further from this line as orbital radius increases.)

It is fair to say that, currently, a monistic view of Solar System formation is supported by most scientists. It should be understood that the details of the process are still open to interpretation, and are matters of research and great debate. There is no one single 'this is exactly how it happened' theory. However, in this chapter we will discuss an overall formation process that best fits the data known today, pointing out where details are uncertain as appropriate.

By the close of the 18th century, the French mathematician Pierre-Simon Laplace had derived the classical monistic model of Solar System formation – an idea that was eventually to be called the **nebula hypothesis**. Laplace demonstrated that a rotating sphere of gas eventually flattens into a spinning disc. As matter begins to contract towards the centre (matter that will eventually form the **protoSun**), the disc starts to spin faster, due to conservation of angular momentum. (See Box 5.7.) Much of the material in the disc eventually goes on to accrete into the planets (the details of which we will discuss throughout this chapter). This is the overall process you looked at in Chapters 1 and 2.

However, this hypothesis was not always accepted. One of the major problems that brought the nebula theory into question, involved the disparity between the present-day distributions of mass and angular momentum in the Solar System. This is normally referred to as 'the angular momentum problem'. Put simply, while only about 0.14% of the total mass of the Solar System is contained in the planets, they nonetheless account for more than 99% of the angular momentum of the system as a whole (i.e. within their orbital motions). In contrast, any simple view of the formation of the Solar System from a nebula would have at the centre, a body containing most of the mass and correspondingly most of the angular momentum. Thus we might expect the Sun to be rotating far more rapidly than it actually is. (The Sun currently rotates on its axis once every 25.5 days; this is entirely typical for stars that are in other ways similar to the Sun.)

The example that we have used before to illustrate angular momentum is that of an ice skater performing a spin. With arms outstretched the skater rotates at one speed, but as he brings his arms inwards towards his body (thereby concentrating mass, like an accreting Sun), he rotates more rapidly (i.e. conserving angular momentum). However, this is not what is observed in the Solar System. The magnitude of the deficit is shown in Figure 8.1.

Let us now look at the formation process in more detail, and see how the nebula theory does, in the end, 'solve' the angular momentum problem, and does indeed form a good model of the origin of our planetary system.

8.2 Physical formation processes

8.2.1 The beginnings of a solar system

In this section we shall examine how the Solar System may have formed, about 4.56 Ga ago. You may recall that you encountered this same value in Section 2.4.3 where it was quoted as the age of the Earth. In fact, the age itself is derived from meteorites which, being small bodies that cooled quickly after their formation, effectively document the age of the Solar System, thus

the age of meteorites = the age of the Solar System = the age of the Earth

We shall start our enquiry before the Solar System formed and then follow the story from the time that the dust and gas cloud around the protoSun began forming into planets. Some very general sense of these processes has already been given in Chapters 1 and 2. Here we supply more detail.

In order to understand the details of how a planetary system forms it will be necessary to understand a few astronomical principles, and to be aware of certain telescopic observations of objects beyond the Solar System. After all, the Sun is merely a star, of which there are some 10^{11} in our Galaxy, the Milky Way. Ideas about how stars form come from observations of stellar objects and their precursors – since there are so many stars potentially available for study, it is inevitable that we will be able to see examples of objects at all stages of evolution, from birth, to quiescent normal life, to terminal decline, and ultimately death.

Stars are born out of clouds of molecular gas and dust (see Figure 8.2) and they come in a complete range of sizes from bodies of mass about one-twelfth times the mass of the Sun (that is $0.08M_\odot$) up to giants of about $90M_\odot$. However, for our purposes we are only really interested in stars that are about the same size as the Sun (a range of, say, 1 to $3M_\odot$). The reason for this is that stars which are very small are unlikely to support planetary systems, while stars that are relatively large have short lifetimes and, furthermore, above $8M_\odot$ are destined to become supernovae, i.e. stars that will undergo a violent explosive death. While stars in their own right are interesting things to study, when considering the origin and evolution of our own Solar System it makes sense for us to concentrate on the workings of stars that are of a similar size to the Sun. In this regard you should note that the Sun is quite average in terms of its size and other characteristics, such as chemical composition, luminosity, age, etc.

Figure 8.2 Spectacular HST image of a star-forming region in the Cone Nebula, which is in the constellation of Monoceros (a distance of 2500 light-years away, i.e. 2.4×10^{16} km). The image shows the upper 2.4×10^{13} km of the nebula. (Space Telescope Science Institute)

■ Since we will be making a number of references to the masses of astronomical objects, check that you are happy with the notion of relating such things to the mass of the Sun. For instance, what is the mass of the Earth expressed in solar masses?

❑ The Earth has a mass of 5.97×10^{24} kg, while that of the Sun is 1.99×10^{30} kg. Hence, the Earth mass, relative to that of the Sun is $(5.97 \times 10^{24})/(1.99 \times 10^{30}) = 3 \times 10^{-6} M_{\odot}$.

By a similar calculation, Jupiter has a mass of about $0.001 M_{\odot}$.

As the process of starbirth advances, any gas/dust that remains gravitationally bound to a young stellar object, while not actually falling onto the star itself, ends up as a swirl of material around the star (known as a **circumstellar disc**). It is within discs like these that planets may be formed. Some of the details of this process will be described in the next section. For now, it is appropriate to marvel at some truly remarkable images obtained using the Hubble Space Telescope (HST) which show discs of material around young stars (see Figure 8.3). The features themselves are revealed by infrared (IR) observations detecting thermal radiation from dust grains that have been heated by absorption of shorter wavelength radiation from the star at the centre of each disc.

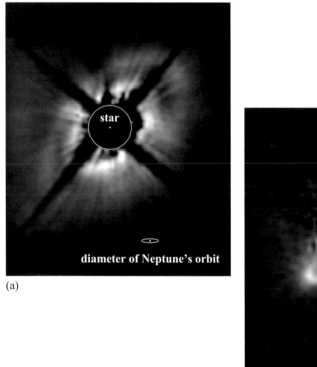

(a)

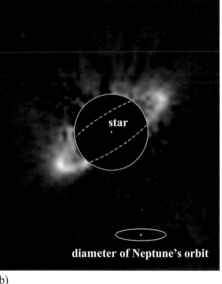

(b)

Figure 8.3 HST images of dusty circumstellar discs around two young stars. (a) HD 141569. (b) HR 4796A. The central dot in each image represents the position of the star; the circles mark the positions of a physical disc positioned so as to restrict the starlight entry to the telescope. The diameter of Neptune's orbit is about 9×10^9 km. (Space Telescope Science Institute)

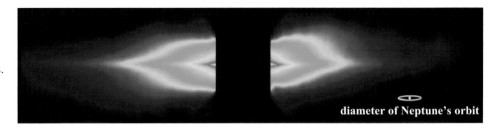

Figure 8.4 HST image of β Pictoris. The diameter of Neptune's orbit is about 9×10^9 km. (Space Telescope Science Institute)

> ■ Contrast the position of the ring around HR 4796A (Figure 8.3b) with the known dimensions of our own Solar System.
>
> ❑ Neptune orbits at 30 AU, while the Kuiper Belt extends to around (at least) 50 AU. It would appear that HR 4796A is of a similar size to our own Solar System.

The mass of dust in discs around young stars, like those shown in Figure 8.3, ranges from $0.001M_\odot$ to $0.1M_\odot$, which is in accord with theories of Solar System formation, as will be described shortly. In detail, extremely tenuous discs extend out to hundreds of AU. That observed around the star known as β Pictoris is believed to contain as little as $10^{-7}M_\odot$ of material (Figure 8.4). This particular star is interesting for many reasons, not least because it is the first ever object that was observed to have a dusty disc around it. At an age of about 20 Ma, it is likely that β Pictoris is at a more advanced stage of evolution than the stars considered previously, and it is likely that a planetary system has already formed.

QUESTION 8.1

In total extent the β Pictoris circumstellar dust disc is 1500 AU across. Light from the star prevents the disc being seen closer in than about 66 AU, but IR observations indicate that its inner edge lies about 20 AU from the star. How well does the suggested extent of this disc correspond with the distribution of planets and other bodies in our own Solar System?

In addition to observable circumstellar discs there is now a substantial body of evidence for planetary systems around stars obtained from observational methods that target the stars themselves. For instance, an apparent wobble in the position of a particular star is taken to imply the presence of a large planet exerting a gravitational pull as it orbits the star. Alternatively, changes in the apparent brightness of a star can be used to infer the passage of a planet between the star and the observer. So far, the kinds of planets determined by these methods are thought to be Jupiter-sized. But be aware that they are seen to orbit their parent star very rapidly (in some cases a matter of days rather than the 11.86 year orbital period of Jupiter). This necessarily means that these extra-solar planets must be close to their respective stars, which has earned them the name 'hot jupiters'. At the time of writing about 100 such planets have been discovered, but progress in this area has been rapid and the number is likely to continue to rise. While these are fascinating discoveries and have attracted much interest from astronomers, they are somewhat less relevant to planetary scientists who ultimately would like to find evidence of systems like our own, i.e. with a range of planetary types from Earth-like to gas giants. However, one extremely important outcome from the discovery of hot jupiters is the realization that in planetary systems as a whole, the orbits of the constituent bodies are subject to change during their lifetimes. This has to be true in the case of the extra-solar planets because there seems to be no logical way to form such large bodies as close to their stars as they appear to be. This finding has alerted people studying the Solar System to the prospect of *planetary migrations* (i.e. changes in orbital distances) affecting our own system.

8.2.2 From dust to protostars

You will now examine some of the details of the formation of **solar systems**, starting with a closer look at the origin of stars. Direct observations of circumstellar discs, achieved with increasing clarity since the early 1980s, support the theory of planetary-system formation proposed originally by Laplace (as was introduced in Chapter 1, Section 1.2). It was suggested by Laplace that the Solar System was formed by the gravitational collapse of a large, initially spherical, rotating mass. According to this theory, the central region grew denser and became the Sun, and the remainder was forced into a disc of gas and dust, called the Solar Nebula, within which the planets formed. This nebula theory has been with us ever since.

Even though most scientists believe that the Solar System formed from a nebula, there is no general agreement as to how it happened. There are many versions of the nebula theory, but it will not be possible to examine them all here. Instead we will present you in this section with a discussion that fits most of the available evidence, and is at least moderately self-consistent. You should *not* regard it as representing a consensus view in all its details, and we will encourage you to be aware of difficulties and uncertainties when they crop up.

Before we consider what happens to the cloud of gas and dust that forms the Solar Nebula it is appropriate to ask how such an entity arose. To tackle this question we need to consider material beyond the Solar System. When we study the distribution of stars within the Milky Way, we find that it is far from even. You can easily see this in Figure 8.5, which shows a panoramic view of the sky. Apart from the general 'splash' of stars across the sky indicating the position of our own Galaxy, the Milky Way, there are, even within the Milky Way, many regions where stars seem to be missing.

Figure 8.5 Full-sky view of the stars that constitute our own Galaxy – the Milky Way. This image captures almost half of the Milky Way and was taken at a time when the Galactic Centre was overhead (although it cannot be seen in visible light because of the obscuring effects of dust and gas in the plane of the disc). (Copyright Axel Mellinger)

Figure 8.6 View of the southern skies showing the Coal Sack region as a dark patch within an otherwise brightly lit region. The Coal Sack itself is a cool dense cloud, located about 500 light-years away; it is situated such that it blocks out the light from the stars beyond it. (Akira Fujii)

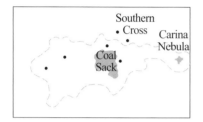

5 degrees

An example of a dark region can be seen in Figure 8.6; the so-called Coal Sack region of the Milky Way. However, we now know that this apparent lack of stars is not what it seems. These gaps arise because our view of the stars is obscured in some directions by intervening clouds of gas (consisting mainly of hydrogen) and dust (about 1% of the mass of the gas and consisting mainly of heavy elements and their compounds). These are the **dense clouds**, which are characterized by a low temperature (typically 10 K), a high density (at least by the standards of the interstellar medium, the material between stars, which in our Galaxy consists of 99% gas and 1% dust) and a rich collection of molecules. Such clouds are relatively common. Within the Milky Way there are about 2000 dense clouds in a ring-like structure about half-way between the position of our Solar System and the Galactic Centre (each cloud has an overall diameter greater than about 10^{15} km). Although the typical number density, n, of gas in the interstellar medium is of the order 10^6 particles m^{-3} (particles/m^3), in the dense clouds n varies from about 10^9 to about 10^{12} particles m^{-3}. Since gaseous hydrogen (H$_2$) is by far the most dominant constituent of these dense clouds, we could also say that n is equivalent to 10^9 to 10^{12} molecules of H$_2$ per cubic metre.

The dense clouds are large and contain a significant amount of material. They make up perhaps 45% of the total mass in the interstellar medium.

■ Assuming a density of about 10^9 particles m^{-3}, with each molecule having a mass of 3.34×10^{-27} kg (that is the mass of a hydrogen molecule), what is the overall mass of a typical dense cloud? (Assume a spherical geometry with a radius of 5×10^{17} m.)

❏ Volume of cloud $= \frac{4}{3}\pi r^3$

$= \frac{4}{3}\pi \times (5 \times 10^{17}\,\text{m})^3$

$= 5.24 \times 10^{53}\,\text{m}^3$

There are 10^9 molecules of H_2 in each cubic metre, i.e. $1 \, m_3$ contains a mass of $3.34 \times 10^{-27} \times 10^9 \, kg = 3.34 \times 10^{-18} \, kg$. So, the overall mass of the cloud would be:

$5.24 \times 10^{53} \times 3.34 \times 10^{-18} \, kg = 1.75 \times 10^{36} \, kg.$

Since the mass of the Sun is $1.99 \times 10^{30} \, kg$ we could say that the cloud had a mass of $8.75 \times 10^5 M_{\odot}$, i.e. about a million solar masses (as an order-of-magnitude estimate). You can see that 2000 clouds of this sort would contain about 2 billion $M_{\odot}$ of gaseous hydrogen.

Figure 8.7 Detailed picture of the Orion Nebula – a feature which is visible to the naked eye, although not as portrayed in the figure. The nebula is made mainly of hydrogen gas, which is glowing as a consequence of four young massive stars in the brightest region. (NASA)

Figure 8.8 The four young bright stars from the Orion Nebula, known as the Trapezium. The image shows other stars at least some of which are also within the Orion Nebula (the others are foreground stars). The region shown corresponds to the inner, bright part of the image shown in Figure 8.7. (Anglo Australian Telescope Board; photograph by David Malin)

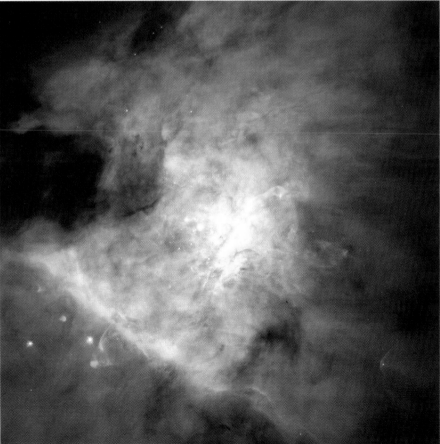

It is the dense clouds where it appears that conditions are particularly favourable for stars to form. Let's look at the evidence that supports the view that dense clouds are the place of starbirth.

Some young star clusters seem to be surrounded by the remnants of the original cloud from which they formed. Figures 8.7 and 8.8 show the Orion Nebula. It is visible to the naked eye as a haze surrounding the star θ Orionis in the sword of Orion. Observations with a telescope show this region to be one of the most visually magnificent in the sky. There is plenty of evidence to suggest that star formation took place here very recently (on the astronomical timescale).

All other evidence points to these stars being very young. What interests us here, though, are the vast clouds of glowing gas and obscuring dark clouds of dust and gas. These are almost certainly the remnants of the dense cloud from which the young stars in the nebula formed. Lit up by the intense radiation from these stars, the hydrogen in the cloud is ionized and glows in its characteristic red light.

Another strand of evidence is that some dense clouds are found to contain a large number of compact IR sources. A good example is shown in Figure 8.7, which is a closer view of the Orion Nebula. It is thought to contain some very young stars, unobservable in visible light but detectable in the IR part of the spectrum. It seems that the IR radiation comes not from the young star itself but from the cocoon of dust still surrounding it. Heated to a temperature of a few hundred kelvin by the recently formed star, the warm dust re-radiates in the IR part of the spectrum.

Theoretical models of star formation all point to regions of low temperature and high density as being the most likely sources of starbirth, i.e. the dense clouds. By way of a brief introduction to these models note that each piece of matter attracts every other piece with the force of gravity. When we are dealing with the gas and dust clouds spread through the vast expanses of the interstellar medium, we cannot ignore this force and must consider the gravitational attraction between different sections of these interstellar clouds. Sir James Jeans (1877–1946) was able to show that, under appropriate conditions, a cloud (or part of one) would start to contract under the influence of the gravitational force. He derived a simple formula for calculating the mass and size that a cloud would have to reach, as a function of its temperature and density, before gravitational contraction could start.

We have pointed out that all atoms, molecules and particles in a cloud are attracted to each other by the gravitational force. However, observations show that many clouds appear to be in a state of equilibrium – in other words, they don't seem to be in a state of gravitational contraction. The constituent particles don't all collapse into a very small volume because their continuous motion means that particles undergo random collisions, creating an outward pressure, which counteracts the tendency of the gas to contract. The basis of the approach used by Jeans was to consider the balance between the two forces (pressure and gravity). He proposed that if the force due to gravity was the greater, then gravitational contraction could occur. Using this simple criterion, Jeans was able to show that, for a given set of conditions of temperature and particle number density, there is a minimum value for the mass of a uniform spherical cloud at which the force of gravitational attraction will overcome the force due to the motion of the particles, and contraction will occur. This critical mass is known as the **Jeans mass**. A cloud which possesses the Jeans mass is said to satisfy the *Jeans criterion* for contraction.

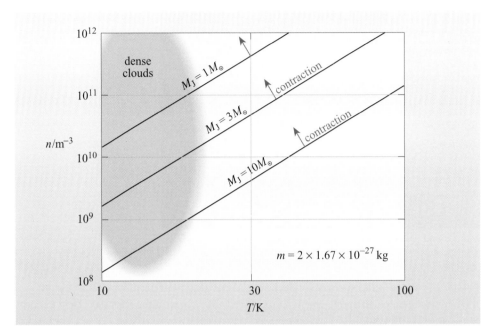

Figure 8.9 The relationship between particle number density, n, temperature, T, and the mass required in a spherical cloud for the gravitational force to cause contraction to occur. This is the Jeans mass, M_J. It is assumed that the particles are hydrogen molecules such that the molecular mass $m = 2 \times 1.67 \times 10^{-27}$ kg.

Figure 8.9 shows that dense clouds of only a few solar masses would contract. Many dense clouds are far more massive than this. Moreover, there are yet denser regions within dense clouds (called cores and clumps), with masses between about 0.3 and $10^3 M_\odot$, that can therefore satisfy the Jeans criterion on their own. Thus, gravitational contraction of dense clouds is to be expected.

The picture that has been painted so far, however, makes simple assumptions.

▪ What are the factors that could complicate this simplified approach to the gravitational contraction of a dense cloud?

▢ We wouldn't necessarily expect a typical cloud to be spherical or to have the same temperature and density throughout.

In addition, no account has been taken of rotation or of magnetic fields, which may exert an influence within the cloud thereby affecting the process of gravitational attraction. In detail these effects may be quite negligible at the start of the process, but become more important as cloud collapse proceeds. There may also be other external influences such as the compressive effects of shock waves accompanying the detonation of a nearby supernova.

Despite its over-simplifications, the Jeans approach is a good starting point for more sophisticated, and perhaps more realistic treatments of the early stages of star formation. Exactly how and why the gravitational collapse is triggered in the first place is perhaps more speculative, but there are plenty of ideas, e.g. a shock wave from a nearby supernova, passage of the cloud through one of the Galaxy's spiral arms, two clouds interacting, and so on.

Figure 8.10 In dense clouds stars form in open clusters of several hundred. This particular example, which has no name (except for its catalogue number, NGC 2264) is about 5 Ma old. The dense cloud, which gave birth to the stars, is still very much in evidence in this visible image. (Anglo Australian Telescope Board; photograph by David Malin)

We know from observations that stars are often found in groups, referred to as clusters (or open clusters). Examples can be seen in Figure 8.10. The stars in each cluster appear to have formed at about the same time. The number of stars in an individual cluster can range from 10 to 1000 (with the volume of space they occupy varying accordingly).

How is the presence of clusters consistent with the picture of gravitational contraction that we've painted so far? The answer is believed to lie in the phenomenon of *cloud fragmentation*. As an individual cloud contracts, it is thought that it starts to break up into fragments, each having the potential to continue collapsing separately from the others. It is quite possible for a breakaway fragment to satisfy the Jeans criterion in its own right and if so, it will continue to undergo gravitational contraction. In this way you can see, at least qualitatively, how a cloud initially with a mass of hundreds or even thousands of solar masses can ultimately produce a large number of small fragments, each collapsing on its own, to yield a cluster.

We have now reached the stage where our dense cloud is contracting, probably in the form of many fragments. Let's now concentrate on the development of a typical fragment. As this contracts, the gravitational energy of the particles is converted, via mutual collisions, into thermal kinetic energy, and the temperature rises. Unfortunately, the process is taking place behind a shield of gas and dust, which effectively screens the fragment from view. For this reason we are forced to fall back on theoretical calculations and computer models of this phase. These seem to show that, after only a few thousand years of gravitational contraction, the edge of the cloud has heated up to between 2000 K and 3000 K. At this stage, the chain of events has started that will lead the fragment, almost inevitably, to become a normal type of star. For this reason, we are now justified in calling the fragment a **protostar**. Virtually all fragments that ultimately become ordinary stars take less than about 10^8 years to pass through the protostar phase – rather short on the astronomical timescale.

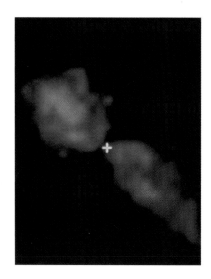

Figure 8.11 Image taken by a radio astronomy technique of the object known as IRS5 (about 520 light-years away). The outflows from a young star (marked with a cross) are clearly visible as two oppositely directed lobes – the one coloured blue is flowing towards us and the one coloured red is flowing away. These lobes together constitute bipolar outflow. (Ronald Snell)

Knowledge of the protostar phase of stellar evolution has been enhanced by observations made by radio astronomers (a technique which allows us to look through the obscuring gas and dust). It transpires that a large number of protostars show evidence of a phenomenon called bipolar outflow – that is, gas flowing at high velocities (in what is known as a *stellar wind*), typically 50 km s^{-1}, in two streams moving in opposite directions. An example is shown in Figure 8.11.

Bipolar outflows are thought to last for only a relatively short time (perhaps 10^4 years), but during this period they eject a significant amount of mass and therefore require a lot of energy to sustain them. The outflows accompany the formation of a circumstellar disc or torus, shown schematically in Figure 8.12. Another class of object that is thought to be relevant to the early stages of evolution is **T Tauri** stars, which are young (typically 1 Ma old) and in the process of losing mass through stellar winds (at speeds of up to 100 to 200 km s^{-1}). These objects can lose as much as $0.5M_\odot$ in the form of a stellar wind during the T Tauri stage. Some T Tauri stars also show evidence of thin discs of circumstellar material, which demonstrates a qualitative relationship with stars having bipolar outflow sources (although in T Tauri stars the outflow is in all directions).

Objects that have bipolar outflows, or are designated as T Tauri stars, are protostars in their final stages of birth, i.e. they are just about to start the process of nuclear fusion reactions. They are generally in the mass range of about 0.2 to $2.0M_\odot$. As a protostar evolves it begins to show a continuous rise in temperature as the protostar contracts. The critical point comes when the temperature and pressure in the centre of the protostar becomes sufficient for nuclear fusion to be triggered in the core. The energy released raises the core temperature (and pressure) sufficiently to halt contraction, and marks the conversion of the protostar into a proper star.

8.2.3 The Solar Nebula

Before we start to consider the details of processes occurring within the Solar Nebula, a word of caution is in order. In the following sections we will deal with topics such as condensation, coagulation, planetesimal formation, planetary embryos and planet growth (as were considered in Chapter 2). While in the most general sense, these processes are part of a time sequence which moves in one direction (from the collapse of the gas–dust cloud towards the formation of planets) it is important to realize that they do not take place one after another. There was not an epoch of condensation, followed by one of coagulation, and so on. The Solar Nebula was a very complicated environment and, indeed, one that we do not yet fully understand. It is likely that small clumps of rocky materials (the building blocks of planets) were forming from dust in certain parts of the nebula, while at the same time there were solid materials condensing from the gas phase in other parts. Furthermore, materials were being moved around by phenomena such as stellar winds and the general turbulent motions of particles in the dust cloud itself. Impacts would have acted to transform clumps back into dust (and gas), and in some regions, dust produced within the Solar Nebula was mixing with the unaltered pre-solar dust which had coalesced in the first place. This pre-solar dust would have had any number of origins – as condensates from other stars, materials formed and processed in the interstellar medium, supernova debris, and so on. That such materials were present is not idle speculation – we have examples of pre-solar grains preserved in meteorites, which represent the fossilized remains of Solar Nebula processes (see Chapter 9).

Meteorites help us to constrain the timescales of events which took place during Solar System formation. In fact, the ages of bodies like the Earth and the Sun are based on data acquired from meteorites. The age that we assign to the Solar System is 4.56 Ga.

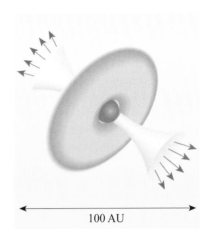

100 AU

Figure 8.12 Schematic representation of bipolar outflow showing a central protostar, a circumstellar disc or torus and a strong stellar wind. The torus confines the wind to flow predominantly in two opposing directions.

The concept of the origin of the Solar System as a single event is not without its critics. However, an age for the Solar System is clearly a useful notion and one which we use extensively herein.

In astrophysical terms, four key stages are recognized during the formation of the Solar System:

1 Dense cloud collapse (an interval of between 0.1 Ma and 0.5 Ma).

2 Disc dissipation, where material falling onto the accretion disc is transported inwards onto the protoSun (an interval of 0.05 Ma).

3 Terminal accumulation of the Sun, during which time the protoSun becomes a T Tauri star and planetary accretion begins (an interval of between 1 Ma and 2 Ma).

4 Gas dissipation, where planetary accretion in the inner Solar System ends and residual nebula gas is largely removed by T Tauri winds (an interval of between 3 Ma and 30 Ma).

Out as far as Uranus and Neptune the process of accretion may actually have continued for another 500 Ma to 1000 Ma as the planets continued to sweep up the remnants of nebula gas not completely dissipated during the T Tauri stage.

With this potted history of the Solar Nebula in mind let us now get back to the collapsing dust and gas cloud of the type observed around HP 4796A (Figure 8.3b). To understand the dynamics of this environment you may need to refresh your memory regarding the conservation of angular momentum (see Box 5.7).

■ If this cloud were originally rotating a little, what would happen to the rate of rotation as the cloud became smaller and denser during collapse?

❏ The cloud would have to spin faster, in order to conserve angular momentum.

Conservation of angular momentum also explains the directions of rotation of the Sun, and the orbits of most planetary bodies in the Solar System. A remarkable aspect of the arrangement of the Solar System is that the orbits of all the planets, and most asteroids, lie close to the same plane (the ecliptic plane), and that the planetary orbital motion and the Sun's axial rotation are in the same direction (prograde).

The general sense of prograde motion close to a common plane would be a natural consequence of formation from a contracting, rotating cloud. The inward gravitational force on matter contracting within the plane of the cloud's rotation would be opposed by the centrifugal force due to rotation and by gas pressure, but the gravitational force on matter contracting from above either pole of the cloud would be opposed only by pressure, which would not become great enough to halt the polewards contraction until the nebula had become dense. Thus collapse would be easier near the poles, and the cloud would be flattened into a rotating disc, as shown in Figure 8.13.

Centrifugal force is an apparent force that pulls an orbiting body away from the centre of its orbit.

The conversion of gravitational energy to thermal kinetic energy during contraction leads to the possibility of heating. As the nebula became progressively denser in its centre, where the protoSun was forming, it would have become more opaque. This would inhibit heat loss by radiation, and cause a substantial temperature rise. The pressure of this hot gas would eventually stop the gravitational contraction of the protoSun and of the innermost part of the disc. Gravitational contraction of the outer part of the disc would continue to be opposed mainly by the disc's rotation.

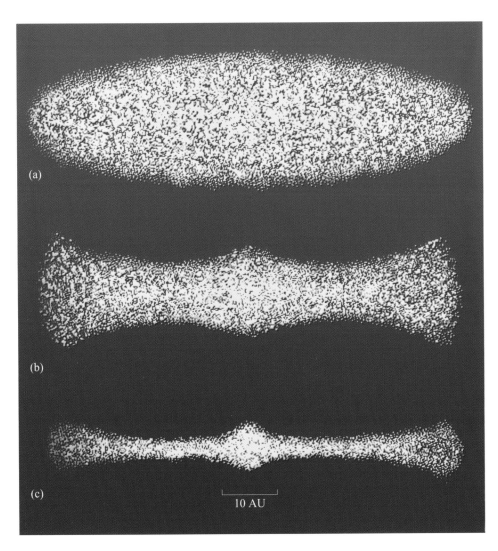

(a)

(b)

(c)

10 AU

Figure 8.13 A cross-section through the Solar Nebula, at three stages in its contraction, as rotation forces it into a disc shape. The time difference between (a) and (c) was probably of the order 10^5 years.

So far we have not considered in any detail the relative masses of the protoSun and the surrounding nebula. Here we touch on one of the more controversial aspects of Solar System formation. There are two major variants of the nebula theory. One is the low-mass or **minimum solar nebula model**, which assumes that the mass of the nebula around the protoSun was equivalent to the present total mass of the planets, with the addition of just enough hydrogen, helium and icy material to make the overall composition the same as the Sun today. This requires a total nebula mass of about $0.01M_\odot$. On the other hand, the **massive solar nebula model** has the mass of the nebula around the protoSun to be of order $1.0M_\odot$. Note that in both solar nebula models the mass of the protoSun itself is considered to be about $1M_\odot$ (in fact, perhaps a little bit more to allow for losses during the T Tauri phase).

Just in case you have calculated the masses of the present day planets (plus asteroids and comets) and realized a figure of $<0.0015M_\odot$, we will briefly explain the discrepancy with the value of $0.01M_\odot$ used in the minimum model. To have enough matter to produce those Solar System bodies that are *not* composed of hydrogen and helium (i.e. the inner planets) it is necessary to have more than $0.0015M_\odot$ of starting materials (which, as you should recall, are mostly hydrogen and helium). It is apparent that a large proportion of the total hydrogen and helium must be

You will see later in this chapter that hydrogen, helium and ice were prevented from being incorporated into planets with increasing efficacy the closer they were to the Sun.

subsequently lost. The minimum model requires that all the suitable nebula material was converted into planets. In contrast, for the massive solar nebula model to be valid, we have to account for how a large proportion of its mass could have been lost. How this may have happened is tied in with the present distribution of angular momentum within the Solar System.

Recall from Section 8.1. what we called the angular momentum problem. The dynamical problem for models of Solar System formation is that if the Sun simply represents a contraction of the central part of a rotating nebula we would expect it to be rotating more rapidly than it actually is. Its present 25.5 day rotational period is far too slow, and although the Sun contains over 99.8% of the mass of the Solar System it possesses only a tiny fraction of the Solar System's angular momentum. Most of this resides in the orbital motions of the giant planets, as you saw in Figure 8.1.

For the Sun to have so little angular momentum, the matter that formed the protoSun must have transferred most of its angular momentum to the surrounding disc. This could not have happened if the gas molecules and dust grains in the Solar Nebula were moving independently in orbits dictated by Kepler's laws. They must have interacted, so that the disc as a whole behaved viscously; that is to say, there would have been **viscous drag** between adjacent elements of the disc. There are several factors that could have contributed to this. Probably the most important factor was **turbulence**, initiated by random motions inherited from the initial contraction.

QUESTION 8.2

Consider a disc of gas and dust around the protoSun.

(a) What would be the relationship between the orbital periods of particles in circular orbits in the inner and outer parts of the disc, if Kepler's third law were applicable, i.e. are the orbital periods the same, or is one much shorter than the other?

(b) What would tend to happen to the orbital periods in the inner and outer parts of the disc if these were linked by viscous drag, so particles could not orbit entirely independently?

Despite this effect, the whole of the disc could not be forced to rotate at a uniform rate unless it were an impossibly strong solid, so even allowing for viscous drag, the outer parts of the nebula would continue to rotate more slowly than the inner parts. However, an important outcome of viscous drag would have been the outward transfer of angular momentum. This occurs because material in the outer part of the disc (having been speeded up) would flow outward, carrying angular momentum with it, but material in the inner part of the disc (having been slowed down) would fall inward onto the protoSun.

Thus, viscous drag within the Solar Nebula can explain the Sun's low angular momentum. The resulting outward flow at the outer edge of the nebula can account for the mass loss that is necessary if we subscribe to a massive solar nebula model. There are two other ways in which the nebula could have lost mass at this stage: loss to space in bipolar outflow (see the previous section), and transfer of mass from the nebula to the protoSun, although neither explains the Sun's low angular momentum. There would have been further losses of both mass and angular momentum later, if the Sun went through a T Tauri phase of violent stellar winds (see the previous section).

It is much easier to explain the Sun's low angular momentum if large quantities of mass were lost from the Solar Nebula. This is a strong argument against the minimum solar nebula model, but we do not have to invoke the full massive solar nebula model in order to account for the present distribution of angular momentum. It seems reasonable to assume that the Solar Nebula began with a mass somewhere between that of the extreme models, say of the order $0.1M_\odot$ (a figure, which you may recall, is at the end of the range of that invoked from observational evidence of circumstellar discs).

QUESTION 8.3

Why would a rotating Solar Nebula assume the shape of a disc, and what opposes the collapse of this disc?

QUESTION 8.4

What processes in the Solar Nebula can account for the Sun's low angular momentum, and at what stage in the formation of the Solar System did each of these probably occur?

8.2.4 Condensation of materials

Until now, we have considered the composition of the Solar Nebula only in terms of abundances of elements. At temperatures below those of stellar atmospheres, most elements will form chemical compounds, including some well-known minerals. The behaviour of these compounds will have controlled the evolution of the Solar Nebula. Compounds that are important in the evolution of the Solar Nebula include water (H_2O), methane (CH_4), troilite or iron sulfide (FeS), corundum (Al_2O_3), and pyroxene ($CaMgSi_2O_6$). We are interested to know whether these compounds were present as gases or in the form of tiny solid particles. The answer to this question depends mainly on the temperature within the nebula, which must have been greatest near the centre, and will have to some extent changed over time (especially during the initial stages of its life). Astronomical observations are much better at determining the sizes of grains in circumstellar dust clouds (typically 1 to 30 μm around T Tauri stars) than their compositions, and unfortunately it is not clear whether such grains are dominated by ices, carbon-rich material or silicate minerals. Let us now pursue the story of the Solar Nebula, by considering how solid material might have begun to be gathered together.

The Solar Nebula must have grown hotter as the nebula became too dense for radiation from the protoSun to escape directly to space. This heating would have been augmented by viscous drag, and continued conversion of gravitational energy to thermal kinetic energy. Evidence from meteorites suggests that the temperature never rose above about 400 K at about 4 to 5 AU from the protoSun. However it is likely that the maximum temperature about 1 AU from the protoSun was around 2000 K, near the time of stage (b) in Figure 8.13. This temperature would have been sufficient to break any pre-existing grains (of virtually all feasible compositions) into their constituent atoms. At this stage, most of the elements within 2 to 3 AU of the protoSun would have been in the gaseous state, as dissociated atoms, although there would have been some simple molecules such as H_2 and CO. Subsequently, as the nebula became more transparent to solar radiation and as the other heat sources ceased to act, it would have begun to cool. This would have allowed new compounds to form, growing as dust grains, by **condensation** (Box 8.2).

BOX 8.2 TWO TYPES OF CONDENSATION

You probably know condensation as what happens when you breathe out on a cold day, and water vapour condenses to form the tiny droplets that make your breath visible. On cooling, molecules of H_2O in the gaseous state lose energy and come together to form a liquid. This type of condensation is characteristic of molecular substances. However, under low pressure (as in the Solar Nebula) gases of molecular substances condense directly to form solids, with no intervening liquid phase (i.e. this is the reverse of sublimation where a solid goes straight to the gaseous state).

Condensation of ionic substances is a different process and usually occurs at higher temperatures (you have already met ionic substances in Chapter 2). In the Solar Nebula, elements that had existed as free atoms in the nebula gas at high temperature would have joined together as the nebula cooled to form chemical compounds, involving two or more elements. Once a tiny grain has condensed in this way, it is easier for individual free atoms to attach to its surface, rather than to start a new grain.

The nature of the starting material from which the planets in the inner Solar System developed was probably dictated by the highest temperatures reached locally. Further out from the protoSun it becomes more likely that pre-existing grains could have survived; we know from studies of meteorites (Chapter 9) that some of these have survived and can be studied in the laboratory today.

The sequence in which solid particles would have condensed while the Solar Nebula was cooling depends on a complex interplay of the *stability* of compounds and the *rate* at which they could form from a gas at the prevailing temperature and pressure. A simplified **condensation sequence** was shown in Table 2.4, showing the temperatures at which some notable constituents of the Solar System probably began to condense.

Minerals and other compounds that condense at relatively high temperatures are described as **refractory**, and those that do so at relatively low temperatures are described as **volatile**. The condensation sequence is the reverse of the evaporation sequence, that is to say if the temperature were to increase, the most volatile compounds would evaporate first whereas the more refractory compounds would be the last to do so.

It is generally believed that the planet-forming process began with the cooling of the Solar Nebula after it had reached its peak temperature. We have to imagine progressively more dust-sized grains forming by condensation within the nebula at about stage (c) of Figure 8.13. As these grains formed, they would have settled towards the mid-plane of the nebula, producing a sheet rich in dust and larger particles. This migration of grains towards the mid-plane occurs because of the continuing influence of gravitational forces; once in the mid-plane, particle–particle collisions act to eradicate those trajectories that would otherwise take grains out of the ecliptic. This settling would be possible only after turbulence within the nebula had declined, and it would have become more rapid as the grains grew.

8.2.5 Coagulation of grains

Now we have to consider how these grains could continue to grow, especially after much of the nebula material had condensed. Gravitational attraction between individual dust grains is too slight to bring grains together. Instead, chance encounters seem to have been the most important factor. In a collision between two particles, any of three things can happen:

- one or both particles can be fragmented (which is the opposite of what we are looking for),
- they can bounce off each other (which would not help us either), or
- they can stick together (which is what we want).

Fragmentation would be common only if the collisions were more violent than the sort of chance encounters likely in the Solar Nebula. One suggestion as to why colliding grains stuck together instead of bouncing apart is that their surfaces were fluffy, perhaps riddled with cavities caused by evaporation of ice that had previously condensed there. Magnetism and small electric charges may also have helped to hold grains together. Whatever the mechanism, this sticking together of grains that had grown by condensation in the Solar Nebula is generally referred to as **coagulation**.

An alternative term for coagulation is cohesion.

The rate at which coagulation could proceed must have depended on the density of material within the nebula. One model suggests that once settling of dust grains towards the mid-plane had begun, it would have taken only about 2000 years to produce particles up to 10 mm in diameter at 1 AU from the Sun, about 5000 years to produce particles up to 15 mm in diameter at about 5 AU (the Jupiter region) and about 50 000 years to produce particles up to 0.3 mm in diameter at about 30 AU (the Neptune region). You should be aware at this point that we are only here considering the details of initial grain growth. The overall timescales for the formation of planets as a whole depends upon what happens after initial growth (as we consider in the following sections). You will have already seen that, at least in some models, Uranus and Neptune may be formed over extended periods. As far as Jupiter is concerned, we can be sure that this is not the case and that we must be looking at a timescale that is compatible with that of the inner planets. The reason for this is that if Jupiter, as a large body, was not present during the formative stages of accretion of material in the inner Solar System, then the materials which are at the present day represented by the asteroid belt, would have accreted to form a rocky planet (at about 2.8 AU, with a mass of about 4 times that of the Earth). Unlike a number of other aspects of Solar System formation we can be absolutely sure that this did not happen!

So, while there are factors which will inevitably lead to slower rates of accretion in the outer Solar System (i.e. Uranus and Neptune) there must also be some counter-balancing processes which promote accretion in the Jupiter–Saturn region. There are three factors that would contribute to the slower growth of grains further from the protoSun. First, **column mass** decreases outwards, so grains might be expected to condense more slowly in the outer reaches of the nebula (in this context, column mass is the mass per square metre of the Solar Nebula, measured perpendicular to its plane). Second, the lower density would result in less frequent collisions and hence slower coagulation. Third, the nebula was probably fatter in its outer reaches (Figure 8.13b and c), and so the majority of grains forming there would have to fall further to reach the mid-plane, taking longer to complete their journeys. Estimates of the rate of coagulation depend on assumptions made about the densities of both gas and dust present in the Solar Nebula. A reasonable model is shown in Figure 8.14, which shows the variation in column mass against distance from the centre. Note that this model has several orders of magnitude more mass present in the form of gas than as dust, especially in the inner Solar System.

You may also find column mass called column density. The two terms mean the same thing. You met this concept in Chapter 5 when considering atmospheres.

Given that Figure 8.14 represents the situation after everything able to condense from the nebula has already done so, can you suggest what remains in the gas?

The gas will be made up of the abundant volatile substances, notably hydrogen (as H_2 molecules) and atomic helium (He), neither of which would have condensed.

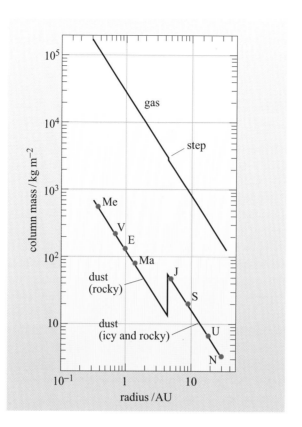

Figure 8.14 The column mass of gas and dust in the Solar Nebula (at a stage represented by Figure 8.13c) plotted against radius from the protoSun. Obvious abbreviations are used to indicate the positions of the planets that ultimately formed. The reason for the increase in column mass of dust near 5 AU is due to the condensation of water (see text). (Smaller steps associated with the condensation of more volatile compounds such as ammonia have been omitted.)

Considering now the processes that might act in opposition to slower rates of accretion, note from Figure 8.14 that the sharp increase in dust column mass near 5 AU is anticipated because the temperature at this distance would have been low enough for water to condense, in the form of ice. In fact, this ice can be thought of as being quite fluffy in texture, (i.e. more like snow) which will encourage individual grains to stick together. Note that there is a downward step in the column mass of the gas at the same radius. Although this downward step seems a lot smaller than the upward step, when you allow for the logarithmic scale, it exactly compensates for the increase in column mass of dust.

Estimating the rates of processes such as condensation within the Solar Nebula is fraught with difficulties, but the icy nature of most of the major satellites of the outer planets demonstrates that water did indeed condense in considerable quantities near Jupiter and beyond.

- Can you suggest how it is possible for water to have been incorporated in material that condensed at this stage considerably *closer* to the Sun than 5 AU?

- It can be incorporated in hydrated minerals, which condense at higher temperatures than that at which water condenses to ice (see Table 2.4). This is probably the origin of much of the water in the Earth.

'Relict' is a term used in meteoritics to describe inherited features, e.g. particles whose formation pre-dates the Solar System.

A possible window back to the time when the processes of coagulation within the nebula had resulted in objects of a few millimetres across, is provided by what may be relicts of these particles in certain meteorites (of which you will learn more in Chapter 9). For now we will just note that these consist of small, irregularly shaped objects comprised of refractory minerals (known as calcium–aluminium-rich inclusions) and globules of silicate minerals a few millimetres across (known as **chondrules**). Inclusions and chondrules are found embedded in a (predominantly) silicate matrix, in meteorites known as **chondrites**, or chondritic meteorites. The important thing here is that calcium–aluminium-rich inclusions and chondrules *could* represent the types of grains that were available for coagulation. We will consider how these materials and the matrix could have been gathered together, in the next section.

In less-primitive chondrites, new minerals have replaced the original ones, showing that entities such as chondrules and matrix were heated after they were assembled.

We have now traced the story of the early Solar System from the origin of the Solar Nebula up to the formation of particles a few millimetres in size. The time required for this appears to have been surprisingly brief. Condensation probably took just a few thousand years. We have not considered in any detail whether or not the processes we have discussed are likely to have been going on simultaneously in all parts of the Solar Nebula, but you have seen examples of both time-dependent and space-dependent variations. For example, in the innermost Solar System the temperature was always too high for water to condense as ice. Also, things will have happened more slowly in the outer Solar System because the lower density of the nebula would give grains less frequent opportunities to collide, and thus coagulate. In the next section we will consider how the grains could come together to form the bigger bodies necessary to create the planets.

QUESTION 8.5

Many chondrules contain small amounts of iron sulfide (FeS) mixed in with the silicate minerals. If we assume that the iron sulfide has not grown within the chondrules subsequently, what constraint (upper limit) does this put on the temperature at which the chondrules formed?

QUESTION 8.6

According to the model in Figure 8.14, what are:

(a) the total proportion of 'icy and rocky' dust expressed as a ratio of the total material in the Solar Nebula?

(b) the ratio between rock-forming and ice-forming material in the nebula?

8.2.6 Planetesimals and embryonic planets

Recall that young stars of around $1M_\odot$ experience a T Tauri phase of violent stellar winds, probably when they are about a million years old. This is known from observations of such stars whose light is reaching us today. At face value, therefore, we can conclude that the Sun itself probably went through such a phase (there is also other circumstantial evidence that we will mention in Section 8.2.8). To appreciate the magnitude of the T Tauri phenomenon consider that during such a process it is estimated that freely orbiting bodies in the inner Solar System of up to 10 m in size would have been 'blown' out of the Solar System. In the outer Solar System the T Tauri effects were nowhere near as intense because otherwise they would have removed all of the gas, (i.e. there would have been no gas to form the giant planets).

■ The density of a typical rock might be around $3 \times 10^3 \, \text{kg m}^{-3}$. For illustration purposes, what is the mass of a hypothetical cube of rock measuring 10 m on its side?

❑ Mass = density × volume, and volume is (10 m) × (10 m) × (10 m) = $10^3 \, \text{m}^3$, so
mass = $(3 \times 10^3 \, \text{kg m}^{-3}) \times (10^3 \, \text{m}^3)$
= $3 \times 10^6 \, \text{kg}$ (or 3000 tonnes)

It is clear that any materials destined to form planets in the inner Solar System must have been gathered into masses of greater than this before the onset of the T Tauri phase. Indeed it is widely believed that the next stage in planetary formation after initial coagulation led to the growth of bodies about 0.1 to 10 km across, known as **planetesimals**. One model for how this happened calls for a continuation of coagulation during collisions in the central dust sheet. Another theory is that once this sheet had contracted to a thickness of less than about 100 km it became gravitationally unstable, breaking into a multitude of turbulent knots each of which developed into a planetesimal. Either way, it appears that coagulated grains of all sizes were gathered together to form planetesimals, probably within a few hundred thousand years after condensation began (within the inner Solar System, at least; we have already stressed the point about things taking longer in the outer regions).

■ Can you suggest what could become of any dust trapped between larger coagulated grains while a planetesimal was forming?

❑ This material could become the matrix between chondrules in a chondritic meteorite.

The complex histories indicated by the mineralogy and texture of most meteorites suggest that they are fragments of planetesimals or larger bodies that were broken apart by collision, but the least-altered varieties are the best relics available of the process of planetesimal formation.

Once a planetesimal has reached about 10 km across, it has become massive enough that its own gravitational attraction is sufficient to perturb the motions of other nearby planetesimals. A larger planetesimal will deflect the trajectories of smaller planetesimals towards it (the phenomenon known as **gravitational focusing**) resulting in more frequent collisions. Provided most of the material in two colliding planetesimals ends up together – the process of **accretion** – instead of fragmenting and dispersing, the bodies that remain will get progressively larger, and the large bodies will become fewer in number.

It would not be possible for colliding planetesimals to stick together *without* fragmenting, because the collisions would be too violent. After a collision therefore, the new, larger, planetesimal is made of a mixture of accreted fragments from the two planetesimals that collided.

It is impossible to model exactly how planetesimals would have interacted because there were far too many of them and there are factors of unknown magnitude, such as the degree of viscous drag due to the remaining gas, that should be taken into account. However one important outcome is clear: because of their greater strength of gravitational focusing, larger planetesimals would have grown more rapidly than smaller ones, while the total mass incorporated in smaller bodies declined. It is probable that a situation of **runaway growth** developed in each neighbourhood, in which the growth of one planetesimal outpaced that of all its rivals, allowing it become a much larger object known as a **planetary embryo**.

QUESTION 8.7

Assume the density of the material forming planetesimals was $3 \times 10^3 \, \mathrm{kg \, m^{-3}}$. How many 10 km diameter planetesimals would it take to form the Earth, which has a mass of approximately $6.0 \times 10^{24} \, \mathrm{kg}$? *Hint*: It is a reasonable approximation to treat these planetesimals as spheres. The volume V of a sphere of radius R is given by $V = \frac{4}{3} \pi R^3$.

8.2.7 Planetary growth in the inner Solar System

The style and speed of runaway growth is likely to have varied with distance from the Sun, slowing drastically in each vicinity as the supply of smaller bodies available for capture became depleted. Limiting masses calculated for planetary embryos at different distances from the Sun are indicated in Figure 8.15.

We will follow events in the inner Solar System first, before considering what happened further out. In the region where the terrestrial planets are now found (from Mercury to Mars) one planetary embryo probably grew about every 0.02 AU outward from the Sun. One of the many models describing the evolution of a swarm of planetesimals into a planetary embryo in this region is illustrated in Figure 8.16. Please study this figure and its caption; it is probably correct in a *general* sense, though its details are not to be trusted.

If the models illustrated in Figures 8.15 and 8.16 are even roughly correct, runaway growth appears able to have produced substantial planetary embryos before the

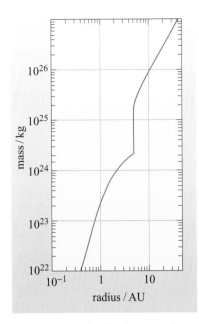

Figure 8.15 The final masses for the largest planetary embryos produced by runaway growth at different distances from the Sun. The step-up at about 5 AU is because of the condensation of water-ice beyond that distance (see Figure 8.14). Note that the curve does not keep on increasing beyond the edge of the graph. In fact, it reaches a maximum and then declines – otherwise there would be very large planets beyond Neptune.

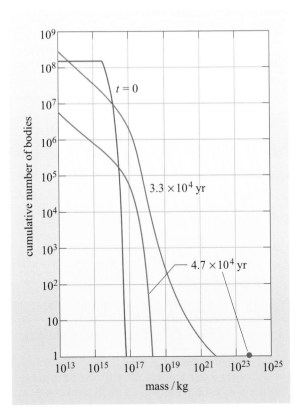

Figure 8.16 The evolution of a swarm of planetesimals orbiting the Sun between 0.99 and 1.01 AU. The curves show the distribution of planetesimals by mass at three different times, beginning at $t = 0$ with almost all the planetesimals between about $10^{16} \, \mathrm{kg}$ and $10^{17} \, \mathrm{kg}$ in mass (about 10 km in diameter). At 33 000 years, the largest planetesimal is just under $10^{22} \, \mathrm{kg}$, and represents the high-mass end of a continuum of planetesimal masses. However, after 47 000 years runaway growth has led to the production of a single planetary embryo between 10^{23} and $10^{24} \, \mathrm{kg}$ in mass. By that stage, no other body in this region of the Solar System has a mass within five orders of magnitude of this. Note that the axes are logarithmic, and that the vertical axis shows the *cumulative* number of bodies greater than the mass plotted on the horizontal axis. (Weatherill, 1989)

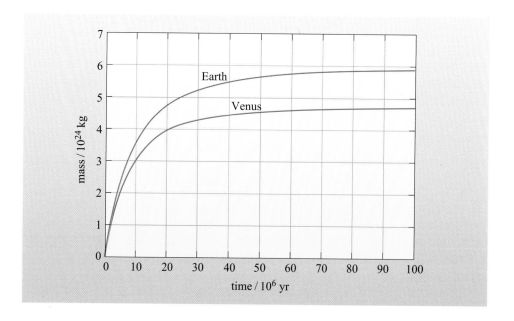

Figure 8.17 A model for the timescale for the growth of the Earth and Venus as a result of collision between planetary embryos. Mercury and Mars probably grew at similar rates. (Weatherill, 1989)

Sun's T Tauri stage. However the supply of planetesimals within range of each embryo's gravitational influence, was sufficient only to allow it to reach about one-tenth of the mass of the individual planets that would eventually form.

After this, planetary growth can have proceeded only at a slower pace, as a result of chance collisions between planetary embryos whose orbits crossed. Whenever two embryos collided, the smaller of the two would have been fragmented and mostly accreted onto the larger one. A collision between two embryos of similar mass would probably have broken both into a swarm of debris, most of which would have rapidly come together (accreted) under its own gravitational attraction. Either way, the result would be the same; fewer and fewer remaining embryos that would become more massive after each collision. Calculations for Mercury, Venus, the Earth and Mars suggest that in each case it would take about 10^7 years for growth to about one-half of the final planetary mass and about 10^8 years for growth to be essentially complete (Figure 8.17). Recall that the curious case of the Moon was examined previously in Section 2.3.

8.2.8 Formation of the giant planets

What we will describe in this section is a traditional view of the formation of the giant planets in which the bodies themselves are assumed to have formed at the orbital distances that are observed today. In fact, you will see in the next section that this is probably not the case and that the orbits of the giant planets have migrated from their original formation positions. However, the overriding principles discussed in the present section are likely to be correct. Furthermore, planetary migration is an active and somewhat controversial subject; there are still many people who reject the idea in the case of the Solar System. As such it seems entirely appropriate to continue with a standard model for the formation of the giant planets.

You can see from Figure 8.15 that in the outer Solar System the expected masses of individual planetary embryos in the Jupiter–Neptune region (5 to 30 AU) are at least two orders of magnitude larger than those at 1 AU. However, the large volume

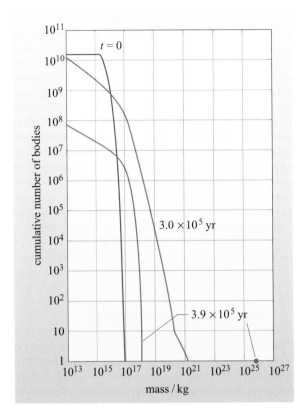

Figure 8.18 The evolution of a swarm of planetesimals near 5 AU leading to the production of a Jupiter kernel by runaway growth. Compare with Figure 8.16. (Weatherill, 1989)

of space occupied by the planet-forming material in the outer Solar System makes it more difficult for runaway growth to begin (indeed, some authorities dispute that it could have happened at all), and when it did get underway it would produce fewer but **larger planetary embryos**. Figure 8.18 shows a simulation of how runaway growth might proceed at 5 AU.

According to this, an embryo (probably a mixture of rocky and icy materials) with a mass somewhat less than 10^{26} kg (about $10M_\oplus$), is produced after 3.9×10^5 years. You may be able to recall from Chapter 5 that Jupiter was estimated to have an icy–rocky core of $5M_\oplus$. (As is often the case in science, different models and methods produce slightly different interpretations so it is probably wise to consider that the planetary embryo mass was about $5M_\oplus$ to $10M_\oplus$.) Note that the timescale 3.9×10^5 years is about ten times as long as it would take to produce a planetary embryo at 1 AU, according to the comparable simulation shown in Figure 8.16. The hypothetical $5M_\oplus$ to $10M_\oplus$ body would act as a **kernel** that would, with the aid of gravitational focusing, sweep up planetesimals, smaller debris and nebula gas, leading to the formation of the planet Jupiter. As this body grew, it would be able to capture a significant amount of gaseous molecular hydrogen (H_2) and atomic helium (He) from the Solar Nebula (a process that would have been enhanced by the large gravitational attraction of the planet). This envelope of gas ultimately became the Jovian atmosphere that we see today.

Runaway growth such as this would take even longer to produce the kernels for Saturn, Uranus and Neptune; probably about 2, 10 and 30 Ma respectively. This increasing timescale with distance from the Sun, is probably the key to understanding the outward trend in composition among the giant planets.

We are using the conventional symbol $M_\oplus$ to indicate the mass of the Earth.

> ▨ Look at the data for Jupiter, Saturn, Uranus and Neptune in Table A1, Appendix A. Can you see any trend in their densities as their distance from the Sun increases?
>
> ❑ No, a trend does not appear to exist.

The reason is that the more massive the planet, the stronger its gravity and the more compressed, and therefore denser, will be the material in its interior. If this **self-compression** is taken into account, then the densities of the four giant planets can be explained by each containing roughly similar amounts of rock and ice ($10M_\oplus$ to $20M_\oplus$) but diminishing amounts of hydrogen and helium further from the Sun. Jupiter is thought to contain about $300M_\oplus$ of H_2 and He, Saturn about $70M_\oplus$, and Uranus and Neptune about $1M_\oplus$ each.

> ▨ Can you suggest an event that would have prevented the more distant planets from gathering as much gas from the Solar Nebula as Jupiter did?
>
> ❑ If their kernels formed later (as we have suggested) then maybe most of the gas in their neighbourhood had been removed before they grew massive enough to attract much gas gravitationally. The most likely explanation for this removal is the T Tauri wind.

This is the circumstantial evidence that the Sun did pass through a T Tauri phase that we have referred to previously. The T Tauri wind blew away the remaining H_2 and He, so that no more gas could be captured by the giant planets. A related argument is that the Earth should have captured an atmosphere of about $0.03M_\oplus$ from the Solar Nebula. This is 3×10^5 times the mass of its present atmosphere, which furthermore has a totally different composition. We can conclude that the Earth's primitive atmosphere (and those of the other terrestrial planets) was probably lost during the T Tauri phase.

QUESTION 8.8

The simulations in Figures 8.16 and 8.18 both begin with a population of planetesimals between about 10^{16} and 10^{17} kg in mass, i.e. about 10 km diameter. The flat tops to the curves for $t = 0$ show that there are no planetesimals less than about 10^{16} kg in mass. In both simulations, at times later than $t = 0$ the low-mass ends of the curves slope down to the right. What does this tell you about the masses of the smallest bodies present at these later times, and to what process do you attribute this?

QUESTION 8.9

What factors are responsible for the giant planets being richer than the terrestrial planets in volatiles such as hydrogen and helium?

8.2.9 Planetary migration

In Section 8.2.1 we alluded to planetary migration in respect of extra-solar planets. By planetary migration we mean a change in the orbital radius (i.e. semimajor axis) of a planet with time. If we just take this at face value for the moment, what this would mean for descriptions of the Solar System is that the orbit of a particular planet, observed today, may not correspond with the orbit of the same body at the time of its formation (this phenomenon is only currently being thought of in respect of the giant planets). If true, there would be no real point in developing a model of planetary formation in which it was implicitly assumed that orbits have remained the same. The gauntlet has now been thrown down, challenging theoreticians who derive models of the formation of the Solar System to contemplate this extra level of complexity.

The issues that have brought this problem to a head can be summarized as:

*   The discoveries of the Kuiper Belt and Oort cloud (see Section 7.4) have shown that the Solar System does not end at Neptune (or Pluto). Between about 20 and 60 AU there are now thought to be 100 000 icy minor planets of 100 to 1000 km diameter mostly orbiting in the plane of the ecliptic.

*   The detection of extra-solar planets of Jupiter size orbiting at distances of less than 0.5 AU is problematic because, from theoretical considerations, there would not be enough material in a nebula disc to form large planets so close to a star.

*   Recent measurements of volatile elements (e.g. noble gases, carbon, nitrogen and sulfur) in the atmosphere of Jupiter show enrichments which suggest that the planet formed at colder temperatures than previously thought. One way of interpreting this is that Jupiter formed much further away from the Sun (e.g. 30 AU) than its present position (about 5 AU).

Without getting into any of the details of planetary migration you should be aware that in a general sense we could be talking about orbital distances that either decrease or increase, depending upon circumstances. Clearly in the case of 'hot jupiters' the orbits are thought to have moved closer to the central star (the original planets having formed at several AU). For the Solar System it has been proposed that the orbits of the giant planets may all have moved further away from the Sun with time. The details of this process are complicated, but in essence the theories are constrained by the presence of Kuiper Belt objects. Theories will undoubtedly become refined as more of these objects are detected. To illustrate the magnitude of the process, the orbital radius of Neptune may have increased by 30%, while those of Uranus, Saturn and Jupiter may have increased by 15%, 10% and 2% respectively. The migration itself is presently considered to have taken place over a timescale of less than 0.1 Ga straight after the formation of the Solar System. Because of uncertainties in ideas regarding the length of time required for the formation of the giant planets, it is difficult to state whether migration accompanied

the tail-end of planet formation, or took place early in the history of the Solar System. Planetary migration is a hot topic. Just a few years ago the idea of this having taken place within a Solar System context would have been considered extremely dubious. And yet, should we have been surprised by the phenomenon? It has been known for a long time that the orbits of many planetary satellites have changed since their formation. Taking the example closest to home, the Moon is believed to have formed within 30 000 km of Earth and has subsequently moved out to its present distance of 384 000 km as a consequence of gravitational tidal forces exerted by our planet. A billion years ago the Moon was 100 000 km closer to the Earth; next year it will be 3 cm further away. It is ironic that the object most readily visible in the night sky should be helping to open our minds to the notion of generalized processes taking place in solar systems across the Galaxy.

8.2.10 Origin of the asteroids and comets

Jupiter is much the most massive body in the Solar System after the Sun, and its gravitational field is likely to have been an influential factor in the history of the Solar System ever since its formation. The most obvious sign of Jupiter's influence can be seen in the absence of a substantial planet between Mars and Jupiter, even though simulations of runaway growth suggest the appearance there of planetary embryos in excess of 10^{24} kg within 4×10^5 years. Today this region is populated only by the asteroids, all less than about 10^{21} kg in mass, and amounting to a total present-day mass of no more than about 3×10^{21} kg (about $5 \times 10^{-4} M_\oplus$). The asteroids exist because the gravitational influence of the growing Jupiter kernel perturbed the orbits of planetesimals orbiting near 3 AU, thus increasing their eccentricities. This would have ensured that when these planetesimals collided with each other, the velocity (and hence the kinetic energy) of the impact was so great that fragmentation and dispersal became just as likely as growth by accretion.

As a result, runaway growth never took place, and the orbital perturbations probably led to much of the original material being scattered. By now most of this debris has collided with other bodies, forming the impact craters that characterize old surfaces throughout the Solar System (see Chapter 4). The irregular shapes of the surviving asteroids – which are essentially fragments of the original planetesimals – attest to their violent pasts.

Jupiter is also likely to have played a significant role in the origin of comets. Long-period comets could originally have been icy planetesimals that were flung outwards (or indeed, inwards) as a result of a close encounter with Jupiter or one of the other giant planets. Such encounters would send planetesimals off in very elongated orbits, extending out to form the Oort cloud.

8.2.11 Satellite systems

We have described processes that can account for the origin of all bodies in the Solar System except planetary satellites. We will now take a look at the major satellites of the giant planets. With the exception of Neptune's large satellite Triton, these are in orbits close to their planet's equatorial plane, travelling in the same direction as the planet rotates. It is unlikely that these satellites could have formed elsewhere and been captured later by their planets, because in that case their orbits would not be so well-ordered. Instead, the satellites probably condensed from a disc of gas and dust about each primitive planet, rather like the Solar Nebula in miniature. One fundamental difference between the satellite systems of the outer planets and the Solar System as a whole is that, whereas the latter suffers from what we have called 'the angular momentum problem', the planetary satellite systems do not. While the physical mechanics of the satellite systems are readily understood there is still a debate about whether a **protosatellite disc** is more likely to have been shed by a young planet as it contracted, or to have been gathered around it by scavenging of stray planetesimals and other material from the remains of the Solar Nebula.

Whatever the origin of a protosatellite disc, we can imagine it developing into individual satellites in much the same way as the planets had formed. Small particles would have aggregated into larger ones that would, over time, have collided. Eventually the largest bodies would have swept up all the debris in their vicinity to form the surviving satellites. The protosatellite disc would have been rotating in the same direction as the Solar Nebula, and this explains the satellites' consistent orbital directions.

We can learn more about the origin of planetary satellites by examining the four large satellites of Jupiter, known as the **Galilean satellites**, after Galileo who discovered them in 1610 using one of the first telescopes.

QUESTION 8.10

(a) Data on the Galilean satellites (Io, Europa, Ganymede and Callisto) are included in Appendix A, Table A2. Can you see any systematic trend in their densities?

(b) The density of ice is about $1.0 \times 10^3 \, \text{kg m}^{-3}$, whereas that of any rock within these bodies is likely to be about $3.0 \times 10^3 \, \text{kg m}^{-3}$. What does this extra information tell you about the probable compositions of each of these satellites?

Figure 8.19 A close-up view of Ganymede. Spectroscopic studies show that the surface is icy, despite the craters and other features that would seem to suggest a rocky surface. (NASA)

Spectroscopic studies of sunlight reflected from the surfaces of Ganymede (Figure 8.19) and Callisto demonstrate that their surfaces at least, are dominated by water-ice. Similar data show that ice is prevalent over the whole of Europa too (Figure 8.20), although Europa's high density demonstrates that its coating of ice can be no more than a few tens of kilometres thick, and its interior is probably mostly rock. Io is an essentially rocky body with no surface ice (Figure 8.21). Its high density shows that its interior contains even denser material, probably in the form of an iron-rich core.

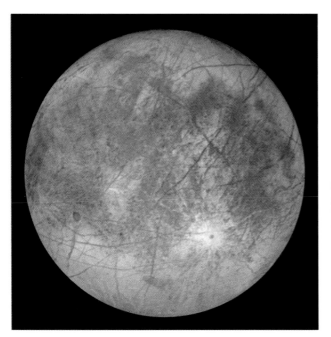

Figure 8.20 Image of Europa taken by the Galileo spacecraft. (NASA)

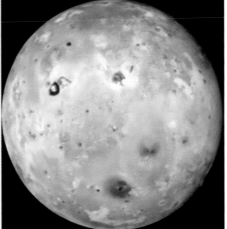

Figure 8.21 Jupiter's volcanically active moon, Io, as observed by the Galileo spacecraft. Io's surface is covered with volcanic deposits that are thought to contain ordinary silicate rock, along with various sulfur-rich compounds that give the satellite its distinctive colour (see Section 3.4.6). Dark areas are regions of recent or current volcanic activity. (NASA)

Thus the Galilean satellites resemble the Solar System in miniature. The inner ones are less massive and denser, whereas the outer ones are more massive but less dense.

▉ Can you suggest a condition within the protosatellite disc around Jupiter that could account for this density trend?

❏ Perhaps it was too hot in the inner part of this disc for ice to exist, that is above about 180 K, according to Table 2.4.

It is now accepted that heat – generated by the conversion of gravitational energy to thermal kinetic energy as material in the protosatellite disc fell in towards the primitive Jupiter – probably raised the temperature in the inner part of the disc. This either prevented ice from condensing there, or vaporized any ice that had already condensed.

There is no such density trend within the satellite systems of Saturn and Uranus. All their major satellites consist of a rock plus ice mixture, and none are essentially rocky like Io and Europa. However, this cannot be used as evidence that they did not form in a protosatellite disc, because, being less massive, the primitive Saturn and Uranus would have caused less heat to be imparted to infalling material than Jupiter. An additional factor that could be responsible for the lack of consistent density trends is that most of these satellites are much smaller than the Galilean satellites, so the inner ones in particular would have been more vulnerable to fragmentation by impacts involving large comets or other bodies. After fragmentation, but before re-accretion to form the satellites we see today, material from different satellites could become mixed. Although they are smaller, the satellites of Saturn and Uranus are similar in composition to Ganymede and Callisto, except that, having formed further from the Sun and consequently in a lower temperature environment, they were probably able to incorporate methane and ammonia into their ices (see Table 2.4).

QUESTION 8.11

Can you think why inner satellites are more vulnerable to impacts than outer ones, and why smaller ones are more likely to break up as a result of an impact?

Satellites less than about 200 km in radius are mostly irregular bodies (e.g. Figure 8.22), and may be fragments resulting from such collisions. Collisions may also be the origin of the rings of debris that are now known to encircle each of the giant planets (see Figure 8.23). Rings could also result from the break-up of a satellite that strayed so close to its planet that it was pulled apart by tidal forces.

In fact, all bodies less than about 300 km are likely to be irregular because their internal gravitational fields cannot overcome their intrinsic strengths and make them spherical.

Figure 8.22 Two of the largest irregular-shaped satellites in the Solar System. (a) Amalthea, an inner satellite of Jupiter (radius dimensions 135 km × 80 km × 75 km). (b) Hyperion, an outer satellite of Saturn (radius dimensions 135 km × 120 km × 75 km). (NASA)

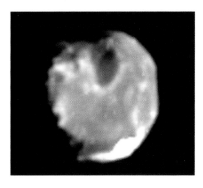
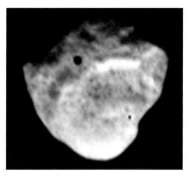

Figure 8.24 Phobos, the larger of the two tiny satellites of Mars, both of which are almost certainly captured asteroids. (Dr Michael H. Carr, National Space Science Data Center)

Figure 8.23 Detailed view of the rings of Saturn as seen by the Voyager 1 spacecraft. The rings consist of icy debris orbiting in the plane of Saturn's equator. The rings are probably the remains of a broken up satellite, and most pieces are only a few centimetres in size. (US Geological Survey)

Other small satellites, notably those in retrograde orbits (i.e. travelling in the opposite direction to their planet's rotation), are more likely to be captured bodies, in which case they probably began life either as asteroids or comet nuclei. The two satellites of Mars, Phobos (Figure 8.24) and Deimos, are almost certainly captured asteroids.

Neptune's only large satellite, Triton (Figure 8.25), has a retrograde orbit, which suggests that it was captured by Neptune after formation, instead of forming in a protosatellite disc. Triton is actually somewhat larger than the planet Pluto, which, so far as we can tell, resembles Triton rather closely in being an ice–rock mixture partly covered by condensed nitrogen. Triton and Pluto may represent stunted planetary embryos, intermediate in size between the Neptune kernel and the more distant bodies in the Kuiper Belt.

QUESTION 8.12

State four possible origins of planetary satellites, with examples.

QUESTION 8.13

If a planetary system is forming at β Pictoris, what stage (if any) do you think it has reached in the sequence suggested in this chapter for the development of our own Solar System, and why?

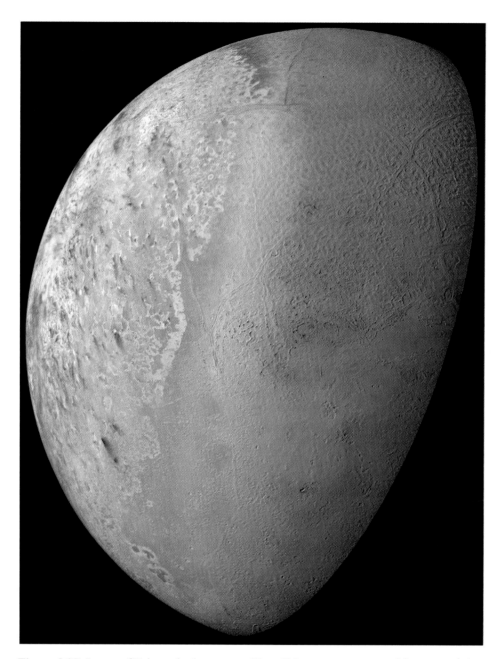

Figure 8.25 Image of Triton, the largest satellite of Neptune, constructed from a mosaic of the highest resolution images taken by the Voyager 2 spacecraft in 1989. (NASA)

8.3 Summary of Chapter 8

- The currently favoured paradigm for the origin of the Solar System involves the formation of a protoSun surrounded by a disc of material that rotates, flattens and separates to form the planets. This so-called nebula hypothesis is now corroborated by observations of other solar systems in the making.

- Within the rotating disc of the early Solar Nebula, angular momentum was evidently transferred outwards from the protoSun into the nebula.

- Much of the original dust in the Solar Nebula may have been vaporized, especially in the inner Solar System, principally as a result of heat released by loss of gravitational energy within the nebula and within the protoSun. Nebula material would have begun to condense as the temperature dropped, with the refractory substances appearing first and the more volatile substances condensing later, according to the condensation sequence. Within about 5 AU of the Sun the temperature never dropped low enough for ice to condense, although some water will have become trapped within hydrated minerals that condensed.

- Grains of dust that condensed from the Solar Nebula settled towards the central plane of the disc. When they came into contact they tended to do so gently, and stuck together (coagulated) to form larger particles. This sticking together may have been encouraged by fluffy surface textures, magnetism and electric charges.

- Within a few hundred thousand years of the creation of the Solar Nebula, grains in the dust disc that had formed near its central plane became gathered into planetesimals about 0.1 to 10 km in size, either because of gravitationally-driven turbulence or as a result of continued coagulation.

- Once a planetesimal had reached about 10 km in size, its gravitational field would have become sufficient to encourage collisions, by means of gravitational focusing. Eventually, the growth of one body would have outpaced all the others in the vicinity (runaway growth), producing a planetary embryo. At about 1 AU these were probably about one-tenth of the mass of the Earth. The creation of planetary embryos probably took of order 10^4 or 10^5 years in the inner Solar System, but the timescale goes up by roughly an order of magnitude in turn for each of the giant planets.

- Each terrestrial planet grew by a series of collisions between planetary embryos that took about 10^8 years to complete. Runaway growth took longer to get established in the outer Solar System, but the planetary embryos that grew there became sufficiently large that they could act as kernels around which large masses of gas were captured directly from the Solar Nebula.

- The onset of a powerful solar wind as the Sun entered its T Tauri phase swept the remains of the Solar Nebula outwards into space, and this evidently happened early enough to forestall Saturn, Uranus and Neptune from capturing as much gas as the earlier-formed Jupiter was able to. Loss of the primitive atmospheres of the terrestrial planets can be attributed to the T Tauri wind, but the giant planets were able to hold onto most of the gas they had already captured.

- The sub-planetary size of the asteroids can be attributed to gravitational perturbation of the orbits of planetesimals in this region by the young Jupiter, which ensured that mutual collisions were too energetic to lead to runaway growth. Long-period comets are icy bodies that may have been flung outwards after passing close to Jupiter or another giant planet to form the Oort cloud.

- The orbits of the major satellites of Jupiter, Saturn and Uranus suggest that these bodies grew out of a protosatellite disc about each planet, rather like the Solar Nebula in miniature.

- Smaller satellites may be collisional fragments, or captured asteroids and comets.

CHAPTER 9
METEORITES: A RECORD OF FORMATION

9.1 Introduction

Meteorites are pieces of extraterrestrial material (meteoroids) that fall to Earth from the sky having survived the retardation (deceleration) caused by their impact with the atmosphere, and which then survive to be collected as some sort of solid entity. While there are some meteorites that represent small pieces of the Moon or Mars (ejected from their respective surfaces during impact events), the majority are fragments of asteroids. In the present Chapter we concentrate solely on asteroidal meteorites. You should recall from Chapter 7 the nature and basic properties of asteroids, and from Section 8.2.10 the formation of asteroids.

Even today meteorites are the subject of much curiosity, but in the past some samples were revered and worshipped, while others were collected and used for making tools. Because meteorites are now considered collectable items, some individuals are prepared to pay large sums of money for specimens. This has had the effect of (a) pushing up prices, and (b) increasing the number of spurious claims regarding meteorite discoveries. Scientists try to work around these problems by maintaining a network of trustworthy individuals, both professional and private, who provide meteorite samples for all kinds of investigations.

The collection of a meteorite sometimes takes place after it has been observed to travel through the Earth's atmosphere. This is known as a **meteorite fall**. Events like this are rarely witnessed – only about 6 to 8 falls are recorded each year. A relatively recent event, the fall of the Peekskill meteorite in New York, USA, has become a rather celebrated case for two reasons. Not only was the incoming meteorite witnessed by many individuals (Figure 9.1), but the body itself struck a car which has since been displayed around the world (Figure 9.2).

In contrast to observed falls, several hundred meteorite fragments are collected each year without being seen to fall; these are the meteorite **finds**. An example of an impressive meteorite find can be seen in Figure 9.3, which shows the Hoba meteorite (Namibia) being excavated in the 1930s. Note that Hoba is the largest single meteorite ever found, with a mass of about 5.5×10^4 kg, and is estimated to have fallen to Earth 80 000 years ago (see Box 9.1). It is still possible to see Hoba in the place it was originally found (Figure 9.4).

The reason we stress the notion of meteorite survival here is that there may be whole classes of potential meteorites which we do not know about because they are completely pulled apart within the atmosphere, or they effectively disappear after contact with the ground. Consider, for instance, a meteoroid composed largely of ice.

Figure 9.1 The Peekskill fireball. (Copyright S. Eichmiller)

Figure 9.2 The Chevrolet Malibu car that was struck by the Peekskill meteorite on display at a minerals exhibition in Munich, Germany. (Copyright © R.A. Langheinrich)

Figure 9.4
The Hoba meteorite as it is displayed today. (Copyright South Africa Online Travel Guide)

Figure 9.3 Excavation of the Hoba meteorite.

Figure 9.5 Mr. Arthur Pettifor holds a meteorite that struck a tree in his garden in Cambridgeshire, UK, in May 1991. If he had not been outside at the time of its fall, it would, like many thousands of other meteorites, have gone unnoticed. (Copyright Natural History Museum)

Meteorites are generally given the name of their collection site, regardless of whether they represent a fall or find – hence Peekskill and Hoba.

How many meteorites arrive at the Earth annually? It is known from continuous observations of the night sky that 20 000 bodies with masses in excess of 0.1 kg should reach the Earth's surface each year. Even of those that fall on land, it is clear that the vast majority of these meteorites are never recovered.

In principle, meteorites can be found anywhere on dry land (e.g. Figure 9.5), but various factors affect the chances of recovery. For instance, areas of dense vegetation are not good places for finding meteorites. In contrast, meteorites that fall in relatively barren environments, such as hot deserts, have a greater potential for collection. Antarctica is an even better place for finding meteorites, since glacial processes sometimes act to concentrate the samples (see Figure 9.6). Dedicated collection parties frequently visit Antarctica in order to collect materials for scientific research. Figure 9.7 shows some examples of meteorite collection.

The term meteorite is used to denote a sample of material that has fallen to Earth and has subsequently been collected as a coherent mass. As we saw in Chapter 7, while still in space, the same body is referred to as a **meteoroid**. The term meteoroid refers to fragments of asteroids, or dust/ice particles ejected from comets (to form meteoroid streams, as discussed in Chapter 7). However, as ejected cometary meteoroids are made of relatively 'fluffy' dust and ice, the bodies will not generally be strong enough to survive their journey through the atmosphere (i.e. they will simply burn-up as meteors). Thus, the relatively large meteoroids that subsequently survive atmospheric entry to become meteorites, will almost certainly be asteroidal in nature (i.e. relatively strong rocky/metallic bodies). Direct evidence of this comes from three well-documented cases, where meteoroids (that survived to become

BOX 9.1 ASSESSING HOW LONG AGO A METEORITE FELL TO EARTH

For observed meteorite falls we can be certain of when the body fell to Earth. For finds, however, this is a more difficult proposition. In the extreme, where meteorite remains are found within geological formations, the date of fall is gauged from the age of the rocks in which they are found. For finds of more recent vintage we use another method. The so-called *terrestrial age* of a meteorite (the period of time that the sample has been resident on Earth) is estimated by measuring the abundances of specific short-lived radionuclides. The nuclides in question are formed when the body is out in space, via the interaction with high-energy particles (cosmic rays – mainly H$^+$ ions with high kinetic energies – which, themselves, originate beyond the Solar System). In space, an equilibrium is set up between the production of the radionuclides (through cosmic-ray irradiation) and radioactive decay. A recently fallen meteorite shows this equilibrium concentration. Most cosmic rays do not penetrate the Earth's atmosphere, so the short-lived radionuclides in a meteorite start to decay as soon as a sample arrives at the Earth's surface since they are no longer being replenished. Knowing the half-lives of the radionuclides in question, determination of the extent of the decay gives a measure of the terrestrial age of a meteorite.

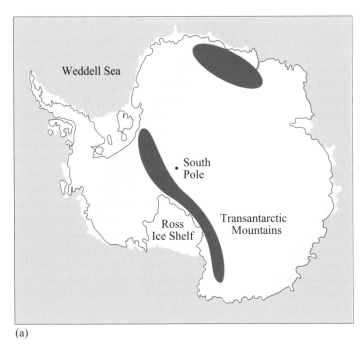

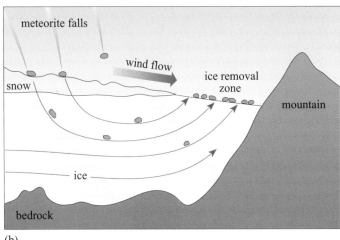

(b)

Figure 9.6 (a) Map of Antarctica showing locations of meteorite finds (red areas). (b) Schematic cross-section through the Antarctic ice sheet showing how ice movement acts to concentrate meteorites from a wide area in which they fall. Where a glacier abuts a mountain range the ice flow (red arrows) is diverted upwards. Strong winds blowing from the South Pole cause the ice to be removed, leaving the meteorite samples exposed.

(a)

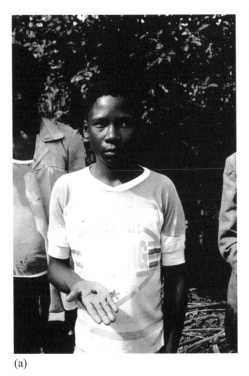

(a)   (b)

Figure 9.7 Meteorites are collected from a variety of environments. (a) The Mbale meteorite fell in Uganda in 1992 as a shower of several hundred fragments, the largest of which had a mass of more than 27 kg. Shown in the picture is one of the smaller fragments, of mass 3 g, which bounced off the leaves of a banana tree and hit a young Ugandan boy on the head. (b) A meteorite being collected in Antarctica. It is being put in an ultra-clean Teflon bag using stainless steel tongs in order to minimize contamination. ((a) University of Leiden; (b) Dr I. A. Franchi/Open University)

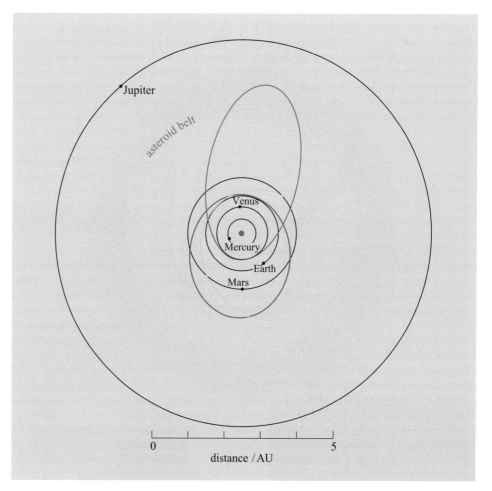

Figure 9.8 The orbits of three meteorites (shown in red, yellow and green) have been determined from data recorded by photographic networks (one in Czechoslovakia in 1959, one in the USA in 1970 and one in Canada in 1977). In these three cases the orbits can be traced back to the asteroid belt. More recently, evidence from video recordings has been used to reconstruct the orbits of two other meteorites (Peekskill and Tagish Lake) and both of these can also be traced back to the asteroid belt. (Wasson, 1985)

meteorites) were photographed during their fall, using networks of cameras. From the photographic evidence it has been possible to assess the meteoroids' orbits, which can be traced back to the asteroid belt in each case (Figure 9.8).

For those asteroidal meteoroids that become meteorites, the brief atmospheric passage is quite eventful. People who have witnessed falls can attest to the spectacular visual display and accompanying loud detonations. At an altitude of about 120 km, frictional heating by the atmosphere begins to melt the surface of an incoming meteorite. For a body of >10 mm in diameter, only the outer layers at any instant are heated sufficiently to cause melting and vaporization. This heating of the surface results in the formation of a glassy **fusion crust**, which is a further diagnostic indicator of a meteorite. During the fall, molten material is lost from the surface by a process called **ablation**. For those groups of meteoroids which have ultimately produced meteorites that are well represented in our collections on Earth, perhaps 50% of the total mass is lost this way.

During their fall to Earth the very smallest meteoroids, which are often called **micrometeoroids**, become (not surprisingly) **micrometeorites**. Unlike larger samples, micrometeoroids, which are of roughly micrometre- to millimetre-sized dimensions, become *totally* molten during their fall. Some of these vaporize completely while others survive to form **cosmic spherules**, which can be collected at the Earth's surface. Ordinarily these spherules go unnoticed. However, they are found in abundance in oceanic sediments (they were first identified in Pacific Ocean clays during the Challenger expeditions of the 19th century). Figure 9.9 shows a picture of cosmic particles extracted from a deep-sea sediment. Curiously, the very smallest micrometeoroids (<0.01 mm in size), which can be cometary in nature, do not melt during their fall to Earth. This is because they are decelerated at sufficiently high altitudes, where the atmosphere is more rarefied, and so avoid melting by frictional heating.

As a meteoroid travels down through the atmosphere it eventually reaches a level at which atmospheric gases surrounding the incoming body become ionized and form a plasma. As the body continues to move, this produces a streak of light known as a meteor (i.e. the process discussed in Chapter 7 for both cometary and asteroidal meteoroids), or, if bright enough, a **fireball**. Typically, meteors appear at an altitude of about 90 to 120 km and become extinguished at about 10 to 30 km (in other words, most of this activity takes place in a layer above the stratosphere). For large meteoroids the fireball can be seen in daylight, as can the accompanying trail of ablated material (which has the appearance of smoke). For smaller bodies, a meteor trail is only visible at night.

As a meteoroid descends to Earth, atmospheric drag causes the body to slow down, and so frictional heating is reduced. Eventually, surface melting ceases. For a meteoroid of >0.1 m diameter, the interior remains relatively cool during atmospheric entry; in fact, immediately following their fall, meteorites that have been broken open have sometimes been observed to form frost on their inside surfaces. Furthermore, the outer layers of a modest-sized meteorite will be cooled in the lower regions of the atmosphere, such that by the time it lands, the outer crust is also usually cool.

250 µm

Figure 9.9 Cosmic spherules, collected from the floor of the Pacific Ocean at a depth of about 5 km. (D. Brownlee, University of Washington)

9.1.2 The meteorite–asteroid connection

Meteorites can be broadly classified as irons, stones or stony–irons, reflecting their parent asteroid composition. As their names might suggest, irons (or **iron meteorites**) are predominantly composed of iron metal while stones (or **stony meteorites**) are composed mostly of silicate minerals. **Stony–iron meteorites** are mixtures of metals and silicates. Figure 9.10 shows a variety of different meteorite types.

In order to learn something of the origins of meteorites it is important to know the relative proportions of irons, stones and stony–irons that arrive at the Earth. For instance, which meteorites are the most common and thus most typically represent the source? Studying the meteorite finds cannot answer this question. This is because a piece of metal weighing several kilograms is obviously unusual and more likely to be collected as a find than a silicate meteorite, which may look like any other piece of rock. In order to properly assess which sorts of meteorites are most common it is necessary to consider only those samples *observed to fall*. Using this criterion, irons constitute about 5% of all meteorites and stony–irons amount to about 1%. Note that if we had considered only the meteorite finds, irons would appear to constitute 40% of all meteorites.

Stony meteorites are thus the samples that fall to Earth most frequently. Of these, more than 90% are **chondrites**, i.e. samples that generally contain **chondrules**.

■ What are chondrules?

❏ Chondrules are globules of silicate minerals, up to a few millimetres in size (see Box 2.3).

Figure 9.10 Different types of meteorites. (a) A cut and polished surface of an iron meteorite that has been etched with acid. (b) A carbonaceous chondrite, representing some of the most primitive material in the Solar System. (c) A stony meteorite of basaltic composition, formed by melting processes on its parent asteroid. (d) A stony-iron meteorite, or pallasite, that comes from deep within an asteroid, that is from the boundary of an iron-rich core and silicate-rich mantle. (Copyright Natural History Museum)

(a)

(b)

(c)

(d)

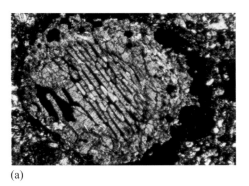

(a)

(b)

Figure 9.11 Meteorites under the microscope. (a) A relatively large, mm-sized chondrule from the Cold Bokkveld meteorite (carbonaceous chondrite). Individual crystals are clearly visible within the chondrule. The dark matrix represents a fine-grained silicate dust containing water-bearing minerals and organic compounds. (b) A collection of different types of chondrules, including fragments and one that has been squashed to some extent, from the Sharps meteorite (ordinary chondrite). The scale of both images is the same. (Copyright Natural History Museum)

In detail, chondrules are small spherules of silicate materials (olivine, pyroxene and glass, with some minor amounts of iron–nickel metal and iron sulfide) which experienced melting *before* they were incorporated into meteorites. Their generally spherical shapes attest to the formation from molten droplets within low gravitational fields. Laboratory experiments have shown that chondrules would have formed in minutes to hours, implying that they formed in local, transient heating events (and not, for instance, as a result of the entire solar nebula cooling at once). The exact formation mechanisms of chondrules are the subject of much activity and debate. While they are mainly considered to have formed by condensation, or re-melting of dust, in the solar nebula, it is also possible that some of them are formed at the surfaces of planetesimals as the result of impact melting. Some examples of chondrules are shown in Figure 9.11.

Most chondrites belong to a group known as the **ordinary chondrites**. The term 'ordinary' is used because they are the most common sort of meteorite. Ordinary chondrites contain varying proportions (5 to 15%) of iron–nickel metal, a factor which, along with the presence of chondrules, makes them distinctive from terrestrial rocks. Another important chondrite group is represented by the **carbonaceous chondrites**; these are carbon-rich meteorites that contain organic compounds such as amino acids. Overall, the carbonaceous chondrites have a *primitive* chemical composition (see Box 2.3). The so-called **achondrites** are stony meteorites that do not contain any chondrules; these have generally been heated to their melting temperatures, and formed by crystallization of magmas on their respective parent bodies. Some achondrites are mechanical mixtures of melted fragments and pieces of chondrites. These meteorites formed at the surface of a parent body where, early in the history of the Solar System, impact processes assembled a variety of materials.

A typical collected stony meteorite has a mass of 1 to 10 kg (diameter 0.1 to 0.2 m), while the mass of an iron is generally in the range 10 to 40 kg. Although meteorites come in all shapes and sizes, there are many more small samples than large ones. Meteorites with masses greater than 100 kg are rather rare (recall that Hoba has a mass of 5.5×10^4 kg). The smallest micrometeorites (i.e. micrometre-sized dust) have masses of around 10^{-15} kg. Samples of this nature are constantly falling to Earth. Indeed, a person who spends a few hours outside during the day is likely to be hit by at least one extraterrestrial particle (of this sort of size). This constant rain of material to Earth amounts to at least 10^7 kg per year. This sounds a lot, but the *total* annual flux of extraterrestrial material, which includes bodies larger than micrometeorites, is estimated to be about 10^8 kg per year. It should be noted that since the mass of the Earth is 5.98×10^{24} kg (that is more than 16 orders of magnitude larger) the yearly input of extraterrestrial material does not cause any obvious changes to the physics or chemistry of our planet as a whole.

QUESTION 9.1

Assuming that the flux of extraterrestrial material has remained constant over the last billion years, calculate the mass and volume that this extra matter has added to our planet during this time. If all the extraterrestrial materials were available now as a surface layer spread uniformly across the whole of the planet, how deep would this be? (Assume a density for the incoming material of $1.5 \times 10^3 \, \text{kg m}^{-3}$.)

What does it feel like to be hit by a small meteorite? Let's consider how fast these objects are travelling. In space they travel at speeds typically of 20 to $40 \, \text{km s}^{-1}$ (for comparison, the speed that a rocket needs to escape from the Earth's gravitational field is about $11.2 \, \text{km s}^{-1}$; the velocity of a bullet from a gun is of the order of $1 \, \text{km s}^{-1}$). Fortunately the Earth's atmosphere acts to retard projectiles with masses of less than a few kilograms (equivalent to a body of 0.1 m diameter). These smaller bodies fall to Earth under the influence of gravity opposed by atmospheric drag and have impact speeds of less than $5 \times 10^{-2} \, \text{km s}^{-1}$. In fact, for sub-millimetre meteorites, the final fall speed is nearer $10^{-5} \, \text{km s}^{-1}$. The smallest micrometeoroid dust particles (micrometre-sized) are essentially totally decelerated in the atmosphere, and gently 'float' to the ground over a period of days or even months. Much larger objects, on the other hand, arrive at the Earth's surface with a significant fraction of the speed they had in space.

- ■ What physical effects will large meteorites exert on our planet?

- ❏ You saw from Chapter 4 that, following the collision of large cosmic debris with various planetary bodies, impact craters are produced which may be many kilometres in diameter.

You have seen how meteorites can be distinguished, on the basis of their principal constituents, as irons, stones etc., and you have learned that meteorites are thought to originate in the asteroid belt. In Chapter 7 you learned about different sorts of asteroids. If we could relate the various meteorite types to asteroid groups we would effectively establish their origin, and be able to study the asteroids in great detail using conventional laboratory-based analytical techniques.

- ■ How are asteroids classified?

- ❏ Asteroids can be classified (among others) as C-type and S-type, using observational astronomy (see Chapter 7) which entails measuring the brightness of reflected sunlight at different wavelengths, to obtain reflectance spectra.

Meteorite surfaces can also be analysed using a similar sort of measurement. In this way we find that the C-type asteroids most closely resemble carbonaceous chondrites, while the S-type asteroids represent stony–irons. A further group, the M-type asteroids, look like iron (metallic) meteorites. The agreement between results obtained from the observations of asteroids and analyses of meteorites is strong circumstantial evidence for a relationship between the two. Unfortunately we have not yet found a direct match between the most common group of meteorites, the ordinary chondrites, and any asteroid taxonomic class. This is because the very surfaces of asteroids (i.e. the parts that are observed) are altered by the phenomenon of 'space weathering' which, in the case of the rocky bodies, has resulted in subtle changes that have acted to confound the observational data.

Micrometeoroids do not have a single origin. Some are formed by collisional processes in the asteroid belt, just like their larger counterparts. Others are ablation products formed in the Earth's atmosphere during the passage of the parent meteoroid. However, the majority of small (sub-mm) micrometeoroids represent fresh debris liberated by comets near perihelion (see Chapter 7).

In addition to asteroids and comets, a few meteoroids have origins on planetary bodies. Several meteorites are known to have come from the Moon. These were ejected when the Moon was itself bombarded by a projectile (asteroid, or comet). How can we be so sure about their lunar origin? One way is to compare the chemistry of these samples directly with materials returned from the Moon (by US and Soviet spacecraft during 1969–76). A further group of meteorites (currently represented by more than 20 specimens) has been recognized as unusual for over a century – these samples are now thought to have originated on Mars.

QUESTION 9.2

A meteorite has been found which is composed of a mixture of iron and silicates. Within the silicate portion there are fragments of materials recognizable as chondrites and achondrites. The meteorite has obviously been subjected to intense shock. Consider how it might have formed.

QUESTION 9.3

20 000 meteoroids of mass >0.1 kg arrive at the Earth's surface each year. On average, what is the frequency with which materials of this size would be encountered in a town with a radius of 5 km?

9.2 The forensic record

9.2.1 Cosmic sediments

The meteorite known as Allende (pronounced eye-endy) fell in 1969. Like many other meteorites, Allende fragmented during descent through the atmosphere, resulting in a shower of stones. More than 2000 kg of material was subsequently gathered over an area of about 300 km², in the vicinity of Pueblito de Allende in Mexico. Murchison, a shower of 500 kg, also fell in 1969, in Victoria, Australia. Meteorites like Allende and Murchison are carbonaceous chondrites and have been studied intensively. You will see in this section how some of the components of these meteorites appear to document changes in formation conditions within the solar nebula. You will also learn that some of their organic compounds pre-date the formation of the Solar System; by studying these materials it is possible to gain an insight into interstellar chemistry.

You may recall from Chapter 2 that, when all but the most volatile elements are considered, the CI carbonaceous chondrites have chemical compositions similar to the Sun.

■ Why are Earth rocks not solar in composition?

❏ Rocks at the Earth's surface are unrepresentative of the Earth as a whole because of geological processes which have acted to segregate elements on a planetary scale (a simple example is the concentration of iron and nickel in the Earth's core).

The carbonaceous chondrites enable us to undertake laboratory studies of materials that are fully representative of the chemical composition of the solar nebula (remember that we called this a *primitive* composition). Recall the data in Figure 2.2. There is clearly a very wide range of abundances of different elements in the Solar System.

■ How much less abundant is silver (Ag) than silicon (Si)?

❏ The abundance of Si is (by convention) 10^6, whereas the abundance of Ag is about 10^{-1} in both the Sun and in carbonaceous chondrites. Thus the abundance of Ag is about seven orders of magnitude less (i.e. seven powers of ten less) than the abundance of Si, which means there are about 10^7 atoms of Si for each atom of Ag.

Despite the wide range of abundances, elements that are abundant in carbonaceous chondrites (e.g. Si, Mg, Fe) are also abundant in the Sun, and those that are rare in carbonaceous chondrites (e.g. Ag, Eu, Rh) are also among the rarest in the Sun. This can be seen by the fact that the elemental abundances in Figure 2.2 fall fairly close to the straight line representing equal abundances in the Sun and carbonaceous chondrites. The relative abundances of the elements in the carbonaceous chondrites are often referred to as representing **chondritic composition**.

Thus, so far as we can tell by examining primitive meteorites, they have pretty much the same composition as the Sun (gas-forming elements such as H and He excepted), and so appear to reflect the composition of at least that part of the solar nebula in which they formed.

We have already said that Allende and Murchison are both carbonaceous chondrites. Thus, we would expect them to be very primitive materials, which have not had their chemistry changed by secondary processes. In a general sense this is true, but when we look in detail at carbonaceous chondrites we find that there are various groups that can be distinguished on the basis of slightly different chemical compositions. Allende is a type CV3 carbonaceous chondrite, whereas Murchison is a CM2 (see Box 9.2). The meteorites which most closely resemble the Sun are the CI carbonaceous chondrites. The meteorite Ivuna is a CI1 chondrite and chemically one of the most primitive of all; a progression away from primitiveness is observed in samples of successively higher types (i.e. from CI1 to CM2 to CV3). This can be understood by looking at Figure 9.12; here are plotted the abundances of certain elements found in the silicate and oxide minerals of Murchison and Allende. The elements are arranged from left to right according to volatility. It would be meaningless to plot absolute concentrations in Figure 9.12 since carbonaceous chondrites contain variable amounts of water, as hydrated minerals and organic compounds. Instead, the elemental data are normalized to a non-volatile element – in this case silicon (so, for instance, the Ca/Si mass ratio is used). These are then plotted with respect to the corresponding ratios found in CI meteorites. This may all seem a little complicated until you realize that in Figure 9.12 the chemical composition of a typical CI meteorite would plot as a horizontal line, at a value of 1.0.

All but one of the CI chondrites are of type CI1. Only recently has a CI2 chondrite been found. Thus CI chondrites are often referred to as a single group (see Box 9.2).

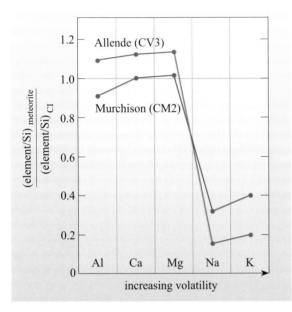

Figure 9.12 Plot of elemental concentrations of aluminium, calcium, magnesium, sodium and potassium (all normalized to silicon) in Allende (CV3) and Murchison (CM2), relative to data from CI meteorites. The most primitive samples are represented by the CI data, i.e. a horizontal line at a value of 1.0.

BOX 9.2 TYPES OF CHONDRITES

Chondritic meteorites, those that contain chondrules, comprise the bulk of the meteorite specimens available for study. You do not need to understand the details of how chondrites are classified, that is beyond the scope of this book, however you will come across the abbreviations that are used to refer to different types of chondrites and this box outlines how these are derived.

Chondrites have been classified on the basis of their mineralogical and chemical composition into 3 main types and a series of letter and number abbreviations are used to refer to them:

1   The carbonaceous chondrites, which can contain up to 5 wt% carbon, are denoted by the letter C. This important, although relatively rare class of meteorites, has been further sub-divided on the basis of chemical and mineralogical composition into about 7 sub-groups. These sub-groups include the CI chondrites such as the Orgueil meteorite, CM chondrites such as the Murchison meteorite and CV chondrites such as the Allende meteorite. The second letter refers to what is considered to be the characteristic meteorite from each sub-group. For the CI chondrites this is the meteorite Ivuna, for the CM chondrites it is the meteorite Murray and for the CV chondrites it is the meteorite Vigarano.

2   The enstatite chondrites that contain the mineral enstatite ($Mg_2Si_2O_6$), a form of pyroxene, are denoted by the letter 'E' and are therefore referred to as the E chondrites.

3   The ordinary chondrites, which comprise the majority of observed falls. The initial letter designation for ordinary Chondrites, O, is usually deleted. Instead, this group is subdivided into meteorites that are relative high in iron, the H chondrites, those with a low iron content of 5 to 10% metallic iron, the L chondrites, and those with very low iron contents (less than 2%) the LL chondrites.

When examined in detail, many classified chondrites are given a further number designation to indicate the state of preservation of the chondrules they contain. For example, the Allende meteorite is classified as a CV3. This number is called the petrologic type of the meteorite and a designation of 3 indicates that the meteorite contains essentially unaltered chondrules. Higher numbers indicate progressively greater amounts of thermal metamorphism, for example the Glatton meteorite found by Mr Pettifor (Figure 9.5) is an L5 indicating that its chondrules have been fairly extensively altered by thermal metamorphism. A designation of 6 would indicate that the meteorite's chondrules had been almost entirely obliterated by thermal metamorphism. Numbers of 1 or 2 indicate that the meteorite's chondrules have been altered by aqueous processes with a designation of 1 denoting that the chondrules have been completely altered by water. Thus the Murchison meteorite, a CM2, contains evidence that liquid water existed for periods of time on the asteroid from which it was derived.

You will find in the scientific literature and in this book that the CI chondrites are usually referred to as a single group without an indication of petrologic type, hence we refer to data being normalized to CI chondrite elemental abundances (see Box 2.3). In fact, all but one of the CI chondrites that have been studied in detail are of type CI1, the exception being the Tagish Lake meteorite that fell in Canada in 2000, which is thought to be of type CI2. You should note that data on Tagish Lake are not included in the average CI chondrite chemical compositions used to compare the elemental concentrations between meteorites and other samples.

■ Refer back to Table 2.4 and assess the volatilities of aluminium and calcium relative to other elements.

❑ Corundum (Al_2O_3) and perovskite ($CaTiO_3$) are some of the first minerals predicted to condense (at temperatures in excess of 1600 K) from a cooling gas of solar composition. Al and Ca are thus refractory elements (the least volatile). For comparison, the volatile elements Na and K are incorporated into alkali feldspars and appear in the condensation sequence at <1000 K.

You can see from Figure 9.12 that the elements Al, Ca and Mg are somewhat more abundant (relative to silicon) in Allende (CV3) than in the other two types. These elements also lie on roughly horizontal lines (other refractory elements also display this trend, although for simplicity these have been left off the figure), showing that refractory elements occur in approximately the same relative proportions to each other in CI–CV3 carbonaceous chondrites (the Ca/Al ratio is virtually constant, for instance). The volatiles in Murchison and Allende (Na and K) are depleted with respect to the refractory elements. Note that the depletion is greater in the CV3 sample than the CM2. Thus, in going from CI to CV3 meteorites, the chemical compositions become progressively less like that of the Sun, i.e. less primitive.

Although the CI meteorites are the most primitive in a chemical sense, it transpires that they have suffered the complication of **hydrothermal alteration**. CI meteorites were endowed with a large complement of volatiles; they contain about 20% (by mass) of H_2O, for instance. Transient heating events on the parent bodies of these meteorites (during impacts, for instance) have acted to mobilize volatiles such as water, which in turn have reacted with the more refractory components. The effect of this alteration is so intense that recognizing primary physical features is difficult – the bulk of the meteorites have been converted into complex hydrated minerals somewhat similar to those found in terrestrial clays.

QUESTION 9.4

The chemical composition of CI meteorites is considered to be the same now as it was when they were formed, even though the original minerals have been altered. What can you say about the nature of the alteration process that has affected CI meteorites?

A detailed look at carbonaceous chondrites shows that the effects of hydrothermal alteration are most intense in CI samples, less so in CM2s and are virtually absent in CV3s. Thus, although a CV3 meteorite like Allende is not the most chemically primitive of samples, it is representative of the least-altered of carbonaceous chondrites. It is therefore a good place to commence the study of meteorites.

Allende is composed of rounded chondrules and large irregular-shaped white inclusions set in a fine-grained, dark matrix. Some examples of chondrules were shown in Figure 9.11. This collection of materials was assembled from nebular dust, upon the outer layers of a planetesimal. This action can be likened to the deposition of a sediment – for this reason we think of carbonaceous chondrites as **cosmic sediments**. Be sure that you understand that there were no liquids present during the meteorite formation process. The water which was imparted to CI and CM2 meteorites settled out with the refractory dust as solid ice grains.

QUESTION 9.5

Consider the refractory elements Ca and Al shown in Figure 9.12. What differences in chemical composition exist between CI, CM2 and CV3 meteorites? How could this be explained?

9.2.2 Refractory bits and pieces

The white inclusions in Allende contain refractory elements. Because of their chemical compositions they are frequently referred to as **calcium–aluminium-rich inclusions**, or **CAIs** for short. The mineralogy of these refractory inclusions matches very well with what is predicted to form at high temperatures in the condensation sequence (see Table 2.4). It would appear, therefore, that the inclusions contain some of the first minerals to form in the cooling solar nebula. Equally well, however, they may represent evaporation residues (i.e. aggregations of minerals that lost all traces of volatile materials through heating and evaporation). Yet another way of looking at the nature of CAIs may be to consider that since they are so refractory, then perhaps they represent lumps of unmelted pre-solar material. In principle a material which condenses from the cooling gas cloud associated with the solar nebula could also have formed by condensation processes operating around other astronomical objects. The chemistry and mineralogy of such materials might be identical in both cases.

Is there a way that we can try to distinguish between the competing theories for the origins of refractory inclusions? Two measurements are of relevance here: age-dating and stable isotope compositions (see below). You have already met radiometric age-dating in Section 2.4.3. Earth Scientists use techniques of this sort in order to date the timing of geological processes. Since the very earliest rocks formed on Earth have all been reprocessed by the action of geology we use meteorites to date the formation of the Earth. A range of different meteorite types give ages that are around 4.5 or 4.6 Ga. For most geologists this is all they need to know. But for meteoriticists they want to understand the detailed chronology of meteorites in order to understand the sequence of events that took place in the solar nebula, and which led to diverse entities such as chondrules, CAIs, primitive melt residues, and so on. And for this purpose a different type of radiometric dating is used, one which uses short-lived radionuclides.

We will not go into the details of the technique here, but in many ways the principles are similar to that of the normal dating techniques. Consider a radioactive isotope like ^{26}Al, which has a half-life of 7.2×10^5 years, and decays to non-radioactive ^{26}Mg by a process known as positron decay. From various lines of evidence we know that ^{26}Al was present at the time when the Solar System formed – we say that the ^{26}Al was 'live' at this time. Where the ^{26}Al came from is another matter and one of continued research – the essential fact is that live ^{26}Al was brought in from the pre-solar environment and then, once in the nebula, began to decay (i.e. ^{26}Al was not being produced within the solar nebula). Consider that if the radioactive aluminium was to become incorporated into minerals (one of those present in the CAIs, for instance) then it would be observable today as a small amount of ^{26}Mg. On the other hand, if the timescales involved in the solar nebula were such that before any Al-bearing minerals could form, all of the ^{26}Al had decayed, then CAIs measured today would contain no ^{26}Mg. Measuring the precise abundance of ^{26}Mg therefore constrains the formation times of minerals *relative* to the time after which the nebula became cut off from its supply of ^{26}Al. The technique can be used to obtain relative ages equivalent to a few half-lives of the radiometric system concerned (at which point scientific instruments become insensitive to the small amounts of decay products that would have accumulated). In the case of the Al–Mg system this means relative ages of a few million years at most. To look at longer or shorter timescales we use different isotopic systems.

Figure 9.13 shows the chronology of the early Solar System, assessed using a different age-dating technique: the ^{53}Mn–^{53}Cr system (half-life of ^{53}Mn = 3.7 Ma). The reason that we have shown the ^{53}Mn–^{53}Cr system is that this can be easily applied to two groups of igneous meteorites (known as HEDs and angrites), which can also be dated using normal (long-lived) radiometric methods. It is not possible to use the ^{26}Al–^{26}Mg system in this case because the mineralogy of these samples is not appropriate. Using a combination of ages derived by the short-lived ^{53}Mn–^{53}Cr system and absolute ages determined in the normal way it is possible to construct a highly detailed chronology of events in the early Solar System. You will learn more about an HED meteorite in Section 9.2.5; the age of 4.563 Ga shown in Figure 9.13 represents the onset of differentiation of the HED meteorite parent body (actual formation ages of HED meteorites span a range from this figure up to about 4.53 Ga, i.e. a range of about 4.53 to 4.56 Ga). Angrites are a very distinctive group of igneous meteorites which all show a similar formation age (of 4.558 Ga).

The acronym 'HED' stands for Howardite–Eucrite–Diogenite.

The exact interpretation of information from radiometric chronometers is the subject of much discussion. Indeed, as you can see from Figure 9.13, there are two possible interpretations when the ages are considered in absolute terms. In contrast, the relative ages are far less ambiguous – as such, we can say with certainty that the time interval represented in Figure 9.13 represents a period of 13 Ma. The processes, as recorded in meteorites, began with the formation of CAIs, which formed by direct condensation

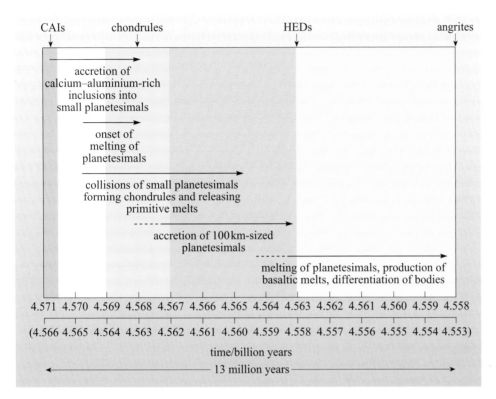

Figure 9.13 Chronology of the early Solar System – an interpretation based on the short-lived radionuclide ^{53}Mn, which decays to ^{53}Cr. The relative ages obtained by the ^{53}Mn–^{53}Cr system are converted into absolute ages by measurements of long-lived radionuclides found in two particular meteorite groups (HEDs and angrites) – these are the ages given in the upper scale. The scale enclosed in brackets (which is displaced from the above by 5 Ma) represents yet another attempt to combine data from short- and long-lived radionuclides.

from a nebular gas of solar composition (a process which lasted less than a million years). Once formed these CAIs began to assemble into small planetesimals; as these bodies grew they started to melt because of impact processes and the effects of heat generated from the decay of short-lived radionuclides (of which the decay of ^{26}Al to ^{26}Mg was the most important). Some 2 Ma after the first CAIs had formed, collisions of CAI-bearing planetesimals were implicated in the formation of chondrules (and at the same time, examples of primitive melts were injected back into the nebula). The chondrule-forming period lasted for about 2 Ma. As this episode began to wane, planetesimals up to about 100 km in size started to form. And as these grew to larger sizes, melting processes operating on planetesimal scales resulted in the differentiation of bodies – i.e. the formation of iron-rich cores, and basaltic mantle rocks (the first examples of geological processing within the Solar System).

You should note that there are some disagreements in detail regarding the absolute ages recorded from meteoritic components. It is for this reason that we assign the age of the Solar System to be a rather non-committal 4.56 Ga. Within the context of Figure 9.13 this represents a time period in which planetesimals were accreting and melting. For the purists the age of the Solar System is considered to be that time when CAIs started to form – some scientists would say that this was 4.571 Ga ago, while others would have this at 4.566 Ga (and, be advised, there are others who would advocate slightly different dates). For our purposes, 4.56 Ga is an honourable compromise!

Meteorites like Allende contain a vast number of CAIs and we have age-dating measurements for just a few of them. Thus, confidence in our belief that CAIs formed in the solar nebula needs to come from other measurements. In this regard **stable isotope compositions** can be used to provide additional information. Stable isotopes are introduced in Box 9.3 using oxygen as a pertinent example.

BOX 9.3 STABLE ISOTOPE COMPOSITIONS

You saw above that certain elements have radioactive isotopes. Aluminium, for instance, has a radioactive isotope, ^{26}Al, which decays to give ^{26}Mg. But aluminium also has another isotope, ^{27}Al, which is not radioactive. Furthermore, the decay product itself, ^{26}Mg, is also not radioactive. ^{27}Al and ^{26}Mg are called **stable isotopes** because they do not decay. Most elements consist of mixtures of stable isotopes (in this regard aluminium is unusual). Oxygen has three stable isotopes: ^{16}O, ^{17}O and ^{18}O. All the isotopes of oxygen contain 8 protons – this is the atomic number. However, the isotopes differ in their number of neutrons; ^{16}O, ^{17}O and ^{18}O have 8, 9 and 10 neutrons respectively.

In astronomical entities, such as stars, the variations in oxygen isotope compositions can be quite large and so the differences are described in terms of their $^{17}O/^{16}O$ and $^{18}O/^{16}O$ ratios. An example of the radial variation of $^{18}O/^{16}O$ ratios of stars within our Galaxy is shown in Figure 9.14 (note that, as yet, $^{17}O/^{16}O$ ratios cannot be measured with sufficient precision to provide meaningful results). What we call the 'solar value' of $^{18}O/^{16}O$ is shown as the horizontal line in Figure 9.14 (in fact this is the isotopic composition of seawater on Earth, of which more below).

It transpires that those meteorites we describe as cosmic sediments contain, in addition to materials formed within the solar nebula, small isolated grains of unmelted materials which pre-date the Solar System. These µm-sized pre-solar grains are remnants of dust from the protostellar disc; their extraction from meteorites, and their ultimate explanation, was one of the major triumphs of meteorite research towards the end of the 20th century. Detailed study of individual grains, which is entirely possible with appropriate laboratory equipment, allows us to investigate nucleosynthetic products from a range of astronomical objects.

1 µm = 10^{-6} m.

Figure 9.14 Plot of $^{18}O/^{16}O \times 10^{-3}$ versus radial distance (i.e. outwards from the centre) for stars in our Galaxy. Note that the distance scale is given in kpc, or kiloparsec (where 1 kpc = 3.084×10^{19} m, or 3260 light-years). On this scale the Solar System would plot at 8.5 kpc. You may be able to detect a trend of gradually decreasing $^{18}O/^{16}O$ with radial distance – this actually represents a Galactic evolutionary phenomenon (which we will not study further here). The figure also includes a scale for oxygen isotope measurements as $\delta^{18}O$, which is described in the text. (Adapted from Wannier, 1980)

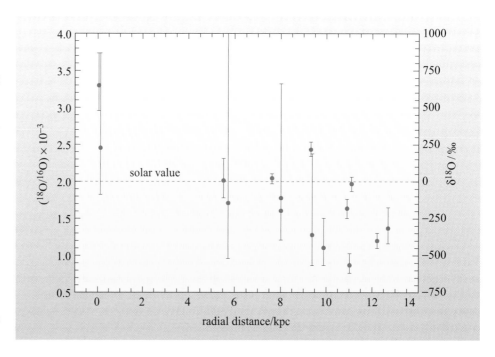

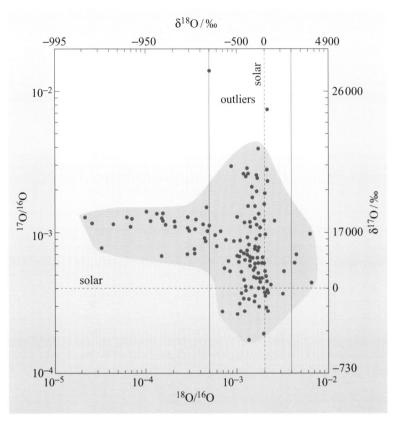

Figure 9.15 Plot of $^{17}O/^{16}O$ versus $^{18}O/^{16}O$ for pre-solar Al_2O_3 grains extracted from meteorites. The figure includes scales for oxygen isotope measurements as $\delta^{17}O$ and $\delta^{18}O$, which are described in the text. The majority of the data for the grains plot within the shaded field. The two outliers probably represent some unusual example of stellar processing. The two solid vertical lines constrain the range of $^{18}O/^{16}O$ values shown in Figure 9.14. While most data for the grains are included within these lines there are clearly points outside this range. Again this represents an interesting aspect of the workings of stars, a phenomenon which can now be studied in great detail by making measurements in the laboratory. The vertical dashed line represents the solar value for $^{18}O/^{16}O$; similarly the horizontal equivalent is for $^{17}O/^{16}O$. Their intersection represents where the Solar System would plot. (Adapted from Nittler, 1996)

Figure 9.15 shows the $^{17}O/^{16}O$ and $^{18}O/^{16}O$ ratios measured from individual grains of meteoritic corundum, Al_2O_3 (we should stress here that these isolated grains of Al_2O_3 are relatively rare and have nothing to do with the bulk of the aluminium and oxygen which is found in CAIs). There are a couple of things to note here. Firstly, the laboratory measurements are much better than those made by observations of stars (Figure 9.14) and, of course, we have measurements of ^{17}O as well. In order to compare the meteorite data with those from stellar measurements, the two vertical solid lines in Figure 9.15 represent the limits of $^{18}O/^{16}O$ represented by the boundaries of the plot in Figure 9.14. The two points shown as outliers in Figure 9.15 have error bars within the individual points – compare those with the data in Figure 9.14. Secondly, the range of $^{18}O/^{16}O$ values recorded from the pre-solar grains is much larger than those measurements of stars – it is information like this that is allowing astrophysicists to refine their models of nucleosynthesis.

For meteorite samples (or indeed, terrestrial rocks) plotting oxygen isotope ratios as shown in Figures 9.14 and 9.15 is not practical because the variations in oxygen isotope composition are relatively minor – tremendous significance may be placed on deviations of less than 1%. In order to describe these small variations it is convenient to express the *differences* in isotope compositions, rather than the absolute ratios. This is done by relating the sample ratio to a reference point. In the case of oxygen, the water which makes up the Earth's oceans is usually used as a reference since this is essentially homogeneous with respect to isotope composition, having $^{18}O/^{16}O$ of 0.002 0052 (which is approximately equivalent to 1/499, i.e. a $^{16}O/^{18}O$ = 499) and $^{17}O/^{16}O$ of 0.000 372 (1/2690, i.e. $^{16}O/^{17}O$ = 2690). On the other hand, the oxygen that makes up the rocks of the Earth's crust has a somewhat variable isotope composition (dependent upon many factors, including formation conditions, temperature, etc.). A typical basaltic rock may have $^{16}O/^{18}O = 496$ and $^{16}O/^{17}O = 2682$. Be sure that you take note of whether we are using, for instance, $^{16}O/^{18}O$ or $^{18}O/^{16}O$ ratios. Astronomers tend to quote isotope ratios as high/low abundance (e.g. $^{16}O/^{18}O$), whereas stable isotope chemists would use $^{18}O/^{16}O$ (there are good reasons for this but we do not have space to go into them here). The net result is that one has to become accustomed to switching between the systems. Colloquially, at least, most people prefer to talk in terms of ratios like $^{16}O/^{18}O$, but as you will see below, data manipulation relies on $^{18}O/^{16}O$. That's life!

The difference between the oxygen isotope composition of ocean water and rocks is relatively small but easily and reproducibly measurable using an instrument known as a mass spectrometer.

It is a convention to relate the stable isotope composition of an unknown sample to that of a reference material. Thus if R_s is the isotope ratio of the sample and R_r is the corresponding ratio in the reference, the difference is $R_s - R_r$, which when expressed as a fraction of R_r, becomes

$$\frac{R_s - R_r}{R_r} = \frac{R_s}{R_r} - 1$$

In this formula, we use the ratio of the rare to the abundant isotopes since, in effect, we are interested in variations of the rare isotopes. Because these differences are very small, they are usually quoted in parts per thousand (known as parts per mil), abbreviated to ‰.

> Comparing this with parts per hundred, that is per cent (%), you can see that $1\% = 10‰$.

The calculated result, δ (delta, in ‰), is called the *differential isotope composition*, or more commonly 'the δ value'. Thus

$$\delta(\text{in ‰}) = \left(\frac{R_s}{R_r} - 1\right) \times 1000 \qquad (9.1)$$

Since oxygen has three stable isotopes there are two different delta values, $\delta^{18}O$ and $\delta^{17}O$, defined relative to the abundant isotope, ^{16}O. Thus, for a $\delta^{18}O$ value, R_s and R_r refer to $^{18}O/^{16}O$ ratios. Similarly, for a $\delta^{17}O$ value, R_s and R_r refer to $^{17}O/^{16}O$ ratios. For the basaltic rock referred to above, inserting values into Equation 9.1 we get

$$\delta^{18}O = \left(\frac{1/496}{1/499} - 1\right) \times 1000 \approx 6‰$$

Since δ values can be either positive or negative numbers, we explicitly give each of them a sign, thus, in the above example, $\delta^{18}O = +6‰$, rather than just 6‰. It can be shown that for this basaltic rock, $\delta^{17}O$ is H $\approx +3‰$, equivalent to about one-half of the value of $\delta^{18}O$. The factor of about 0.5 difference between $\delta^{17}O$ and $\delta^{18}O$ holds true for almost all oxygen-containing materials on Earth. This relationship, which can be related to the difference in mass between the isotopes, is a consequence of the process known as *isotope fractionation*, which occurs when isotopes become separated from each other, or fractionated. (Isotope fractionation can happen during chemical reactions, or during physical processes such as the movement of a gas or liquid through a porous medium.)

What are the $\delta^{17}O$ and $\delta^{18}O$ of ocean water? What are the $\delta^{17}O$ and $\delta^{18}O$ of pure ^{16}O?

On a plot of $\delta^{17}O$ and $\delta^{18}O$, the oxygen isotope composition of samples from the Earth lie on the line called the **terrestrial fractionation line (TFL)** which has a slope of about 0.5 *and* passes through the point (0, 0) where ocean water would plot.

No matter what oxygen isotope composition an individual body (planetary or otherwise) has initially, *secondary* processes that result in isotope fractionation produce data that fall on lines of slope 0.5. When this is observed it is a good indicator that materials are related in some way. Two lines of slope 0.5 that are *not* coincident imply different *initial* oxygen isotope compositions.

■ What do you think would be the slope of an oxygen isotope fractionation line for Mars (i.e. an MFL)?

❑ The MFL value would be 0.5. Indeed, this has been measured for different constituents of Martian meteorites. For information note that the MFL is offset from the TFL, showing that the two planets formed from materials that were slightly different in terms of their oxygen isotopic composition (which means that even though a particular rock from Mars might look identical to one from Earth, it can be distinguished on the basis of its oxygen isotopic composition).

A plot of $\delta^{17}O$ and $\delta^{18}O$ for whole-rock meteorites is shown in Figure 9.16. Compare the scales of this plot with those of Figure 9.15; while astronomical entities display astronomical variations in oxygen isotopic compositions, those of materials formed within the solar nebula are much less variable (although the small differences which can be measured are of immense importance). Note the line of slope 0.5 representing the TFL.

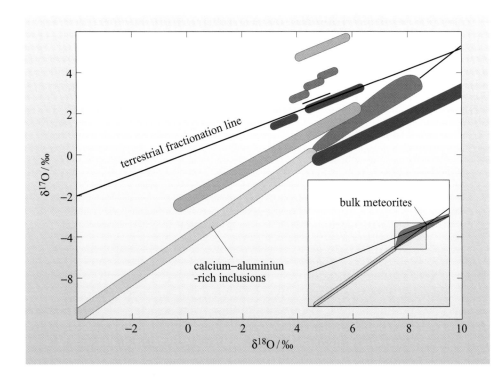

Figure 9.16 Plot of $\delta^{17}O$ and $\delta^{18}O$ for whole-rock meteorite samples. Each different coloured field represents a different type of meteorite group (included on the plot are fields for different groups of carbonaceous chondrites, different types of ordinary chondrites, Martian meteorites, HEDs, and so on). The inset shows the range in oxygen isotopic composition of all bulk meteorite samples and covers a range in $\delta^{18}O$ values from −40 to +20‰ and in $\delta^{17}O$ values from −40 to +30‰. The region of the main graph is shown by the shaded box in the inset. (Franchi *et al.*, 2001)

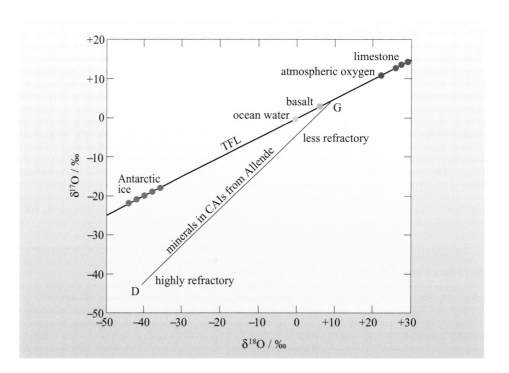

Figure 9.17 Plot of oxygen isotope compositions ($\delta^{17}O$ versus $\delta^{18}O$) of different minerals from refractory inclusions in Allende. The Allende data fall on a line (DG) with a slope of about 1. For reference, a line of slope about 0.5 and labelled TFL is also shown. Almost any sample from the Earth will have an oxygen isotope composition that falls somewhere on this TFL line (a few examples are shown).

Returning to the CAIs, if the inclusions represent samples of pre-solar dust, we may expect them to have a variety of different oxygen isotope compositions, like the isolated Al_2O_3 grains which we encountered in Figure 9.15. This is because the oxygen isotopes would have been synthesized in various ways, with materials from different stars having different isotopic characteristics. In contrast, if CAIs were formed during condensation in the solar nebula we might expect a fairly straightforward relationship between the oxygen isotope compositions of the different inclusions. This is because a hot, turbulent nebula environment will act to homogenize the isotope compositions of pre-solar gas and dust.

A plot of $\delta^{17}O$ versus $\delta^{18}O$ for data from the Allende CAIs is shown in Figure 9.17. In contrast to samples from the Earth, the oxygen isotope composition of different minerals from CAIs define a line labelled DG which has a slope of about 1. That the data fall on a line, rather than at random across the plot, tends to suggest that the minerals were all formed in a common event. As it happens, refractory minerals, like spinel (Table 2.4), fall close to the lower end of the line (D), while less refractory minerals plot close to G.

▪ Remember that refractory minerals condense early in the condensation sequence. How can we explain the oxygen isotope data for the different minerals in CAIs?

☐ The oxygen isotope composition for spinel must have differed from that of less refractory minerals, in the reservoir from which the minerals condensed.

The data have been explained in terms of a dust reservoir (D) and gas (G). The solar nebula started as a mixture of gas and very fine dust. It is proposed that of the minerals that we find in the CAIs, the first to form, were somehow derived from the dust and reflect its initial oxygen isotope composition. This involved reprocessing of dust material, and so it is appropriate to refer to a dust *reservoir* rather than dust particles. As mineral formation proceeded, the isotope composition of the dust reservoir became modified by a process known as **isotope exchange** with the gas reservoir. The gas reservoir was much the larger, and so this exchange led to the dust reservoir becoming more like the gas reservoir. In consequence, the less refractory minerals that formed later, and now constitute part of the CAIs, lie closer to point G.

QUESTION 9.7

Consider for a moment that the dust and gas reservoirs in the solar nebula *both* had oxygen isotope compositions at G in Figure 9.17. Where would the isotope compositions plot if we mixed small quantities of pure ^{16}O into the system? (Remember the δ values of pure ^{16}O from Question 9.6 and consider where this would fall on a plot like Figure 9.17.)

There are two important conclusions from the study of Allende CAIs. Firstly, not all of the oxygen in the solar nebula had the same initial isotope composition. This means that the solar nebula was not completely homogenized before condensation began (if this had happened the dust and gas reservoirs would be related to each other along a line of slope 0.5 in Figure 9.17, in the same way that different components from the Earth are related). Thus, meteorites contain some relicts of the pre-solar gas and dust which coalesced to form the solar nebula. Secondly, the oxygen isotope compositions of minerals in CAIs show a relatively simple pattern, i.e. evidence for only two major components with isotope exchange between them. This simple pattern was almost certainly established within the solar nebula, rather than before the solar nebula formed.

If we were to plot oxygen isotope data for the chondrules in Allende in Figure 9.17, we would find that they fall on a different line from that shown by CAIs. Although one end of the line would be coincident with D, the other would be slightly displaced from G. Thus it would seem that in between the formation of CAIs and chondrules, the oxygen isotope composition of the solar nebula had changed.

QUESTION 9.8

What are the consequences for planetary bodies in general for a variation with time in the oxygen isotope composition of the solar nebula?

QUESTION 9.9

Different sorts of matrix materials from Murchison have oxygen isotope compositions which, on a plot of $\delta^{17}O$ versus $\delta^{18}O$, fall on a line with a slope of about 0.5. This line is parallel to, although displaced from, the TFL. How can the results be interpreted?

9.2.3 Riding the x-wind

It seems appropriate here to delve back into the details of solar nebular dynamics to show how data from meteorites and other samples are helping to constrain models of Solar System formation (Box 9.4). The idea we describe here, known colloquially as the 'x-wind model', is very much a crossover from astrophysics to the planetary sciences. Here a generalized theory of the formation of low-mass stars has been extended to incorporate mechanisms that would allow production of the two main high-temperature features of chondritic meteorites, namely chondrules and CAIs. The formation model starts off with a relatively dense protostar forming at the core of a molecular cloud following the loss of the supporting magnetic field. Rotation involved with the cloud collapse produces, not only a disc around the central protostar, but also a strong stellar wind operating outwards from the polar regions of the star.

■ What is this phenomenon called?

❑ A bipolar outflow (Chapter 8).

Dust and gas accrete onto the central star via the disc, although at the same time the outflow begins to push away the outer envelope of the cloud. Eventually the outflow blows away the surrounding gas and dust entirely and the young stellar object and associated disc becomes visible (and hence observable by telescopes).

■ What name do we give to these kinds of objects?

❑ T Tauri stars (Chapter 8).

It is known that protostars emit more x-rays than young Sun-like stars in more advanced stages of their formation. This is taken as evidence for enhanced magnetic activity on the surfaces of protostars. It is the detailed evaluation of the magnetic fields around the protostar, and how these affect the flow of gas and dust, that results in the identification of the so-called x-wind. In Figure 9.18 we show an overall schematic of the magnetic field lines associated with a typical Sun-like protostar. The x-region is effectively the boundary of the inner edge of the accretion disc (a distance of perhaps 0.075 AU or so, i.e. close to the protostar's surface, but

BOX 9.4 THE GENESIS MISSION

The origin of the oxygen isotopic variations in solar system materials has been a major puzzle for planetary scientists for nearly 30 years. To help solve it, the Genesis spacecraft was launched in August, 2001. The spacecraft contains a number of instruments to monitor the stream of atoms ejected from the Sun into space, i.e. the solar wind, Several different collectors can also be exposed, allowing collection of samples of the solar wind for return to Earth. The abundance of approximately 70 different elements in the periodic table are scheduled for measurement in this way, with the isotopic composition determined for almost 20 of these elements, including the isotopes of oxygen. By returning the samples to the laboratory, high precision measurements can be made at a level unattainable remotely.

It is hoped that a more accurate determination of the oxygen isotopic composition of the Sun, and hence the solar nebula, will provide a test for different theories for the origin of the oxygen isotopic variations among CAIs, chondrules, asteroids and planets. The samples will be returned via a re-entry capsule, which separates from the main portion of the spacecraft and returns to Earth in September 2004.

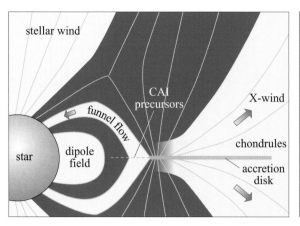 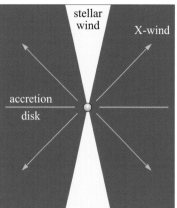

Figure 9.18 Schematic diagram of the magnetic field lines near to a Sun-like protostar. The accretion disc extends inwards to a point which is close to, but not touching, the surface of the protostar. This inner edge of the accretion disc marks the position of the x-region. Dust spiralling towards the protostar, from within the disc, is either accreted onto it (via funnel flow) or blown upwards and outwards (via the x-wind). (Shu *et al.*, 1997)

not touching it). Materials heading towards the central regions of the solar nebula, i.e. spiralling inwards within the accretion disc, arrive at the x-region and either get channelled towards the protostar (along a trajectory labelled as the 'funnel flow' in Figure 9.18), or blown back out of the system by the x-wind. About two-thirds of the material accretes onto the star, the rest is carried off by the x-wind.

Dust grains arriving near the x-region can become heated and melted by the radiation streaming off the protostar, ultimately forming CAIs and chondrules. These are then either accreted onto the star or sent back out of the system by the x-wind. In this way CAIs and chondrules become dispersed across the entire solar system. It is these materials which ultimately aggregate to form planetesimals and planets.

There are some interesting testable aspects to this hypothesis. Firstly, by direct study of CAIs and chondrules it should be possible to infer some of the characteristics of the x-wind all of which have long since disappeared. Secondly, since CAIs and chondrules are essentially the fundamental building blocks of Solar System objects, they should be present, for instance, in comets. Note also that if the x-wind model is a truly generic process then chondrules and CAIs will be a universal feature of solar systems. In other words, we should not be surprised to see chondrite-like debris if we ever get to visit another solar system. Additionally, we have to keep an open mind when considering the characteristics of interstellar dust. Clearly some of the materials which originally coalesced to form our own solar nebula may have included chondrule-like objects from previous generations of solar systems.

9.2.4 Low-temperature materials

In addition to chondrules and CAIs, Allende and Murchison also contain a dark, fine-grained **matrix** – effectively, everything else. The matrix is dark in colour, partly because it is fine-grained, but also because it contains carbonaceous materials. In Allende the matrix is predominantly an assemblage of mineral components such as olivine and sulfides (e.g. troilite, FeS). These minerals formed as a logical progression in the condensation sequence (see Table 2.4). The fine-grained nature of the constituent grains may be due to fragmentation processes occurring in the solar nebula. In contrast to Allende, the matrix of Murchison contains a large proportion of hydrated minerals. You can see from the condensation sequence that in a cooling gas of solar composition, hydrated minerals condense directly in the nebula at temperatures of 330 to 550 K.

■ What can you deduce from the presence of hydrated minerals in the matrix of Murchison (and their absence in Allende)?

❑ The temperature of the nebula was in the region of 330 K to 550 K when the matrix of Murchison was formed. That of Allende, on the other hand, must have formed at temperatures in excess of 550 K.

Not all of the features displayed by the matrix of Murchison can be explained by processes which took place during the condensation sequence. For instance, water-soluble substances, such as carbonates and sulfates, are found deposited as veins running through the fabric of the meteorite. It seems that Murchison, like CI meteorites, has been affected by the action of fluids some time after it formed. Unlike the CI meteorites, however, the alteration in Murchison is not that extensive – the primary constituents such as chondrules and CAIs are still recognizable. On the basis of geochemical evidence it can be deduced that the hydrothermal alteration took place at about 300 K.

Turning now to the carbonaceous materials, the matrix of Allende contains about 0.25% by mass of carbon, mainly in an amorphous form. Murchison, on the other hand, contains about an order of magnitude more carbon (i.e. about ten times more), not in amorphous form but as *organic compounds*, i.e. materials made up from the elements carbon and hydrogen, with the possible involvement of nitrogen, oxygen, sulfur, etc. Most of the organic material in Murchison exists as a complex macromolecular substance, meaning a component comprising many hundreds of atoms which has no clearly defined structure. This is analogous in many ways to the material that on Earth occurs in some sedimentary rocks and is responsible for generating oil. Of the organic compounds that are not part of macromolecular structures, many different constituents are present. Some examples of compounds in meteorites are shown in Figure 9.19. Benzene is a hydrocarbon, notorious because of its carcinogenic properties. Adenine and alanine are, respectively, examples of a purine and an amino acid. On Earth, purines are found in nucleic acids and amino acids are found in proteins. That is not to say that there are nucleic acids or proteins in meteorites! What is present in Murchison are the building blocks of these materials. These same building blocks were presumably brought to the surface of the Earth early in the history of the Solar System, and *may* have triggered the development of life.

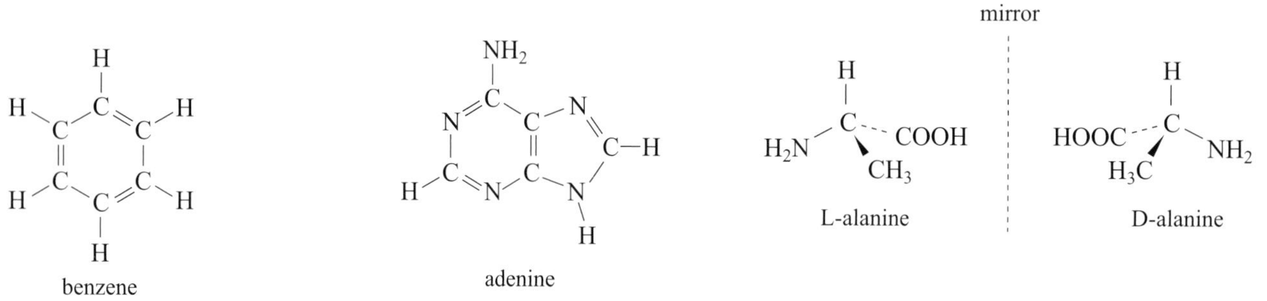

> Amorphous means having no definite shape or form. Thus, amorphous carbon is an occurrence of the element in no obvious crystallographic form – here it implies something that is intermediate between graphite (an obvious crystal form) and a complex of macromolecular organic compounds. Soot and lamp black are good examples of amorphous carbon found on Earth.

Figure 9.19 Three examples of organic compounds from Murchison. Benzene is a hydrocarbon (a compound made only of hydrogen and carbon). Adenine is a purine, a more complicated molecule than benzene, which includes nitrogen in its structure. Alanine is an amino acid; it can exist in two different structural forms, L and D, which are mirror images of each other. (The dashed lines denote bonds directed behind the plane of the paper, and the wedges are bonds coming up out of the plane of the paper.)

Given that organic compounds abound in the Solar System beyond the Earth (not only in primitive meteorites but also in comets, and probably within some of the satellites of the giant planets), mechanisms which might lead to the large-scale production of them in the solar nebula are of great interest. It transpires that there are many different ways in which organic materials can be formed. It is not possible to describe all of these here; we will consider just one possibility. If we continue with the notion of a condensation sequence, then at temperatures below 400 K, gases such as CO (carbon monoxide) and H_2 undergo reactions on the surfaces of earlier-formed mineral grains to form organic compounds. This mechanism, along with others proposed, produces organic materials abiotically, i.e life-forms are *not* involved. Can we be absolutely certain of this? Let us consider the case of amino acids, since these are so closely linked with life. There are many different amino acids; and some of the ones found in Murchison do not form biologically on Earth. At very least, we can be sure that these are not simply terrestrial contaminants introduced to Murchison by micro-organisms.

■ Based on your experience with the study of refractory inclusions, how could we discern whether all the organic compounds in Murchison had a common origin?

❏ One possibility would be to determine the isotope compositions of the constituent elements (carbon, hydrogen and nitrogen) to see if these show any obvious relationships.

The carbon and nitrogen isotope compositions of organic compounds in Murchison are generally fairly uniform, being similar to values found in organic compounds on Earth. This similarity in isotope compositions is taken by some to imply that the majority of organic compounds found in Murchison formed within the solar nebula. However, the hydrogen in some of the organics in Murchison is found to be highly enriched in the isotope deuterium (D, or ^2H). It has been calculated that the magnitude of the observed enrichment cannot be attained in the solar nebula, even under the most extreme conditions imaginable. On the other hand, large deuterium enrichments have been detected in the interstellar medium, using infrared astronomy. The enrichments are found in organic molecules present in dense clouds and are the result of reactions between molecules and ions. It seems, therefore, that some of the organic compounds in Murchison were produced in the interstellar medium. These have survived the processes of heating and mixing in the solar nebula to end up in the organic material found in carbonaceous chondrites.

QUESTION 9.10

If the majority of organic compounds in Murchison were formed in the solar nebula, as a consequence of condensation, how can we interpret the survival of interstellar organic molecules in this meteorite?

QUESTION 9.11

Allende contains carbon mainly in an amorphous form but also as diamonds. How might we try and assess if these elemental forms of carbon are related?

9.2.5 Asteroidal melting processes

Juvinas is an achondrite which, as you should remember, means that it is a stony meteorite containing no chondrules. In detail it is one of the HED meteorites which made an appearance in Figure 9.13. It is thought to have been formed near to the surface of its parent body by an igneous process, i.e. a high-temperature melting event which involved the creation of a magma, of which Juvinas is a fragment. The Cape York meteorite is also the product of melting, although its parent body was different from that of Juvinas. Cape York is an iron meteorite and represents part of an asteroid's core.

Because of melting, Juvinas and Cape York no longer retain their primary chemical and physical characteristics. They are referred to as differentiated meteorites.

■ What is differentiation?

❑ It is the process by which a hot, possibly molten, planetary body segregates into compositionally distinct layers of different density (Chapter 2).

Consideration of iron will help to explain differentiation. When the parent bodies of meteorites were first formed they probably all had a chondritic composition. We know that in primitive chondrites, iron constitutes 20 to 25% by mass and is present in silicates and sulfides, and as the metal itself. However, in Juvinas, iron is present at about 14% by mass, and is almost entirely in silicate minerals – so clearly there is a deficiency of iron in this meteorite. The depletion can be explained by the effects of partial melting (Chapter 2), whereby the magma that solidified to form Juvinas became separated from its source region, which was left with a relatively higher proportion of iron. The extent of partial melting was fairly limited in this case – thus, although the source region became enriched in iron it probably did not form into a metallic core. In contrast, Cape York is >90% iron metal, which represents almost the end-point of total melting. In this case the silicates have been completely removed, leaving a residue of iron–nickel metal.

The observation of meteorites with a variety of iron content led early workers to propose that all of these materials came from a single disrupted planet which once occupied an orbit between Mars and Jupiter. The irons were thought to represent the core of this planet and the igneous samples were considered to be derived from its crust or mantle. The chondrites were assumed to represent samples of unmelted material which had survived unaltered at the planet's surface. This single-parent model has long since been untenable as detailed chemical and isotope evidence demands many different parent bodies for meteorites, e.g. recall the plot of $\delta^{17}O$ versus $\delta^{18}O$ for whole-rock meteorites shown in Figure 9.16. It is now considered that the larger asteroids are planetesimals which were large enough to have undergone a variety of geological-type processes, but (as noted in Chapter 8) were *prevented* from accreting into a single planetary body by the gravitational influence of Jupiter.

Juvinas is one of a group of about 60 related meteorites. The parental material of Juvinas and its associates must have been extruded onto the surface of an asteroid as a thin layer of magma which then cooled relatively quickly. By studying the chemistry and mineralogy of Juvinas we can learn about the magma itself and also about the source region from which it was extracted.

Looking at a specimen of Juvinas shows it to be a basalt, i.e. comprised mostly of the silicate minerals pyroxene and feldspar. From the mineralogical evidence it is possible to conclude that the sample crystallized within the temperature range 1300 K to 1500 K. Be sure that you understand this is the magmatic temperature, and not a measure of temperature during condensation from the nebula.

As already stated, the heating and melting episode that produced Juvinas has acted to change the original chondritic composition. But how can we be absolutely sure that the parent body of Juvinas had a chondritic composition to start with? All we have to work with is a sample of solidified magma, and we have already seen that the chemical composition of this has been changed by the melting process. Consider two more ways in which the chemistry has been changed. Firstly, you saw in a previous section that carbonaceous chondrites contain carbon – however, Juvinas contains hardly any carbon. This is because any organic materials that were once present in the parent body have been destroyed by heating and lost as H_2O and CO_2. Secondly, refractory elements, such as osmium, rhenium and iridium, are depleted relative to chondrites; like iron, these elements were left behind in the source region during partial melting. Fortunately there is a way to assess the original chemical composition of Juvinas. This is by consideration of the **rare earth elements (REE)**, a group of 15 related elements spanning the Periodic Table from atomic number 57 (lanthanum) to 71 (lutetium).

The REE are not present in any meteorite samples at high enough concentrations to form minerals in their own right. Instead they are found as substitutions for other elements in host minerals. The different mechanisms by which REE become incorporated into minerals tend to produce characteristic abundance patterns. The abundances of REE in Juvinas are shown in Figure 9.20, where it can be seen that

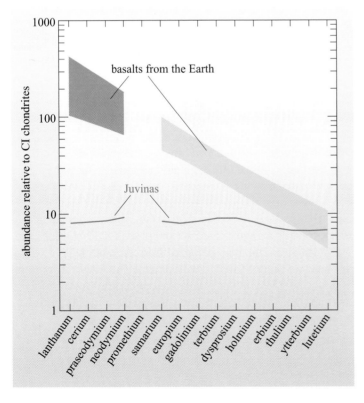

Figure 9.20 Rare earth element abundances in Juvinas and in basalts from the Earth. The data are related to the abundances present in CI chondrites (which, if plotted, would fall on a horizontal line at a value of 1). Since different rock types show a wide range of REE abundances it is useful to plot the data on a logarithmic scale. Note that there are no data for promethium since this element is entirely radioactive.

when compared to CI chondrites, they are uniformly enriched by a factor of about eight. The reason for the enrichment is that during partial melting the REE have become concentrated in the magma rather than the residual solid. Note that the REE in Juvinas plot on what is roughly a horizontal line. This is known as a chondritic pattern since chondrites also plot on a horizontal line (at a value of 1). In contrast, data from basalts found on Earth plot on a steeply sloping line, which demonstrates that these rocks have formed by a more complex series of processes than the simple melting experienced by meteorites like Juvinas.

Let us now consider how the parent body of Juvinas became heated to temperatures great enough to produce magmas.

■ How might the parent body have been heated?

❏ As outlined in Chapter 2, accretion, impacts, core formation, radioactive decay and tidal heating are all possible sources for the thermal energy that causes melting in planetary bodies.

For various reasons, we can discount all but one of these heat-producing mechanisms as causes of melting in asteroids.

- A small body, such as an asteroid, has a large surface-to-volume ratio (compared to a large body) and so thermal energy within the body is rapidly radiated away from its surface. Thus, although kinetic and gravitational energy can be converted to heat during accretion, for small bodies this source is insufficient to cause melting.

- Heating by core formation is discounted since, as you have already seen, a metallic core was probably not formed in the case of the Juvinas parent body.

- Radiogenic heating by long-lived radioactive isotopes, such as ^{40}K, ^{232}Th, ^{235}U and ^{238}U, can produce high temperatures in large planetary bodies, but is again insufficient in small ones.

- Tidal heating may be an important process for the satellites of the giant planets, but not for asteroids.

- Heating through asteroid collisions may be a possibility, but if the process is too violent the bodies will be totally disrupted instead.

So, how *did* asteroids melt? There are two possibilities. The first involves heat production by electromagnetic induction, which you have not met before in this Course. This could arise as the bodies moved through a dense solar wind-like plasma, as might have been produced by the young Sun (during its T Tauri phase). However, there is still some debate about the details of this period of evolution and so it is impossible to assess the significance of this mechanism. A second plausible way of producing heat in asteroids is by short-lived isotopes, such as ^{26}Al. Since the half-life of ^{26}Al is only 7.2×10^5 years, the heat potential from this source is confined to the very early stages of Solar System development. The $^{26}Al/^{27}Al$ ratio in the interstellar medium is about 10^{-5} (recall that ^{27}Al is the only stable isotope of aluminium). If aluminium with this isotope composition was present in chondritic materials, the heat energy released by the radioactive decay of ^{26}Al would be sufficient to melt an asteroid of a few kilometres diameter.

QUESTION 9.12

(a) ^{26}Al decays by positron emission to give stable ^{26}Mg. How could you tell whether a mineral component from a meteorite had once contained ^{26}Al? (b) In what sorts of minerals would the occurrence of previous ^{26}Al radioactivity be most easily recognized?

Unfortunately, as far as Juvinas is concerned, it is impossible to discern absolutely whether ^{26}Al was responsible for melting. However, we do know that ^{26}Al was present in the solar nebula since evidence of its decay product ^{26}Mg has been found in CAIs from meteorites like Allende and Murchison (Section 9.2.2). That the chondritic meteorites did not melt shows that, in those samples at least, the overall concentration of ^{26}Al was not high enough. Thus, while it is a commonly held belief that ^{26}Al was a heat source in the early Solar System, the exact influence is still being actively researched.

Notwithstanding the lack of a consensus regarding the heat source for asteroidal melting processes, the timing of this event can be reliably established by using a radiometric dating technique.

■ What are the formation ages of HED meteorites?

❏ 4.53 Ga to 4.56 Ga ago.

■ If the ages obtained from the oldest meteorites represent the age of the Solar System, what constraints do radiometric ages of meteorites like HEDs put on formation conditions in the solar nebula?

❏ Melting processes on meteoritic parent bodies took place only a short time after the Solar System was formed.

The reflectance spectra properties of meteorites like Juvinas resemble those of the asteroid (4) Vesta, a roughly spherical body of 250 km radius (which makes it one of the largest asteroids). As far as we can tell, the composition of Vesta is uniform across its surface and so, we consider it to be a largely intact body. That is not to say that it necessarily represents an entire, original planetesimal. There is every possibility that Vesta has been subjected to collisional processes that have resulted in fragmentation (we can only speculate about the parent of Vesta itself). However, if Juvinas does come from Vesta then we can constrain the age of any fragmentation events to a period very early in the history of Solar System evolution. And so, it follows that what we see today as Vesta has survived largely intact for at least 4.53 Ga. Of course, in detail, the outer layers will have been eroded to some extent by impacts – one such impact will have removed the fragment which ultimately fell to Earth as the body we now know as Juvinas. But since Vesta is intact, we cannot expect to have samples of its interior. However, we do have several hundred specimens of iron meteorites that are thought to represent the cores of other asteroids. Cape York is one such sample.

Cape York is part of one of the largest meteorite showers from which we have samples. The meteorite fell to Earth over 1000 years ago in Greenland; eight individual specimens have so far been collected, totalling 5.9×10^4 kg. In historical times, fragments of the meteorite were used extensively by the Inuit (Eskimos) in the manufacture of tools. When tools were traded between the Inuit and European explorers, it became apparent to the scientific community that there must be an iron meteorite somewhere in Greenland. In 1897 the largest known fragment of the meteorite (3.1×10^4 kg) was found and taken by sea to New York; another large mass (2.1×10^4 kg), which was unknown to the Inuit, was removed in 1965 by a Danish expedition.

Iron meteorites originate from asteroidal bodies – the processes which contributed heat were similar to those that produced meteorites like Juvinas. However, iron meteorites are not the result of partial melting. They represent the end-product of differentiation and segregation of metal. In other words, the parental magmas of iron meteorites were, at some point in their history, completely molten. During this process, high-density metallic elements percolated down towards the centre of their parent bodies while low-density silicates floated outwards. As such, most iron meteorites represent samples of the cores of asteroids. We can be certain that a number of different parent bodies are represented among all the iron meteorites, since the concentrations of trace elements, such as iridium, gallium and germanium, vary in ways which could not be produced in a single body.

QUESTION 9.13

List the mechanisms that are likely to have contributed heat to asteroids.

Iron meteorites are alloys of iron and a relatively small proportion of nickel. Cape York, for instance, contains about 7.5% nickel. When a meteorite like Cape York is sawn open, polished and then etched with a mild acid, a characteristic pattern can be observed. This is known as the **Widmanstätten pattern** and while it can be reproduced to a very limited extent in the laboratory it is not a normal feature of terrestrially produced iron. An example of an iron meteorite displaying the Widmanstätten pattern can be seen in Figure 9.10a. The pattern is caused by the segregation, during extremely slow cooling, of nickel-rich and nickel-poor alloys. Cooling rates, calculated from the different scales of pattern that have been found in iron meteorites, vary from 0.5 to 500 K Ma^{-1}, and these differences reflect parent bodies of different sizes – larger bodies cooled more slowly. Some iron meteorites, which have related chemical compositions but different cooling rates, are considered to have formed at varying depths within an individual body, i.e. samples closer to the centre cooled more slowly. If this is true, the parent asteroids cannot have had single cores of iron, which has a high thermal conductivity, but must have been composed of individual pods of metal surrounded by silicates. Such parent bodies are said to have a *raisin-bread* structure.

Most iron meteorites are not composed of pure iron–nickel metal. They contain various quantities of phosphorous-containing minerals, sulfide nodules and graphite. Other irons contain inclusions of silicates, which have chemical compositions similar to chondrites.

■ Does the presence of silicate inclusions pose any constraint on the formation conditions of iron meteorites?

❏ The presence of silicate inclusions shows that these samples were never completely molten.

Thus not all iron meteorites were formed by processes that produced total melting. Those with silicate inclusions are considered to be the result of impact melting and mechanical mixing of iron-containing materials with silicate-rich bodies. This happened early in the history of the Solar System when newly-formed bodies were colliding with each other. Thus, not all iron meteorites can be thought of as being the cores of asteroids.

QUESTION 9.14

Some of the observable asteroids, classified as M-type, have a metallic composition. What does this tell us about processes within the asteroid belt?

QUESTION 9.15

Kodaikanal is an unusual iron meteorite which has a radiometric formation age of 3.8 Ga. This is within the range recorded in the majority of impact-produced rocks from the Moon (3.8 Ga to 4.0 Ga). What can we infer here?

9.3 Summary of Chapter 9

- Meteorites may come from asteroids, the Moon or Mars, and small micrometeorites (dust particles) can come from comets.

- Meteorites can be broadly classified into stones, irons and stony–irons. The stones are by far the most common variety.

- Juvinas and Cape York are examples of meteorites which have been subjected to melting processes on asteroidal bodies. We refer to these meteorites and their parent bodies as differentiated.

- Meteorites like Juvinas are comprised of materials which are the result of primary igneous activity, i.e. magmas produced via the partial melting of a chondritic source. They may all come from a single asteroid (Vesta).

- Most iron meteorites are the result of the total melting which produced layering within asteroids. In the case of Cape York, the iron–nickel body formed the core of an asteroid. However, iron meteorites could also arise from asteroids which have a raisin-bread structure, in which there are discrete pods of iron–nickel metal.

- Some iron meteorites were not formed by processes that produced total melting but were instead produced by impact melting and mixing at the surfaces of asteroidal bodies.

- The short-lived radioactive isotope ^{26}Al may have been responsible for heating and melting asteroids.

- The formation ages of meteorites can be assessed by radiometric dating.

- Allende and Murchison are representatives of a chemically primitive group of meteorites known as carbonaceous chondrites. By chemically primitive we mean having a composition which approximates that of the Sun.

- The carbonaceous chondrites are referred to as cosmic sediments because they represent a collection of materials, formed in different thermal regimes, which settled out from the solar nebula onto the surfaces of parent bodies.

- Refractory inclusions, composed of minerals containing elements such as calcium and aluminium, are present in Allende and Murchison. These materials formed early in the solar nebula.

- The oxygen isotope compositions of refractory inclusions show that the solar nebula was never completely homogenized. In other words, these entities retain a signature of the pre-existing materials which became incorporated into the solar nebula.

- Carbonaceous chondrites also contain materials formed at low temperatures, such as carbonates, sulfates and hydrated minerals. The carbonates and sulfates were formed during secondary aqueous activity on the parent bodies.

- Murchison contains 2.5% by mass of carbon in the form of organic compounds. The exact formation mechanism of these materials is unclear. A small fraction of the organic compounds is demonstrably pre-solar in origin and clearly shows that such material was able to survive the conditions prevailing in the solar nebula. For the rest of the material, which is isotopically similar to organic materials on Earth, there are plausible mechanisms which suggest formation in the solar nebula.

ANSWERS AND COMMENTS

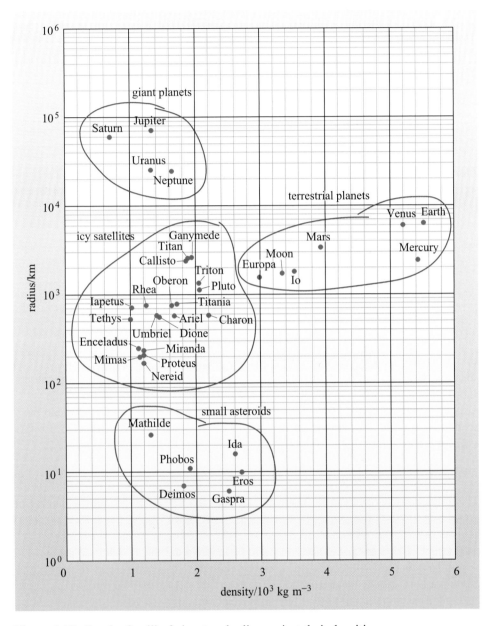

Figure 1.52 Graph of radii of planetary bodies against their densities.

QUESTION 1.2

From the information in Figure 1.52, it appears that planetary objects fall into several groups. The giant planets lie together and have radii greater than 10 000 km and low densities. Another group is apparent, clustered towards the left-hand side of the graph. These are the (icy) satellites. The terrestrial planets (and terrestrial-like bodies) form another group which straggles off towards the high-density area at the right-hand side of the graph. The small asteroids that you plotted form another group.

QUESTION 2.1

(a) Volume of a sphere is $\frac{4}{3}\pi r^3$ where r is the radius. Density $= \dfrac{\text{mass}}{\text{volume}}$

Hence, bulk density of Earth:

$$= \frac{5.9737 \times 10^{24}\ \text{kg}}{\frac{4}{3}\pi(6.371 \times 10^6)^3\ \text{m}^3}$$

$$= \frac{5.9737 \times 10^{24}\ \text{kg}}{1.0832 \times 10^{21}\ \text{m}^3}$$

$$= 5.51 \times 10^3\ \text{kg m}^{-3}\ (\text{or }5.51\ \text{g cm}^{-3})$$

(b) The rocks sampled in the continental areas and from the ocean floors have only about half the density of that of the bulk Earth value. To account for this discrepancy, the Earth must become denser deeper down, and contain materials more dense than the average value to balance the less dense materials near the surface. Even the dense xenoliths, which provide the deepest material that can be examined, are only about two-thirds the bulk Earth density. This means that Earth's densest material must lie at depths much greater than 100 km.

QUESTION 2.2

Using the density of ocean floor crustal rocks ($3.0 \times 10^3\ \text{kg m}^{-3}$), and the density of rocks found within the mantle at 100 km ($3.5 \times 10^3\ \text{kg m}^{-3}$), the linear increase in density per km depth can be calculated at $0.005 \times 10^3\ \text{kg m}^{-3}\ \text{km}^{-1}$ (i.e. an increase of $0.005 \times 10^3\ \text{kg m}^{-3}$ for every kilometre increase in depth). If this increase is extended down to 2900 km as a progressive increase, then the density of material in the outer core should be about $17.5 \times 10^3\ \text{kg m}^{-3}$ (i.e. $3.0 \times 10^3\ \text{kg m}^{-3} +$ (2900 km $\times 0.005 \times 10^3\ \text{kg m}^{-3}\ \text{km}^{-1}$)). In fact, calculations based on seismic studies (Section 2.2.2) indicate that the density of materials in the outer part of Earth's core is actually about $13 \times 10^3\ \text{kg m}^{-3}$. The discrepancy between the two values arises because this simple calculation assumes that density increases are proportional to increases in pressure. This is unlikely to be true because changes in chemical and mineralogical compositions are also known to occur.

QUESTION 2.3

(a)

Table 2.10 Chondrite normalized values of oceanic and continental crust and mantle materials (completed).

	mantle/chondrite		oceanic crust/chondrite		continental crust/chondrite	
SiO_2	1.31	enriched	1.39	enriched	1.72	enriched
TiO_2*	2.29	enriched	20.00	enriched	11.14	enriched
Al_2O_3	1.30	enriched	4.70	enriched	4.86	enriched
Fe_2O_3	0.27	depleted	0.35	depleted	0.25	depleted
MnO	0.74	depleted	0.89	depleted	0.58	depleted
MgO	1.52	enriched	0.38	depleted	0.18	depleted
CaO	1.27	enriched	4.14	enriched	2.41	enriched
Na_2O	0.58	depleted	6.72	enriched	6.42	enriched
K_2O*	0.60	depleted	1.40	enriched	38.20	enriched
Ba	2.90	enriched	4.98	enriched	293.36	enriched
Cr*	0.75	depleted	0.093	depleted	0.030	depleted
K*	0.33	depleted	1.10	enriched	28.94	enriched
Nb	2.85	enriched	10.00	enriched	52.00	enriched
Ni*	0.15	depleted	0.008	depleted	0.003	depleted
Rb	0.28	depleted	1.29	enriched	26.29	enriched
Sr	2.91	enriched	18.73	enriched	69.28	enriched
Th	2.80	enriched	6.67	enriched	190.00	enriched
Ti*	2.16	enriched	18.88	enriched	10.51	enriched
U	2.10	enriched	10.00	enriched	130.00	enriched
Y	2.90	enriched	22.29	enriched	8.92	enriched
Zn	0.06	depleted	0.15	depleted	0.16	depleted
Zr	2.89	enriched	22.74	enriched	32.04	enriched

* Elements that behave as a major element in one group of rocks, and as trace elements in others.

(b)

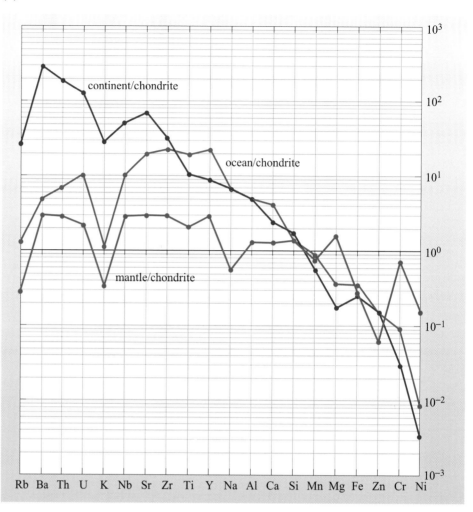

Figure 2.28 Spidergram showing chondrite normalized peridotite, oceanic and continental crustal rocks (completed).

(c) Assuming that primitive meteorites, planetesimals and planetary embryos all condensed from similar nebula materials prior to assembling to form the Earth, then both meteoritic and Earth materials should display similar bulk compositions. However, if terrestrial rocks did have a chondritic composition, then the chondrite normalized value of every element should be 1. Clearly this is not the case, since the three rock types you have plotted show different patterns of enrichment and depletion when chondrite normalized. For instance, elements Rb to Si are enriched in crustal rocks, whilst Mn to Ni are depleted. Moreover, Rb to Zr are much more enriched in continental crust than they are in oceanic crust (remember the vertical axis is a log scale). By contrast mantle (peridotite) rocks are much more similar to chondrite composition (i.e. nearer to $conc_{rel} = 1$), but nevertheless display depletion in Rb, K, Na, Mn, Fe, Zn, Cr and Ni. To summarize, the compositional data in Table 2.2 indicates that there must have been a segregation of elements at a later stage into different layers of the Earth, or else some of the original elements have been 'lost' in another manner (Section 2.3.3).

QUESTION 2.4

From Table A1:

Mass of Mercury = 0.330×10^{24} kg; mass of Earth 5.97×10^{24} kg.

Number of planetary embryos required to assemble Mercury:

$$\frac{\text{mass of Mercury}}{\text{mass of planetary embryo}} = \frac{0.33 \times 10^{24} \text{ kg}}{5 \times 10^{22} \text{ kg}} = 6.6 \text{ (i.e. 6–7 planetary embryos)}.$$

Number of planetary embryos required to assemble Earth:

$$\frac{5.97 \times 10^{24} \text{ kg}}{5 \times 10^{22} \text{ kg}} = 119.4 \text{ (i.e. 119–120 planetary embryos)}.$$

QUESTION 2.5

Collision of planetary embryos during the assembly of Earth would have led to the development of a largely molten body. The nickel–iron component of the original chondritic composition (from which the planetesimals, and hence the planetary embryos formed) would have been much denser and have a higher melting point than the silicates. In this molten state, these denser components would have sunk towards the centre of the newly formed planet, eventually coalescing to form the nickel–iron-rich core. In effect, this segregation process would remove both iron and nickel from the outer layers (i.e. mantle) of the Earth so producing the relative depletions of Fe and Ni shown on the spidergram. Later on, after the last embryo–embryo collision, segregation of the mantle material itself to form oceanic or continental crust would have led to even greater relative depletion of iron and nickel in crustal materials because these elements were preferentially retained in mantle minerals.

QUESTION 2.6

The number of seconds in a year $(s\,yr^{-1})$ is about 3.2×10^7 so the rate of heating is:

$$\frac{3 \times 10^{19} \text{ J yr}^{-1}}{3.2 \times 10^7 \text{ s yr}^{-1}} = 9.4 \times 10^{11} \text{ J s}^{-1} = 9.4 \times 10^{11} \text{ W}.$$

The rate of tidal heating generated within the Earth $= \dfrac{9.4 \times 10^{11} \text{ W}}{6 \times 10^{24} \text{ kg}} = 1.6 \times 10^{-13} \text{ W kg}^{-1}.$

QUESTION 2.7

(a) The age of the Earth is 4.6×10^9 yr. To calculate the proportion of the original radiogenic isotope remaining today, you first need to determine how many half-lives have occurred since the Earth was formed:

$$\text{Number of half-lives} = \frac{\text{age of Earth}}{\text{isotopic half-life}}$$

The proportion remaining after one half-life is $\left(\dfrac{1}{2}\right)^1 = 0.5$, after 2 half-lives is

$\left(\dfrac{1}{2}\right)^2 = 0.25$, and after three half-lives is $\left(\dfrac{1}{2}\right)^3 = 0.125$.

For ^{40}K the number of half-lives $= \dfrac{4.6 \times 10^9 \text{ yr}}{1.3 \times 10^9 \text{ yr}} = 3.54$ half-lives.

So the proportion remaining today is $\left(\dfrac{1}{2}\right)^{3.54} = 0.086$

Note: on many calculators this function can be performed by the key labelled x^y, which is often operated by the SHIFT key. Your calculator may differ in detail, but for many common calculators this calculation would be performed by keying:

0.5 ⌷SHIFT⌷ ⌷x^y⌷ 3.54 ⌷=⌷

For ^{232}Th the number of half-lives $= \dfrac{4.6 \times 10^9 \text{ yr}}{13.9 \times 10^9 \text{ yr}} = 0.33$ half-lives.

So the proportion remaining today is $\left(\dfrac{1}{2}\right)^{0.33} = 0.80$.

(b) The amount of radiogenic heating is determined by utilizing the proportion of the different radiogenic isotopes remaining today (you need to calculate the values for ^{235}U and ^{238}U as above) to calculate the original heat budget.

If only 0.086 of the original ^{40}K remains today and its present rate of heat generation is 2.8×10^{-12} W kg^{-1} (from Table 2.6), its original heat budget

$$= \frac{2.8 \times 10^{-12} \text{ W kg}^{-1}}{0.086} = 3.26 \times 10^{-11} \text{ W kg}^{-1} \text{ (or } 32.6 \times 10^{-12} \text{ W kg}^{-1}\text{).}$$

Likewise:

$$\text{Initial heating by } ^{232}\text{Th decay} = \frac{1.04 \times 10^{-12} \text{ W kg}^{-1}}{0.80} = 1.30 \times 10^{-12} \text{ W kg}^{-1}.$$

$$\text{Initial heating by } ^{235}\text{U decay} = \frac{0.04 \times 10^{-12} \text{ W kg}^{-1}}{0.011} = 3.64 \times 10^{-12} \text{ W kg}^{-1}.$$

$$\text{Initial heating by } ^{238}\text{U decay} = \frac{0.96 \times 10^{-12} \text{ W kg}^{-1}}{0.49} = 1.96 \times 10^{-12} \text{ W kg}^{-1}.$$

By adding the four answers, the total initial heating was therefore about 4×10^{-11} W kg^{-1}, compared with modern total of about 5×10^{-12} W kg^{-1} (obtained by adding the four present-day heating rates given in Table 2.6), in other words at least eight times greater (we have not accounted for other short-lived isotopes such as ^{26}Al and the heat from their decay). This initial decay would have originally provided a major source of internal heating even in relatively small planetary bodies, and so would have contributed significantly to early planetary differentiation.

QUESTION 2.8

(a)

Table 2.11 How layers within Earth are defined (completed).

Layer	How layer is defined
lithosphere	on the basis of its strength and the nature of heat transfer
asthenosphere	on the basis of its strength and the nature of heat transfer
crust	Seismically (note also that the Moho corresponds to a *compositional* change)
mantle	Seismically (note also that the Moho corresponds to a *compositional* change)
core	seismically

(b) Earth's core, mantle and crust are all defined on the basis of changes in seismic speed, which reflect differences in composition and mineralogy. The core is iron-rich, whereas the mantle and crust are silicate-rich. The crust is less dense and more silicate-rich than the underlying mantle. Interfaces between all three of these layers are marked by sharp changes in the speed at which seismic waves are transmitted. By contrast, the lithosphere is distinguished from the asthenosphere on the basis of strength and the differences by which they each transfer heat. The lithosphere comprises Earth's crust and the rigid, non-convecting uppermost part of the mantle. Heat passes through this zone by conduction and advection, and about two-thirds of the global surfaceward heat transfer through the lithosphere is accomplished by plate recycling, involving the creation and destruction of lithospheric plates. The asthenosphere is effectively all of the mantle below the lithosphere, and it loses its heat through solid-state convection.

QUESTION 2.9

(a) The lithosphere must thicken over time since the rate of radiogenic heating will decline.

(b) A similar argument regarding lithospheric thickening will not apply to tidally heated bodies because tidal heating does not decline in the same manner. Whilst a satellite remains in the same type of orbit, the tidal heating effects will remain. (Note: in systems involving several satellites, a satellite can experience several interludes of orbital resonance, each resulting in a heating episode. In such a situation, the lithosphere would continually thicken and thin until such a time when the nature of the orbital interactions changed.) In general, orbital distances between satellites and planets are likely to increase over time (Section 2.3.3), and so the effects of tidal heating will decline. However, this decline is likely to occur over a different time-scale to that of the common radioactive decay systems.

QUESTION 3.1

(a)

(i) Since this planet has a stabilized crust, it must have already passed through the period of intense bombardment, but it will still retain heat from accretion and these large impact events and also from self-compression and segregation of the core. Another major source of heat will come from radiogenic decay. This would be greater in a young planet due to the initial concentrations of radiogenic isotopes. Furthermore the passage of only a relatively short period of geological time since formation would mean few half-lives would have elapsed.

(ii) For an ancient planet, the planet-forming heating processes (i.e. those created by impacts and the core-forming event) would have dissipated significantly, leaving radiogenic heating as the major source of heat. Given this is an ancient planet, only half the size of Earth, it would have contained considerably smaller volumes of radiogenic mantle and crust compared with Earth. Moreover, its antiquity will mean much of the radiogenic material will have already decayed (i.e. many half-lives will have elapsed). Therefore, the planet will produce only a fraction of the radiogenic heating currently occurring within the Earth. A greatly thickened lithosphere would suggest that any heat loss is largely by conduction, rather than convection or associated plate tectonic movement. In summary, the diminished radiogenic decay would be the only significant source of internal heating.

(iii) Since this satellite contains only a very small rocky core, the amount of heating generated by radioactive decay would be small despite its relatively young age. Primordial heat remaining from accretion, self-compression, and internal differentiation may still be present.

The dominant heating would be due to the massive tidal effects resulting from its close elliptical orbit and the enormous gravitational attraction of the nearby giant planet. However, the relatively large size of this satellite (its 3000 km diameter is similar to that of Europa) means it would also retain heat better than a smaller body because it has a smaller surface area to volume ratio (Box 2.8). It would therefore have the potential to heat up internally to such a point where melting of the icy mantle could begin. This melting seems to have occurred since the extensive recent resurfacing and is likely to be the result of cryovolcanism.

(b) The most volcanically active bodies are likely to be (i) and (iii) despite their very different heating processes. However, we don't have enough information to determine which of these is the more volcanically active, except to say that the young Earth-like planet will exhibit silicate volcanism and lava eruptions which are likely to be more frequent and intense than those of present-day Earth. The young icy satellite moon will be characterized by active cryovolcanism and associated resurfacing. By contrast, the old, small rocky planet will certainly be the least volcanically active. Its thick lithosphere and much reduced radiogenic decay means it may have been volcanically and tectonically dead for a considerable part of its later history.

QUESTION 3.2

(a) The major element compositions of lunar and Martian basalts differ, but comparison with peridotite using the tabulated data reveals that both exhibit similar degrees of relative enrichment in Al, Ca, Mn and Fe, and depletion in Mg and Cr (Table 3.4). This is similar to the pattern of element enrichment and depletion observed between mantle peridotite and ocean basalt (MORB) on Earth (see also the answer to Question 2.3b for spidergrams of peridotite compared with oceanic basaltic crust). Since partial melting of mantle peridotite is thought to be the source of basaltic lavas on Earth, the similarity in the pattern of element enrichment and depletion in both lunar and Martian lavas compared with mantle compositions probably indicates that they were also originally derived from a peridotite-like source. These data are shown graphically in Figure 3.43 (a spidergram normalizing the basalt types against peridotite composition, i.e. 'mantle normalized'), from which it can be seen that the Martian basalt and MORB patterns are remarkably similar. Lunar basalt differs slightly in that it is relatively depleted in Na and enriched in Ti compared with MORB and Martian basalt, probably because it is derived from mantle material that had previously experienced loss of volatiles and retention of refractory elements during the Moon's formation (Section 2.3.3).

Table 3.4   Element enrichment and depletion of basalt types compared to mantle peridotite.

	MORB/mantle		Deccan basalt/mantle		Komatiite/mantle		Lunar basalt/mantle		Martian basalt/mantle	
SiO_2	1.06	enriched	1.07	enriched	1.00	–	0.94	depleted	1.10	enriched
TiO_2	7.00	enriched	12.50	enriched	1.00	–	13.00	enriched	4.50	enriched
Al_2O_3	3.58	enriched	3.19	enriched	0.86	depleted	1.79	enriched	1.56	enriched
Fe_2O_3	1.32	enriched	1.80	enriched	1.49	enriched	2.88	enriched	2.54	enriched
MnO	2.00	enriched	2.00	enriched	2.00	enriched	3.00	enriched	5.00	enriched
MgO	0.25	depleted	0.16	depleted	0.84	depleted	0.38	depleted	0.25	depleted
CaO	3.26	enriched	3.12	enriched	1.56	enriched	2.38	enriched	2.94	enriched
Na_2O	11.33	enriched	8.00	enriched	1.33	enriched	0.67	depleted	4.33	enriched
K_2O	2.33	enriched	10.00	enriched	3.33	enriched	3.33	enriched	6.67	enriched
Cr_2O_3	0.12	depleted	0.05	depleted	0.51	depleted	0.58	depleted	0.47	depleted

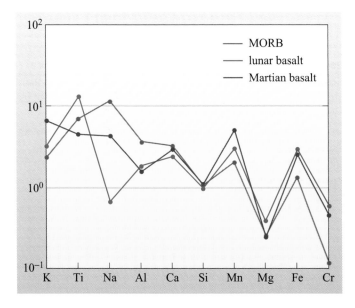

Figure 3.43 Spidergram normalizing basalt types against peridotite mantle composition.

(b) According to the data in Table 3.1, mantle peridotite is characterized by relatively low SiO_2, CaO, Na_2O and K_2O, and relatively high MgO. The example that most closely matches this compositional pattern is komatiite lava (Figure 3.44). This probably represents large degrees of partial melting of a peridotite-like parent, with relatively little modification during magma migration from melt source to surface eruption (i.e. they are 'primitive' compositions). By contrast, the Deccan continental flood basalt provinces and the Martian lava contain much less MgO, and higher CaO, Na_2O, K_2O and SiO_2. These data are also shown graphically in Figure 3.44. Smaller degrees of partial melting lead to significant increases of incompatible elements in the melt (e.g. CaO, Na_2O and K_2O), whilst MgO remains behind in the mantle residue. Thus, smaller degrees of partial melting, together with other modification processes (e.g. crystallization and separation of mineral fractions) acting upon the magma during its migration from source to eruption, will produce more MORB-like compositions. This suggests processes similar to those producing MORB-like magmas on Earth apparently also operated on Mars (and by analogy, may have operated in other terrestrial-like bodies).

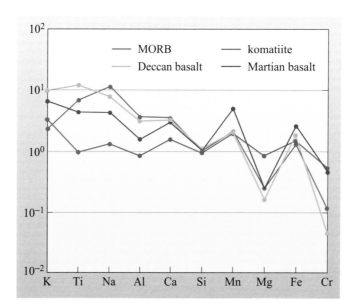

Figure 3.44 Spidergram normalizing basalt types against peridotite mantle composition.

QUESTION 3.3

Mars is a small, silicate planet, lacking major satellites. Its only present-day source of heat must be from the decay of long-lived radioactive isotopes, plus whatever remains of its primordial heat. Since it is small (its mass is only about one-tenth that of Earth), its active volcanic lifetime is likely to have been relatively short, but not as short as that of the Moon. Like the Earth and Moon, abundant evidence of lava eruptions might be expected, particularly in the early history of the planet. Moreover, because Mars is rich in volatiles, there should be evidence of pyroclastic eruptions and, because Mars has an atmosphere (albeit a relatively low density one), convecting eruption columns would also have been possible.

QUESTION 4.1

Because the pit is so tiny, the meteorite was clearly a very small one. Its small size means that its speed must have been severely limited by atmospheric drag. Thus, it more nearly resembles a pebble thrown into mud rather than an explosive hypervelocity impact. The asymmetrical distribution of ejecta suggests that the meteorite was travelling from right to left, and it probably struck the road at an oblique angle.

QUESTION 4.2

Aside from the actual crater, a heat pulse and shock wave will affect the immediate environment. Large impacts will have regional, or even global effects that may kill off many organisms, or even cause mass extinctions of species. Re-entering ejecta overheat the atmosphere causing forest fires, ozone depletion, possibly greenhouse warming depending on the composition of the target, followed by 'nuclear winter' cooling caused by material lofted into the stratosphere. Impacts into ocean present an additional danger. Impact-generated tsunamis could devastate huge areas, and since oceans are large targets, this may represent the most substantial impact-related hazard to human populations.

QUESTION 4.3

The proportion of elliptical craters is negligible. Given the wide range of possible impact angles, this shows that the process of cratering favours the production of circular craters over elliptical ones.

QUESTION 4.4

Starting in the centre of the image, a small cluster of yellow/green can be seen. Moving out from there, most of the base of the crater is red, with blue dominating in the terraces and beyond the rim, and more yellow in the bottom part of the image. In this case, blue is probably relatively fresh ejecta, or fresh material exposed in the terraces. Bright red is associated with deposits of impact melt in the base of the crater, and the yellow tones in the middle may be iron-rich rocks brought up from depth during the formation of the central peaks.

QUESTION 4.5

Truly diagnostic impact signatures in layers of rock that might be distant from the impact site, are shocked quartz containing planar deformation features, and tektites. Chemical analyses of the suspected impact layer may reveal unusual abundances of certain elements, such as iridium, which are relatively much more abundant in meteorites than in the Earth's crust.

Impact structures may be buried, in which case more local shock features will be absent or eroded (although deep coring can reveal rocks from the structure itself). If the structure is eroded, and the original target rocks are accessible, we might see breccias, shatter cones and pseudotachylites within and around the structure. In some cases we can employ techniques used in geophysical exploration, such as gravity meters and magnetometers, to infer the presence of a buried structure, as illustrated in Figure 4.3.

QUESTION 4.6

Figure 4.24 right (an image of the heavily cratered terrain on the Moon) shows a nearly saturated surface. It would be impossible to fit a large crater on it without overprinting older ones. Figure 4.24 (left) shows an area of the lunar mare. It is clearly not saturated, and must be younger than that shown in the image on the right.

QUESTION 5.1

Mass spectrometry identifies molecules on the basis of their masses, so this method would not distinguish between $^{14}N_2$ (RMM = 28) and $^{12}C_2{}^1H_4$ (RMM = 28). (Instruments with very high resolution could differentiate between the molecules but have not, as yet, been sent on planetary missions.)

Each of these chemically distinct molecules could be identified by gas chromatography, provided that the instrument had been calibrated for each substance.

QUESTION 5.2

Only diatomic molecules with a dipole moment, i.e. heteronuclear diatomic molecules, can be detected by infrared spectroscopy, so H_2 would not be detected but CO and HCl would.

QUESTION 5.3

Venus:

$$P_s = 92 \, \text{bar} = 9.2 \times 10^6 \, \text{Pa} = 9.2 \times 10^6 \, \text{kg m}^{-1} \, \text{s}^{-2}$$

Using Equation 5.3:

$$m_c = \frac{9.2 \times 10^6 \, \text{kg m}^{-1} \, \text{s}^{-2}}{8.90 \, \text{m s}^{-2}}$$

$$m_c = 1.03 \times 10^6 \, \text{kg m}^{-2}$$

Mars:

$$P_s = 6.3 \times 10^{-3} \, \text{bar} = 6.3 \times 10^2 \, \text{Pa} = 6.3 \times 10^2 \, \text{kg m}^{-1} \, \text{s}^{-2}$$

Using Equation 5.3:

$$m_c = \frac{6.3 \times 10^2 \, \text{kg m}^{-1} \, \text{s}^{-2}}{3.72 \, \text{m s}^{-2}}$$

$$m_c = 1.69 \times 10^2 \, \text{kg m}^{-2}$$

The column mass for Venus is about 100 times that for Earth and the column mass for Mars is about one-sixtieth of that for Earth.

QUESTION 5.4

The flux due to radiogenic heating is 5×10^{-12} W kg^{-1}. The mass of the Earth is 5.97×10^{24} kg. The total power from radiogenic heating is thus $(5 \times 10^{-12} \text{ W kg}^{-1}) \times (5.97 \times 10^{24} \text{ kg}) = 2.99 \times 10^{13}$ W.

The total solar power absorbed is πR^2 times the solar flux density multiplied by $(1 - a)$, which is the fraction absorbed. The albedo for the Earth is 0.30 (see Table 5.4). The radius of Earth is 6.371×10^3 km, which is 6.371×10^6 m.

Total solar power absorbed $= \pi R^2 \times$ solar flux density $\times (1 - a)$

$= \pi (6.371 \times 10^6 \text{ m})^2 \times (1.38 \times 10^3 \text{ W m}^{-2}) \times (1 - 0.3) = 1.23 \times 10^{17}$ W.

The ratio of these powers is:

$$\frac{1.23 \times 10^{17} \text{ W}}{2.99 \times 10^{13} \text{ W}} = 4.1 \times 10^3$$

Therefore the Sun delivers about 4000 times as much power to the Earth as does radiogenic heating.

Percentage of energy from radiogenic heating $= \dfrac{2.99 \times 10^{13} \text{ W}}{1.23 \times 10^{17} \text{ W}} \times 100\% = 0.02\%$

QUESTION 5.5

From Wien's displacement law, at a temperature of 288 K

$$\lambda_{\text{peak}} = \frac{2.90 \times 10^{-3} \text{ m K}}{288 \text{ K}} = 10^{-5} \text{ m}$$

Thus the peak wavelength is 10^{-5} m or 10 µm. Electromagnetic radiation of this wavelength is in the infrared, and so the Earth's radiation is most intense in the infrared region.

QUESTION 5.6

A high albedo indicates a highly reflecting surface. This suggests either extensive cloud cover or a nearly white surface. A high albedo and large distance from the Sun are consistent with a low temperature.

The atmosphere of Triton is extremely thin and cloud-free; the surface is composed of highly reflective icy material.

QUESTION 5.7

By absorbing infrared radiation that is not otherwise absorbed by the Earth's atmosphere, the chlorofluorocarbons add to the greenhouse effect. The temperature of the Earth's troposphere will therefore be increased by their presence. This role as a greenhouse gas is different and quite separate from the role of chlorofluorocarbons in depleting the ozone layer.

QUESTION 5.8

The partial pressure of CO_2 at the surface of Mars will be close to the total surface pressure, 6.3×10^{-3} bar. A horizontal line representing this pressure on Figure 5.32 would intersect a vertical line at the surface temperature of 223 K, in the area labelled 'gas only' (see Figure 5.51). Condensation of CO_2 will not occur on Mars at this temperature.

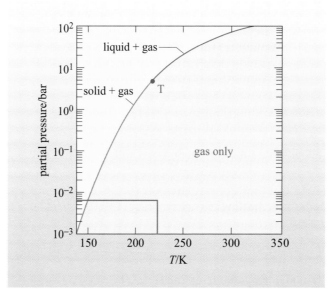

Figure 5.51 Saturation vapour pressure diagram for CO_2 showing the state of CO_2 at 223 K and 6.3×10^{-3} bar.

This question only considered the average temperature, 223 K, but at the poles during winter the temperature can be much lower, below 150 K. At these temperatures, CO_2 will condense as a solid according to Figure 5.51.

QUESTION 5.9

The initial speed of this piece of atmosphere at the equator will be the planet's surface rotation speed, which is 0.997/243 of that of the Earth. The speed of the piece of atmosphere at 30° N will be 0.997/243 of the speed of a piece at the same latitude on Earth. So the gain in speed relative to the surface will also be 0.997/243 of that of the gain on the Earth which we calculated (Section 5.6) to be 134 m s^{-1}. This is

$$\frac{0.997 \times 134 \, \text{m s}^{-1}}{243} = 0.55 \, \text{m s}^{-1}$$

As Venus rotates in a retrograde direction, the direction will be east to west.

QUESTION 5.10

If the rotation of Venus were sufficiently fast, then it would generate a magnetic dipole field (see Chapter 2). The magnetosphere of Venus would then more closely resemble that of the Earth, with the IMF interacting with the planetary magnetic field.

QUESTION 5.11

The atmosphere had only been probed remotely and detection of atmospheric gases would have been through infrared spectroscopy. As N_2 is a homonuclear molecule, it would not be detected in this way and so it would not be possible to say whether it were present or not. It was assumed to be nitrogen because the bulk of the Earth's atmosphere is nitrogen.

QUESTION 5.12

The initial effect of the greenhouse gases is to raise the atmospheric temperature. This would cause evaporation of some carbon dioxide and water ice from the polar ice-caps and from the permafrost. With the added contribution of these gases, further evaporation would occur. If the temperature and pressure rose sufficiently, water could exist in the liquid state, the basis for the far-fetched suggestion in the question.

QUESTION 5.13

No. From Figure 5.32, even at this low temperature the partial pressure is too low for carbon dioxide to condense. The point corresponding to 193 K and 3.6×10^{-4} bar falls below the region shown in Figure 5.32 but, assuming the curve continues smoothly, would be in the gas only region.

QUESTION 5.14

(a) Absorption of UV radiation from the Sun by ozone prevents it reaching ground level (Section 5.4.2). (b) The magnetic field deflects high-energy solar wind particles around the Earth so that few of them reach the surface.

Both short wavelength UV radiation and high-energy solar wind particles are harmful to life.

QUESTION 6.1

For Enceladus, $a = 238\,000$ km and $P = 1.370 \times 24 \times 3600$ s. Using the equation given

$$M_{Sat} = 4\pi^2 \frac{a^3}{GP^2} \text{, we have}$$

$$
\begin{aligned}
M_{Sat} &= \frac{4\pi^2 (238\,000 \times 10^3 \text{ m})^3}{(6.67 \times 10^{-11} \text{ N m}^2 \text{ kg}^{-2})(1.370 \times 24 \times 3600 \text{ s})^2} \\
&= \frac{4\pi^2 (1.348 \times 10^{25} \text{ m}^3)}{(6.67 \times 10^{-11} \text{ N m}^2 \text{ kg}^{-2})(1.401 \times 10^{10} \text{ s}^2)} \\
&= 5.69 \times 10^{26} \text{ m}^3 \text{ N}^{-1} \text{ m}^{-2} \text{ kg}^2 \text{ s}^{-2}
\end{aligned}
$$

Now $1 \text{ N} = 1 \text{ kg m s}^{-2}$ (as given in Appendix B, Table B1), so $1 \text{ N}^{-1} = 1 \text{ kg}^{-1} \text{ m}^{-1} \text{ s}^2$. Therefore

$1 \text{ m}^3 \text{ N}^{-1} \text{ m}^{-2} \text{ kg}^2 \text{ s}^{-2} = 1 \text{ m}^3 (1 \text{ kg}^{-1} \text{ m}^{-1} \text{ s}^2) \text{ m}^{-2} \text{ kg}^2 \text{ s}^{-2}$

$= 1 \text{ kg}$

So the units of M_{Sat} are kg as expected and

$M_{Sat} = 5.69 \times 10^{26}\,\text{kg}$

Density is mass per unit volume. The approximate volume was calculated as $9.0 \times 10^{23}\,\text{m}^3$. Hence the density of Saturn is approximately

$$\frac{5.69 \times 10^{26}\,\text{kg}}{8.3 \times 10^{23}\,\text{m}^3} = 0.69 \times 10^3\,\text{kg}\,\text{m}^{-3}.$$

(This is a very low density. It is lower than the density of water at the surface of the Earth, which is $1000\,\text{kg}\,\text{m}^{-3}$.)

QUESTION 6.2

Liquid mercury, Hg, has a very high density, and a substantial layer of Hg (which would be needed to produce the magnetic field) would make the average densities far higher than those observed.

Hg is not a very abundant element and so a substantial layer of it would be ruled out if we assume roughly solar abundances.

(The form of Hg at the temperatures and pressures in the interior of Saturn is not given, but at such high pressures it is likely that Hg will still be a conducting fluid.)

QUESTION 6.3

(a) For Earth the mean density is $5.51 \times 10^3\,\text{kg}\,\text{m}^{-3}$ and its mean radius is 6371 km. The formula given yields for the pressure at the centre of the Earth

$$P_c \approx \frac{2\pi G}{3}\rho_m^2 R^2$$

$$= \frac{2\pi \times 6.67 \times 10^{-11}\,\text{N}\,\text{m}^2\,\text{kg}^{-2} \times (5.51 \times 10^3\,\text{kg}\,\text{m}^{-3})^2 \times (6.371 \times 10^6\,\text{m})^2}{3}$$

$$= 1.72 \times 10^{11}\,\text{Pa}$$

$$= 1.72\,\text{Mbar}$$

The accepted value is 3.3 Mbar.

(b) For Saturn the mean density is $690\,\text{kg}\,\text{m}^{-3}$ and the mean radius is 58 230 km. The pressure at the centre is calculated as

$$P_c \approx \frac{2\pi G}{3}\rho_m^2 R^2$$

$$= \frac{2\pi \times 6.67 \times 10^{-11}\,\text{N}\,\text{m}^2\,\text{kg}^{-2} \times (690\,\text{kg}\,\text{m}^{-3})^2 \times (5.8230 \times 10^7\,\text{m})^2}{3}$$

$$= 2.3 \times 10^{11}\,\text{Pa}$$

$$= 2.3\,\text{Mbar}$$

(c) For Jupiter the mean density is $1330 \, \text{kg m}^{-3}$ and the mean radius is $69\,910 \, \text{km}$. The pressure at the centre is calculated as

$$P_c \approx \frac{2\pi G}{3} \rho_m^2 R^2$$

$$= \frac{2\pi \times 6.67 \times 10^{-11} \, \text{N m}^2 \, \text{kg}^{-2} \times (1330 \, \text{kg m}^{-3})^2 \times (6.9910 \times 10^7 \, \text{m})^2}{3}$$

$$= 1.21 \times 10^{12} \, \text{Pa}$$

$$= 12.1 \, \text{Mbar}$$

(Because we have only considered self-compression and mean density, the values obtained are much lower than the accepted values. However they do show that the pressure at the centre of Jupiter is considerably larger than at the centre of Saturn.)

QUESTION 6.4

The heat produced by radiogenic decay on Earth is $4.8 \times 10^{-12} \, \text{W kg}^{-1}$. Four Earth masses is $5.97 \times 10^{24} \times 4 \, \text{kg}$, so a core of four Earth masses on Jupiter would produce

$$(4.8 \times 10^{-12} \, \text{W kg}^{-1}) \times (5.97 \times 10^{24} \times 4 \, \text{kg})$$

$$= 1.1 \times 10^{14} \, \text{W}.$$

To obtain the flux per unit surface area, we have to divide this by the surface area of Jupiter. As an approximation we take the emitting layer as the 1 bar layer and the planet as spherical.

(In fact, different wavelengths of radiation are emitted from different depths in the atmosphere, but IR radiation all comes from layers close to 1 bar and the distance of these layers from the 1 bar level will be negligible compared with the radius of Jupiter.)

The approximate surface area is $4\pi R_J^2 = 4\pi \times (6.9910 \times 10^7 \, \text{m})^2 = 6.1 \times 10^{16} \, \text{m}^2$. The flux is thus

$$\frac{1.1 \times 10^{14} \, \text{W}}{6.1 \times 10^{16} \, \text{m}^2} = 1.8 \times 10^{-3} \, \text{W m}^{-2}$$

This is far too small to account for the observed flux ($7 \, \text{W m}^{-2}$).

QUESTION 6.5

This could be evidence *against* the theory of the origin of Saturn's heat excess that we have just described, because helium lost from the metallic hydrogen layer by differentiation would have to be replaced by helium from the molecular layer. Helium lost from the molecular layer would in turn be replaced by helium from the atmosphere.

QUESTION 6.6

Neon like helium is a noble gas and does not form molecules, so in particular, Ne_2 does not exist. It cannot therefore be detected through vibrational spectroscopy. Its atomic spectrum would have to be used.

(The atomic spectrum is in a more accessible region than that of helium.)

The solar abundance of neon is much less than that of helium and so we would only expect a small amount to be present in the atmosphere. This would make Ne harder to detect by its atomic spectrum and rules out the indirect methods used for helium. The Galileo probe detected neon using mass spectrometry.

QUESTION 6.7

When considering how we could estimate the temperature using an ideal-gas model (Section 6.3.2), we rearranged Equation 6.1 to give

$$T = \frac{Pm}{(1.38 \times 10^{-23})\rho}$$

In this question you were given the number density, $n = \rho/m$.

The temperature is thus given by

$$T = \frac{P}{(1.38 \times 10^{-23})n}$$

$$= \frac{0.42 \times 10^5}{(1.38 \times 10^{-23}) \times (2.4 \times 10^{25})}$$

$$= 127\,\text{K}$$

QUESTION 6.8

For methane to condense out, its partial pressure must exceed that of the saturation vapour pressure curve at the temperature of Jupiter's atmosphere. Although relatively abundant, methane does not have a sufficiently high partial pressure to condense out at the temperatures in the atmospheres of Jupiter and Saturn.

QUESTION 6.9

A windless atmosphere would be rotating with the same rotational speed as the core of the planet. At the equator, a piece of atmosphere at the 1 bar level travels the circumference of the equator in 0.412 days. The circumference is given by $2\pi R$ where R is the equatorial radius. Hence the windless atmosphere travels $2 \times \pi \times 71\,490$ km in 0.412 days, which gives a speed of

$$\frac{2 \times \pi \times 71\,490\,\text{km}}{(0.412 \times 24 \times 60 \times 60)\,\text{s}} = 12.6\,\text{km}\,\text{s}^{-1}$$

$$= 12\,600\,\text{m}\,\text{s}^{-1}$$

QUESTION 6.10

A negative velocity corresponds to a piece of atmosphere rotating more slowly than the centre of the planet. A piece of atmosphere moving in the same direction as material at the centre but at a lower speed will have negative velocity.

(A similar effect occurs if you are sitting in an express train and look at a slower train which you are passing. The slower train appears to be going backwards.)

QUESTION 6.11

Jupiter's bow shock will be produced when the field strength of Jupiter's magnetic field is sufficient to resist the interplanetary magnetic field. This is the same mechanism as for the Earth.

QUESTION 6.12

Rocky cores are not essential to produce the average density: as long as it can be shown that the proposed interior is sufficiently dense to reproduce the average density of the planet, then there is no conflict with the observed density.

Gravitational field measurements indicate a dense core but this does not have to be rocky. Again there is no conflict with observation.

The magnetic field originates in a liquid layer outside the core. A solid icy core is compatible with this.

The interior heat is not well understood but the evidence on this does not rule out the model described.

QUESTION 6.13

The oxygen-containing molecule most likely to be found in the atmosphere of Uranus is water, H_2O, as there is a high abundance of hydrogen present. At the atmospheric temperatures in the accessible region of Uranus, water at the abundance expected would be condensed out as ice. The icy layers, presumably, contain a considerable quantity of liquid water which is prevented from reaching the atmosphere.

QUESTION 6.14

Uranus has a substantial atmosphere which would be ionized in the upper reaches to form an ionosphere. Thus in the absence of a magnetic field, the magnetosphere would resemble that of Venus with the ionosphere rather than the magnetic field lines interacting with the IMF. The magnetosphere would be smaller, roughly the size of the planet on the sunward side rather than 18 times the size of the planet, as observed.

QUESTION 6.15

The far greater mass of Jupiter means that self-compression is greater, and were Jupiter and Rhea made of the *same* materials, the density of Jupiter would be far greater. Having the same average density is thus evidence *against* the two bodies being made of similar material and *for* them being different.

QUESTION 6.16

Nitrogen is a homonuclear diatomic molecule and so could not be observed through its vibrational spectrum. (Furthermore, its electronic spectrum lies in the UV and much UV radiation is absorbed by the Earth's atmosphere, making it difficult to observe the electronic spectrum using ground-based instruments.)

QUESTION 6.17

Your list might include the following:

*   Jupiter has an internal heat source and also receives energy from the Sun.
*   The rate of loss of heat to space is low owing to the low temperatures of the radiating levels in the atmosphere.
*   The rate of decline of temperature in the troposphere is close to the adiabatic lapse rate.

QUESTION 6.18

The solar wind is deflected around the planet but there are energetic particles within the magnetosphere stemming from Io and other sources. Thus you would be at danger from high-energy particles even if not from the solar wind.

QUESTION 6.19

Both Uranus and Neptune are observed to have magnetic fields, and so the prediction that Neptune has a solid core and no magnetic field does not agree with observation. The electrically conducting liquid layer is found to lie not in the core but further out. The observed magnetic field data thus do not rule out the model but do not support it as an explanation of the magnetic fields in these planets.

Neptune's internal heat shows that its core is hot but not necessarily molten rock. There is no observational evidence for a hot molten interior in Uranus, although, from our observations on other planets, it would be almost inconceivable that the temperature did not increase towards the centre.

QUESTION 6.20

As on Jupiter and Saturn, the most abundant species is the hydrogen molecule so emissions from H and H_2 would be expected.

QUESTION 7.1

The perihelion distance is given by $q = a(1 - e)$ and so,

$$q = (39.48\,\text{AU}) \times (1 - 0.249) = 29.65\,\text{AU}$$

The aphelion distance is given by $Q = a(1 + e)$ and so,

$$Q = (39.48\,\text{AU}) \times (1 + 0.249) = 49.31\,\text{AU}$$

QUESTION 7.2

As with Question 7.1, for Neptune we obtain

$$q = (30.07\,\text{AU}) \times (1 - 0.009) = 29.80\,\text{AU}$$

The aphelion distance is given by $Q = a(1 + e)$ and so,

$$Q = (30.07 \text{ AU}) \times (1 + 0.009) = 30.34 \text{ AU}$$

QUESTION 7.3

As Pluto's perihelion distance is less than both Neptune's perihelion and aphelion distances, then Pluto must sometimes be closer to the Sun than Neptune. So Pluto is not always the outermost planet – an inconvenience for compilers of general knowledge quizzes!

QUESTION 7.4

Rearranging Equation 7.3 gives $P = \sqrt{ka^3}$ and as $k = 1 \text{ yr}^2 \text{ AU}^{-3}$, then we calculate $\sqrt{0.39^3}$ (which can also be written $0.39^{3/2}$) $= \sqrt{0.0593} = 0.24$. So the orbital period is 0.24 years (about 88 days).

QUESTION 7.5

(a) This satellite will undergo some tidal heating, as the satellite is rotating (like the Earth is rotating while it orbits the Sun), so producing frictional heating due to the movement of the tidal bulge.

(b) This satellite will not undergo tidal heating, as the circular orbit and the synchronous rotation mean that the tidal bulge does not move (and thus no frictional heating is produced).

(c) This satellite will undergo some tidal heating. While there is no contribution solely from the rotation of the satellite, as it has synchronous rotation, the eccentric orbit means that the tidal bulge will move slightly back and forth with respect to the direction of the gravitational force (i.e. libration), producing frictional heating.

QUESTION 7.6

Jupiter has a semimajor axis of 5.20 AU (from Appendix A, Table A1). Using $P^2 = ka^3$ (and remembering that $k = 1 \text{ yr}^2 \text{ AU}^{-3}$ from Section 7.2), Jupiter's orbital period is given by $5.2^{3/2} = 11.86$ years. An object in the $3:1$ resonance has an orbital period one-third of this, i.e. 3.95 years. Using $P^2 = ka^3$ again, we calculate that the value of semimajor axis associated with this period is $3.95^{2/3}$, i.e. a semimajor axis of 2.50 AU. An alternative way to calculate this is to note that Kepler's third law equation can be rearranged such that $k = P^2/a^3$. This is true for any object orbiting the Sun, and the value of k remains constant such that P^2/a^3 for object 1 equals P^2/a^3 for object 2. Thus equating these expressions, we can write

$$\frac{P_1^2}{a_1^3} = \frac{P_2^2}{a_2^3}$$

thus
$$\left(\frac{P_1}{P_2}\right)^2 = \left(\frac{a_1}{a_2}\right)^3$$

where the subscripts, refer to object 1 (e.g. our asteroid) and object 2 (e.g. Jupiter). For our object in the $3:1$ resonance, the ratio of periods (P_1/P_2) is simply $1/3$, and noting that Jupiter's semimajor axis is 5.20 AU we have $(1/3)^2 = (a_1/5.20)^3$ which by solving gives us a semimajor axis of our asteroid $a_1 = 2.50$ AU.

QUESTION 7.7

For an asteroid that has an eccentricity value that is *not* zero (i.e. basically *all* asteroids!), then its aphelion will lie beyond the distance associated with the given Kirkwood Gap, and its perihelion will lie within the distance associated with that Kirkwood Gap. Thus, twice per orbit, the asteroid will be at a distance from the Sun associated with the Kirkwood Gap. Therefore the gaps are not actually regions of space vacant of asteroids. They are just gaps in a diagram of asteroid semimajor axis.

QUESTION 7.8

The craters on Ida are simple bowl shapes, rather than complex craters with central peaks. They are mostly circular with many of the craters (particularly the ones seen at the right-hand side of the image) appearing to have very well defined, and relatively deep bowls, i.e. there is little sign of significant gravitational collapse of the crater walls (as you might expect for a low gravity environment).

QUESTION 7.9

An S-type asteroid would have a higher density (perhaps $2500 \, \text{kg m}^{-3}$) than a C-type (perhaps $1300 \, \text{kg m}^{-3}$) and so, if the 1 km diameter asteroid were an S-type, it would be considerably more massive than if it were a C-type. Thus the S-type would have a greater impact energy, and hence give rise to a somewhat larger impact crater.

QUESTION 7.10

To impact, or have close approaches with, a giant planet, a Kuiper Belt object must have a perihelion distance generally smaller than the semimajor axis of the giant planet. Thus some of the Scattered Disc objects may impact a giant planet (particularly Uranus and Neptune), as Scattered Disc objects can have large eccentricities and relatively small perihelion distances.

QUESTION 7.11

Most comets are thought to actively eject material from only a fairly small area of their surface at any one time (the rest being covered with insulating dust). If however a nucleus were to split apart, then presumably this would expose faces of new 'active' ices, which would then sublimate, producing an outburst of cometary activity.

QUESTION 7.12

The presence of Jupiter, which has the largest gravitational influence of all the planets, can essentially provide some 'shielding' from comets coming from the outer Solar System. Many comets that might otherwise have been destined for the inner Solar System can suffer impacts with Jupiter or be scattered back to the outer Solar System due to a close approach with Jupiter. (Indeed all the giant planets offer some shielding, but Jupiter's effect is greatest.)

QUESTION 7.13

The effect of seeing more meteors after local midnight is a bit like the reason you get more rain drops hitting the front windscreen of a moving car. The Earth gets more meteoroid impacts on its leading side. As the Earth is spinning, a local observer rotates into the leading side after local midnight (and remains on this leading side for 12 hours). Most sporadic meteors are in fact usually seen just before dawn.

QUESTION 7.14

The impact energy is equal to the kinetic energy, which is $\frac{1}{2}mv^2$.

For the car, $m = 1000\,\text{kg}$, and $v = 20\,\text{m s}^{-1}$. So

$$\text{kinetic energy} = 0.5 \times (1000\,\text{kg}) \times (20\,\text{m s}^{-1})^2 = 2 \times 10^5\,\text{J}.$$

For the meteoroid, $m = 1\,\text{g} = 10^{-3}\,\text{kg}$, and $v = 20\,\text{km s}^{-1} = 2 \times 10^4\,\text{m s}^{-1}$. So

$$\text{kinetic energy} = 0.5 \times (10^{-3}\,\text{kg}) \times (2 \times 10^4\,\text{m s}^{-1})^2 = 2 \times 10^5\,\text{J}.$$

So the impact energy of the 1 g meteoroid (which would be about 1 cm in diameter) is the same as that of the car travelling at 50 mph!

QUESTION 8.1

From the Solar System data given in Appendix A you can see that the orbits of Uranus and Neptune lie at 19.2 AU and 30.1 AU respectively. Since the inner edge of the β Pictoris disc is at about 20 AU, any planetary formation in this system could result in bodies that are an equivalent distance from the central star as some of those in our own Solar System. The outer limit of the β Pictoris disc (at a radius of 750 AU) lies at a much greater distance than that of Pluto (39.5 AU), or the anticipated spread of orbits of Kuiper Belt objects (20 to 50 AU), but far less than that of the Oort cloud (10^4 AU and more) where most of the Solar System cometary bodies reside.

QUESTION 8.2

(a) As described by Kepler's third law (Equation 7.3), a particle in an inner orbit would have a shorter orbital period than one in an outer orbit.

(b) The slow outer part of the disc would tend to slow down the faster inner part, at the same time as the fast inner part would tend to speed up the slower outer part.

QUESTION 8.3

The disc shape is the result of conservation of angular momentum, which inhibits collapse in the equatorial plane, but has no effect on matter collapsing from above the poles. In the hot, dense inner part of the disc, collapse is also inhibited by gas pressure. (Collapse may also be inhibited by magnetic forces, which we do not consider in this book.)

QUESTION 8.4

Viscous drag within the solar nebula while the protoSun was forming will have led to outward transfer of angular momentum and inward transfer of mass. The material falling into the protoSun would tend to be that whose speed of rotation had been slowed down by viscous drag. The T Tauri solar wind which occurred later (we will consider how much later, in Section 8.2.8) could also have carried away angular momentum.

QUESTION 8.5

According to Table 2.4, FeS could not have condensed from the solar nebula unless the temperature had dropped below about 700 K, so this value provides an upper limit on the possible temperature at which the chondrules originated.

QUESTION 8.6

The lines on Figure 8.14 are parallel, implying no change in relative composition throughout the solar nebula. The most convenient place for us to read off the proportions of the three components is at the point where the column mass of dust steps up from just rock, to rock plus ice, i.e. at the radius corresponding to Jupiter.

(a) Immediately to the right of the step-up in column mass of dust, the column mass of rock plus ice is about $60 \, kg \, m^{-2}$, whereas the column mass of gas at this point is about $2.8 \times 10^3 \, kg \, m^{-2}$. To two significant figures, this value is also the same as the total column mass of the nebula. The amount of rock plus ice expressed as a fraction of this is given by:

$$\frac{60 \, kg \, m^{-2}}{2.8 \times 10^3 \, kg \, m^{-2}} = 0.021$$

(Your reading of Figure 8.14 is likely to differ slightly from ours, because of difficulties in precisely reading values off the scale, but you should have got a reasonably similar result to ours.)

(b) Immediately to the left of the step-up in column mass of dust, the column mass of rock is about $15 \, kg \, m^{-2}$. The column mass of rock plus ice dust at the same distance (immediately to the right of the step-up) is $60 \, kg \, m^{-2}$, so the ratio of rock to ice is given by

$$\frac{rock}{(rock \; plus \; ice) - rock} = \frac{15 \, kg \, m^{-2}}{60 \, kg \, m^{-2} - 15 \, kg \, m^{-2}}$$

$$= \frac{15}{45}$$

$$= \frac{1}{3}$$

(Your reading of the graph may have been different from ours, but you should have ended up with a fraction with a value of approximately 0.3. Note that this is actually an upper limit because Figure 8.14 shows only the contribution of H_2O to the ice, and does not include other volatiles such as ammonia that add to the amount of ice condensing at greater distances from the Sun. However, H_2O is considerably more abundant than any other ice-forming substance.)

QUESTION 8.7

The volume of a spherical planetesimal that has a diameter of 10 km, and thus a radius of 5 km, is:

$$\tfrac{4}{3} \times \pi \times (5 \times 10^3 \, m)^3 = 5.2 \times 10^{11} \, m^3$$

The mass of an object is given by volume × density, so the mass of a planetesimal with the specified density is

$$(5.2 \times 10^{11} \, m^3) \times (3 \times 10^3 \, kg \, m^{-3}) = 1.6 \times 10^{15} \, kg$$

The number of such planetesimals necessary to make up the Earth is given by

$$\frac{\text{mass of Earth}}{\text{mass of planetesimal}} = \frac{6.0 \times 10^{24} \text{ kg}}{1.6 \times 10^{15} \text{ kg}}$$

$$= 3.8 \times 10^9$$

$$\approx 4 \times 10^9$$

QUESTION 8.8

These are cumulative curves. As we follow a curve from right to left it climbs, which shows that the total (i.e. cumulative) number of bodies greater than the mass indicated on the horizontal axis, is increasing. At times later than $t = 0$, the slope to the left of about 10^{16} kg shows that bodies of less mass than this have been produced. This can be explained by fragmentation during collision, even though collisions have also resulted in an increase in the maximum size of planetesimals (see bottom right of curves). (In the discussion of asteroid formation in Section 8.2.10 you will see that under certain conditions the rate of fragmentation can stop runaway growth.)

QUESTION 8.9

The terrestrial planets formed within 3 AU of the Sun, where the temperature was too hot for water (the least volatile hydrogen-rich compound in the condensation sequence) to condense (as ice) from the solar nebula; most of their water probably arrived in the form of hydrated minerals. The terrestrial planets are too small to have collected much nebula gas directly by gravitation, and what they (or their planetary embryo precursors) did collect would have been lost during the Sun's T Tauri phase. The giant planets formed beyond 5 AU, so water was able to condense at Jupiter, and other hydrogen-rich substances such as methane and ammonia could condense at greater distances. Thus planetesimals colliding to form the kernels of the giant planets would be rich in these materials as well as the more refractory elements. Furthermore, the kernels of the giant planets were sufficiently massive that each was able to capture and retain a large mass of gas (which would be mainly hydrogen and helium) directly from the nebula. This capture was interrupted by the onset of the T Tauri phase, which explains the decreasing volatile-to-rock ratio outwards from Jupiter (which began to form earliest) to Neptune (which began to form last).

QUESTION 8.10

(a) Density decreases outwards from Jupiter, running from $3.53 \times 10^3 \text{ kg m}^{-3}$ for the innermost large Galilean satellite, Io, to $1.85 \times 10^3 \text{ kg m}^{-3}$ for the outermost large Galilean satellite, Callisto.

(b) The densities in Appendix A, Table A2 are mean densities for each satellite as a whole. The mean densities of Ganymede and Callisto are most simply explained if they are mixtures of rock and ice, with roughly equal proportions of each. Europa can be explained by being composed almost entirely of rock, but it would appear that we need to call on a denser component within Io (probably iron or iron and sulfur in its core). (Strictly speaking we should take self-compression into account, but in bodies this small its effects are too slight to affect our argument.)

QUESTION 8.11

The gravitational field of the planet would cause gravitational focusing, whereby the trajectories of incoming projectiles would be deviated towards the planet. As a result, the closer a satellite is to a planet, the more likely it is to be hit. A subsidiary factor is that the planet's gravity would accelerate incoming projectiles, so that those hitting an inner satellite would, on average, be travelling faster than those striking an outer satellite. The smaller the satellite, the less the energy required to break it up, so an incoming projectile carrying a given kinetic energy would be more able to fragment a small satellite.

QUESTION 8.12

The large prograde satellites of Jupiter, Saturn and Uranus probably grew within protosatellite discs around each planet. Neptune's large retrograde satellite, Triton, may be a captured planetary embryo. *Small* satellites in prograde orbits can be formed by collisions involving *larger* ones, but those in retrograde orbits are more likely to be captured asteroids or comet nuclei.

QUESTION 8.13

As Question 8.1 and the associated text discussed, the observable part of the β Pictoris dust cloud extends inwards to about 20 AU (where the orbit of Uranus would be). The dust suggests that the β Pictoris 'solar nebula' has probably completed its condensation phase, and grain growth by coagulation could be well advanced. There is no observational evidence bearing on whether planetesimal formation and subsequent growth of planetary embryos has begun. However, the apparent absence of dust within about 20 AU of β Pictoris could mean planets have already formed in this region. Now, β Pictoris should have already undergone its T Tauri phase, but if this is the case, the survival of dust is perplexing. Probably the best bet is that this is how our own Solar System would have looked while Uranus was growing but before the Neptune kernel had had a chance to form. Section 8.2.8 suggests that there would be an interval of 20 Ma between these events. (This is not to say that we should expect β Pictoris to form planets with the same masses and orbits as those in our own Solar System, because planetary formation must be influenced by the original mass and angular momentum of the nebula, and by subsequent random events; especially during and after runaway growth.)

QUESTION 9.1

The total mass of material added = (yearly flux) × (length of time over which material has been added).

Thus, total mass = $10^8 \, \text{kg yr}^{-1} \times 10^9 \, \text{yr} = 10^{17} \, \text{kg}$.

The volume of material added = $\dfrac{\text{mass}}{\text{density}} = \dfrac{10^{17} \, \text{kg}}{1.5 \times 10^3 \, \text{kg m}^{-3}} = 6.67 \times 10^{13} \, \text{m}^3$.

The surface area of a sphere = $4\pi R^2$.

Thus, the volume of the layer $= 4\pi R^2 \times h$, where R = radius of the Earth and h = the depth of the layer. Therefore

$$6.67 \times 10^{13}\, \mathrm{m}^3 = 4\pi (6.378 \times 10^6\, \mathrm{m})^2 \times h$$

$$h = \frac{6.67 \times 10^{13}\, \mathrm{m}^3}{5.099 \times 10^{14}\, \mathrm{m}^2}$$

$$= 0.13\, \mathrm{m}$$

Therefore the depth of the layer h is 0.13 m.

QUESTION 9.2

It is an impact-produced sample of some description. It is probable that the chondritic, achondritic and metal-rich parts of the meteorite were all formed separately by normal processes. These were then assembled in a regolith during an impact event – for instance, an iron meteorite may have hit the surface of a silicate-rich parent body. (The meteorite described is not a hypothetical example but one member of a well documented, albeit rather rare, class of unusual samples.)

QUESTION 9.3

The surface area of a sphere $= 4\pi R^2$.

The surface area of the Earth $= 4\pi \times (6371\ \mathrm{km})^2 = 5.101 \times 10^8\ \mathrm{km}^2$.

If we assume (reasonably) that meteorites fall evenly over all parts of the Earth, then in one year a single meteorite falls in an area of $\dfrac{5.101 \times 10^8}{20\,000}\ \mathrm{km}^2 = 25\,505\ \mathrm{km}^2$.

The area of the town $= \pi r^2 = \pi \times 5^2\ \mathrm{km}^2 = 78.5\ \mathrm{km}^2$.

In an area of $25\,505\ \mathrm{km}^2$ there could be $\dfrac{25\,505}{78.5} = 325$ towns of radius 5 km.

On average, a meteorite of >0.1 kg size would fall in a town of radius 5 km, every 325 years.

QUESTION 9.4

The alteration has taken place in a closed system. Although fluids have been formed and these have reacted with the original minerals to form new phases, there has been no net change in chemical composition. (Presumably, there was not a great deal of fluid flow through the parent body, otherwise the fluids would have taken up certain elements into solution, or suspension, and removed them, thereby changing the chemical composition.)

QUESTION 9.5

Compared to CI and CM2 meteorites, CV3 samples are enriched in Ca and Al. This can be explained by a relatively higher proportion of Ca- and Al-rich minerals in CV3 meteorites.

QUESTION 9.6

The $\delta^{18}O$ of ocean water is $\left(\dfrac{1/499}{1/499} - 1\right) \times 1000$ which is 0‰. Similarly $\delta^{17}O$ can be shown to be 0‰. In other words, the reference point always has a δ value of 0‰. Relative enrichments of the isotopes ^{17}O and ^{18}O result in δ values which have a positive sign. Pure ^{16}O would have

$$\delta^{18}O = \left(\frac{1/\infty}{1/499} - 1\right) \times 1000 = \left(\frac{0}{1/499} - 1\right) \times 1000 = -1000‰$$

QUESTION 9.7

Remember that pure ^{16}O has both $\delta^{17}O = -1000‰$ and $\delta^{18}O = -1000‰$. Starting at G and adding ^{16}O to the system would produce a line which has a slope of about 1 on Figure 9.17. If you have trouble with understanding this, draw an expanded graph with axes that go from −1000 to 0 for each δ value. Pure ^{16}O plots at (−1000, −1000), while pure G is approximately at (0, 0). A line drawn between these points will have a slope of 1, and any mixture of pure ^{16}O with pure G will plot somewhere along this line. (When the oxygen isotope data from CAIs in Allende were first obtained, the addition of pure ^{16}O to materials with an isotope composition somewhere in the vicinity of G was considered a reasonable explanation of the data (the ^{16}O was assumed to have been injected into the solar nebula at a late stage from a supernova). However, for this theory to have gained acceptance it was considered that further oxygen isotope measurements would have eventually uncovered more extreme values than point D, and ideally all the way down to (−1000, −1000). Thus far, innumerable measurements have failed to exceed the limit set by D which seems to suggest that this was the isotope composition of a large reservoir of dust.)

QUESTION 9.8

The simple answer is that planets and asteroids formed at different times will have different oxygen isotope compositions. (The variation in oxygen isotope compositions displayed by meteorites provides a very valuable classification scheme. Note that some of the differences may be related to formation locale, rather than the timing of formation.)

QUESTION 9.9

That the data from the matrix materials plot on a line of slope 0.5 suggests that the oxygen isotopes in these minerals have been subjected to processes of isotope fractionation, which have resulted in $\delta^{18}O$ being twice that of $\delta^{17}O$. This is evidence that matrix materials were formed on a parent body, rather than in the solar nebula. Since the line of slope 0.5 is not coincident with the TFL it shows that the reservoir from which the carbonaceous chondrites were extracted was not exactly the same as that which formed the Earth.

QUESTION 9.10

The survival of interstellar organic molecules shows us that there must have been places within the solar nebula that were never heated significantly above 400 K.

(Although at the centre of the nebula, in the region of the Sun, temperatures were high, the outer parts, where comets formed, remained relatively cold. Carbonaceous chondrites sample materials from both extremes, demonstrating considerable turbulent mixing in the nebula.)

QUESTION 9.11

A very simple way would be to measure the stable isotope composition of carbon in each form. If there is no difference in $^{12}C/^{13}C$ ratios then the two forms might be related. (In reality there is a distinct difference in carbon isotope composition between the diamond and the amorphous carbon. While it is entirely possible that diamond can be formed from amorphous carbon by shock processes (which result in high pressures) the isotope evidence is at variance with such an origin. Several other lines of evidence also militate against shock.)

QUESTION 9.12

(a) By measuring the content of ^{26}Mg. (This is carried out by measuring the $^{26}Mg/^{24}Mg$ ratio and looking for anomalously high values.)

(b) Because some minerals contain Al and Mg it is difficult in these instances to measure any excess ^{26}Mg. It is desirable to search for the effects of ^{26}Al decay in minerals which contain very little, or no, magnesium.

QUESTION 9.13

Electromagnetic induction and heating through impact events are possibilities. However, the decay of short-lived radionuclides, such as ^{26}Al, is the most likely process to have contributed heat.

QUESTION 9.14

If we assume that all meteorite parent bodies started off with a chondritic composition then in order to obtain a metal-rich body there has to be extensive melting and differentiation. Upon total melting, silicate materials migrate to the outermost parts of an asteroid while metal sinks to the centre of the body. If the metallic asteroids we can observe today are original cores then the silicate crusts and mantles of these bodies must have been removed (most probably by collisional processes in the asteroid belt). Unlike the asteroid Vesta, which is still largely intact, a metallic asteroid represents the remnants of an original parent body.

QUESTION 9.15

With just the knowledge that Kodaikanal has a radiometric formation age of 3.8 Ga we may be inclined to think that this was the age of the meteorite's formation. However, as stated in the text, most meteorite ages fall in the range 4.5 to 4.6 Ga. Perhaps, therefore, we should consider that Kodaikanal formed at this earlier time and that the age of 3.8 Ga represents something that happened subsequently to the meteorite's parent body. Now, the impact-produced rocks from the Moon document a period of Solar System history when planetary bodies were undergoing intense bombardment by meteoroids. Thus, we can infer that the age of Kodaikanal might also document this same period, but in this case the impact event that affected the Kodaikanal parent body occurred in the asteroid belt, rather than on the Moon.

APPENDIX A USEFUL PLANETARY DATA

Table A1 Basic data on the planets (including the Moon).

	Mercury	Venus	Earth	Moon	Mars	Jupiter	Saturn	Uranus	Neptune	Pluto
Mass										
/10^{24} kg	0.330	4.87	5.97	0.074	0.642	1900	569	86.8	102	0.013
/Earth masses	0.055	0.815	1.00	0.012	0.107	318	95.2	14.4	17.1	0.002
Orbital semimajor axis[a]										
/10^6 km	57.91	108.2	149.6	149.6	227.9	778.4	1427	2871	4498	5906
/AU	0.39	0.72	1.00	1.00	1.52	5.20	9.54	19.19	30.07	39.48
Orbital eccentricity	0.206	0.007	0.017	0.055	0.093	0.048	0.054	0.047	0.009	0.249
Orbital inclination /degrees	7.0	3.4	0.0	5.2	1.9	1.3	2.5	0.8	1.8	17.1
Orbital period[b]	88.0 days	224.7 days	365.0 days	27.3 days	686.5 days	11.86 yr	29.42 yr	83.75 yr	163.7 yr	248.0 yr
Axial rotation period[b]/days	58.6	243	0.997	27.3	1.03	0.412	0.444	0.718	0.671	6.39
Axial inclination /degrees	0.1	177.3	23.5	6.7	25.2	3.1	26.7	97.9	29.6	119.6
Polar radius/km	2440	6052	6357	1738	3375	66 850	54 360	24 970	24 340	1137
Equatorial radius/km	2440	6052	6378	1738	3397	71 490	60 270	25 560	24 770	1137
Mean radius[c]/km	2440	6052	6371	1738	3390	69 910	58 230	25 360	24 620	1137
Density/10^3 kg m^{-3}	5.43	5.20	5.51	3.34	3.93	1.33	0.69	1.32	1.64	2.1
Surface gravity/m s^{-2}	3.7	8.9	9.8	1.6	3.7	23.1	9.0	8.7	11.1	0.7
Mean surface temperature/K	443	733	288	250	223					≥40
Effective cloud-top temperature/K						120	89	53	54	
Temperature at 1 bar pressure/K						165	135	75	70	
Rings	0	0	0	0	0	few	many	several	few	0
Satellites	0	0	1		2	≥39	≥30	≥21	≥8	1
Atmospheric surface pressure/bar[d]	≈10^{-15}	92.1	1.01	≈10^{-14}	6.3×10^{-3}					≈10^{-5}
Atmospheric surface density/kg m^{-3}	≈10^{-13}	67	1.293	≈10^{-13}	0.018					≈10^{-4}
Atmospheric column mass[e]/kg m^{-2}	≈10^{-11}	1.03×10^6	1.03×10^4	≈10^{-11}	1.69×10^2					≈1
Atmospheric main components (relatively minor components in parentheses)	O Na H_2 (He)	CO_2 (N_2)	N_2 O_2 (H_2O) (Ar)	Ar H_2 He	CO_2 (N_2) (Ar) (O_2)	H_2 He (CH_4)	H_2 He (CH_4)	H_2 He (CH_4)	H_2 He (CH_4)	N_2 (CH_4) (CO_2)

[a] Semimajor axis also represents the *mean* distance from the Sun.

[b] These are sidereal periods (i.e. referenced to the stars rather than to the Sun) quoted in Earth days or years.

[c] The mean radius is defined as the *volumetric* radius (i.e. the radius the body would have if it were a sphere of the same mass), and is calculated by $(R_e^2 R_p)^{1/3}$. The values quoted here for the gas giants are for the atmospheric layer where the pressure is equal to 1 bar (this also applies to the values for surface gravity).

[d] Although the SI unit for pressure is the pascal, we use bar here for simplicity and easy comparison, as the Earth's surface atmospheric pressure ≈1 bar. Note 1 bar = 10^5 Pa.

[e] Column mass is the mass of atmosphere situated above each unit area (1 m^2) of the planet's surface.

Table A2 Planetary satellites.

Planet	Satellite	Mean distance from planet/10^3 km	Orbital period/days	Mean radius[a]/km	Mass/10^{20} kg	Density/ 10^3 kg m^{-3}
Earth	Moon	384	27.3	1738	735	3.34
Mars	Phobos	9.4	0.32	11.1	0.00011	1.90
	Deimos	23.5	1.26	6.2	0.000018	1.76
Jupiter	Io	422	1.77	1821	893	3.53
	Europa	671	3.55	1565	480	2.99
	Ganymede	1070	7.15	2634	1482	1.94
	Callisto	1883	16.7	2403	1076	1.85
	≥56 others					
Saturn	Mimas	186	0.94	199	0.38	1.14
	Enceladus	238	1.37	249	0.73	1.21
	Tethys	295	1.89	530	6.2	1.00
	Dione	377	2.74	560	10.5	1.44
	Rhea	527	4.52	764	23.1	1.24
	Titan	1222	15.95	2575	1346	1.88
	Iapetus	3561	79.3	718	15.9	1.02
	≥24 others					
Uranus	Miranda	130	1.42	236	0.66	1.20
	Ariel	191	2.52	579	13.5	1.7
	Umbriel	266	4.14	585	11.7	1.4
	Titania	436	8.71	789	35.3	1.71
	Oberon	583	13.46	761	30.1	1.63
	≥16 others					
Neptune	Proteus	118	1.12	209	≈0.5?	≈1.2?
	Triton	355	5.88	1353	215	2.05
	Nereid	5513	360	170	≈0.3?	≈1.2?
	≥9 others					
Pluto	Charon	19.4	6.39	586	19.0	2.2

[a] The mean radius is defined as the *volumetric* radius (i.e. the radius the body would have if it were a sphere of the same mass).

Table A3 Asteroids that have been targets of spacecraft fly-bys or encounters.

Asteroid[a]	Spacecraft	Encounter date	Asteroid size/km	Mean radius[b]/km	Density/kg m^{-3}	Semimajor axis/AU
(951) Gaspra	Galileo	29 Oct 1991	19 × 12	6.1	2500 ±1000?	2.21
(243) Ida	Galileo	28 Aug 1993	58 × 23	15.8	2600 ±500	2.86
(253) Mathilde	NEAR	27 Jun 1997	59 × 47	26.4	1300 ±200	2.65
(9969) Braille	Deep Space 1	29 Jul 1999	2 × 1	0.7	not known	2.34
(433) Eros	NEAR	14 Feb 2000[c]	33 × 13	9.69	2670 ±30	1.46
(5535) Annefrank	Stardust	2 Nov 2002	8 × 4	2.5	not known	2.21

[a] Asteroids are initially numbered, and are then usually named also. We refer to them by (*number*) *name*.

[b] The mean radius is defined as the *volumetric* radius (i.e. the radius the body would have if it were a sphere of the same mass).

[c] NEAR went into orbit around Eros on this date. It remained there for a year and then landed on the surface of Eros on 12 Feb 2001.

Table A4 The largest known minor bodies in the Solar System.

Object	Semimajor axis/AU	Orbital period/yr	Orbital inclination	Orbital eccentricity	Mean radius[a]/km
Largest bodies in the asteroid belt:					
(1) Ceres	2.77	4.61	10.6°	0.079	457
(2) Pallas	2.77	4.61	34.8°	0.230	261
(4) Vesta	2.36	3.63	7.1°	0.090	250
(10) Hygiea	3.14	5.59	3.8°	0.121	215
(511) Davida	3.17	5.65	15.9°	0.180	163
Largest *known* (as of mid-2002) bodies in the Kuiper Belt (excluding Pluto):					
2002 LM$_{60}$ ('Quaoar')	43.2	284	8.0°	0.036	650
2002 AW$_{197}$	47.5	327	24.3°	0.128	400–650?
(28978) Ixion	39.3	246	19.7°	0.245	400–650?
2002 TX$_{300}$	43.3	284	25.9°	0.121	350–600?
(20000) Varuna	43.3	285	17.1°	0.054	450

[a] The mean radius is defined as the *volumetric* radius (i.e. the radius the body would have if it were a sphere of the same mass).

Table A5 Some selected comets.

Comet[a]	Perihelion distance/AU	Semimajor axis/AU	Orbital period/yr	Eccentricity	Inclination	Velocity at perihelion/km s^{-1}
2P/Enke	0.338	2.22	3.30	0.847	11.8°	69.6
46P/Wirtanen	1.059	3.09	5.44	0.658	11.7°	37.3
81P/Wild 2	1.590	3.44	6.40	0.539	3.2°	29.3
26P/Grigg–Skjellerup	1.118	3.04	5.31	0.663	22.3°	36.0
55P/Tempel–Tuttle	0.977	10.3	33.2	0.906	162.5°	41.6
1P/Halley	0.587	17.9	76.0	0.967	162.2°	54.5
109P/Swift–Tuttle	0.958	26.3	135	0.964	113.4°	42.6
153P/Ikeya–Zhang	0.507	51.0	367	0.990	28.1°	59.0
Hale–Bopp	0.925	184	≈2500	0.995	89.4°	43.8
Hyakutake	0.230	1490	≈58000	0.9998	124.9°	87.8

[a] Well observed periodic comets (i.e. short-period comets) are numbered, somewhat like asteroids, and this is indicated by the designation *number* P/, for example 2P/Enke.

Table A6 Major annual meteor showers.

Date of maximum rate	Name of shower	Hourly meteor rate	Parent comet
3 Jan	Quadrantids	130	unknown
12 Aug	Perseids	80	Swift–Tuttle
21 Oct	Orionids	25	Halley
17 Nov	Leonids	25[a]	Tempel–Tuttle
13 Dec	Geminids	90	(3200) Phaethon[b]

[a] This rate is usually what is observed, but every 33 years or so, this shower can display much higher rates.

[b] When discovered, Phaethon was assumed to be an asteroid as no cometary coma was observed. However it is likely that some activity has been present in the past.

Table A7 Some notable Solar System exploration missions.

Mission	Launch	Description
Sputnik 1 (USSR)	4 Oct 1957	First Earth-orbiting satellite. Remained in orbit for 92 days.
Pioneer 4 (USA)	3 Mar 1959	4 Mar 1959: first lunar fly-by (within 60 000 km of Moon's surface).
Luna 2 (USSR)	12 Sep 1959	14 Sep 1959: first spacecraft to land (impact) on the Moon.
Venera 1 (USSR)	12 Feb 1961	19 May 1961: first Venus fly-by. (Contact lost before fly-by.)
Mars 1 (USSR)	1 Nov 1962	19 Jun 1963: first Mars fly-by. (Contact lost before fly-by.)
Venera 3 (USSR)	16 Nov 1965	1 Mar 1966: first spacecraft to land on Venus. (Contact lost before landing.)
Luna 9 (USSR)	31 Jan 1966	3 Feb 1966: First soft landing on the Moon. TV pictures returned to Earth.
Zond 5 (USSR)	14 Sep 1968	First spacecraft to orbit the Moon (18 Sep 1968) and return a payload safely to Earth (21 Sep 1968). Payload included turtles, flies, worms and plants.
Apollo 8 (USA)	21 Dec 1968	First manned mission to orbit the Moon (24 Dec 1968). Returned 27 Dec 1968.
Apollo 11 (USA)	16 July 1969	First manned landing on the Moon (20 July 1969). Crew: Neil Armstrong, Edwin 'Buzz' Aldrin, Michael Collins (orbiter). Returned 24 July 1969.
Apollo 12 (USA)	14 Nov 1969	Second manned landing on the Moon (19 Nov 1969). Crew: Charles Conrad, Alan Bean, Richard Gordon (orbiter). Returned 24 Nov 1969.
Apollo 13 (USA)	11 Apr 1970	Moon mission aborted after onboard explosion on 14 Apr 1970. Crew: James Lovell, Fred Haise, John Swigert (orbiter). Returned 17 Apr 1970.
Luna 16 (USSR)	12 Sep 1970	First robotic sample-return from the Moon. Returned approximately 100 g of lunar material.
Apollo 14 (USA)	31 Jan 1971	Third manned landing on the Moon (5 Feb 1971). Crew: Alan Shepard, Edgar Mitchell, Stuart Roosa (orbiter). Returned 9 Feb 1971.
Mars 3 (USSR)	28 May 1971	2 Dec 1971: first spacecraft to land on Mars. Soft landing. Images returned.
Apollo 15 (USA)	26 Jul 1971	Fourth manned landing on the Moon (30 Jul 1971). Crew: David Scott, James Irwin, Alfred Worden (orbiter). Returned 7 Aug 1971. First lunar rover used.
Pioneer 10 (USA)	3 Mar 1972	First outer Solar System mission. 3 Dec 1973: fly-by of Jupiter. Currently ≈80 AU from the Sun. Will reach the star Aldebaran in 2 million years!
Apollo 16 (USA)	16 Apr 1972	Fifth manned landing on the Moon (21 Apr 1972). Crew: John Young, Charles Duke, Thomas Mattingly (orbiter). Returned 27 Apr 1972.
Apollo 17 (USA)	7 Dec 1972	Sixth (and final) manned landing on the Moon (11 Dec 1972). Crew: Eugene Cernan, Harrison Schmitt, Ronald Evans (orbiter). Returned 19 Dec 1972.
Pioneer 11 (USA)	6 Apr 1973	4 Dec 1974: Jupiter fly-by. 1 Sep 1979: Saturn fly-by.
Skylab (USA)	14 May 1973	First manned orbiting 'space station'. Manned until 8 Feb 1974. Final usage of the Apollo Saturn V rocket.
Mariner 10 (USA)	3 Nov 1973	First (and only) spacecraft to go to Mecury. 5 Feb 1974: Venus fly-by. Mercury fly-bys on 29 Mar 1974, 21 Sep 1974 and 16 Mar 1975.
Viking 1 (USA)	20 Aug 1975	Mars orbiter and lander. 19 June 1976: reached Mars. 20 Jul 1976: lander touched down.

Table A7 continued.

Mission	Launch	Description
Viking 2 (USA)	4 Sept 1975	Mars orbiter and lander. 7 Aug 1976: reached Mars. 3 Sep 1976: lander touched down.
Voyager 2 (USA)	20 Aug 1977	First (only) spacecraft to undertake a tour of all the giant planets. 9 Jul 1979: Jupiter fly-by. 26 Aug 1981: Saturn fly-by. 24 Jan 1986: Uranus fly-by. 25 Aug 1989: Neptune fly-by.
Voyager 1 (USA)	5 Sep 1977	5 Mar 1979: Jupiter fly-by. 12 Nov 1980: Saturn fly-by.
ISEE-3/ICE (USA)	12 Aug 1978	11 Sep 1985: first spacecraft to 'distant fly-by' a comet (Giacobini–Zinner).
Venera 13 (USSR)	30 Oct 1981	1 Mar 1982: Venus landing. Returned colour images from the surface.
Giotto (ESA)	2 Jul 1985	13 Mar 1986: first close (600 km) fly-by of a cometary nucleus (comet Halley).
Magellan (USA)	4 May 1989	Venus orbit insertion 10 Aug 1990. Mapped Venus surface with radar (1990–1994).
Galileo (USA)	18 Oct 1989	First spacecraft to orbit one of the giant planets. 29 Oct 1991: fly-by of asteroid (951) Gaspra. 28 Aug 1993: fly-by of asteroid (243) Ida. 7 Dec 1995: Galileo reaches Jupiter and deployed probe enters the atmosphere of Jupiter. 21 Sept 2003: Galileo impacts Jupiter.
Ulysses (ESA)	6 Oct 1990	First spacecraft to leave the ecliptic plane and orbit around the Sun, passing over the north and south poles. 8 Feb 1992: Jupiter fly-by.
Near Earth Asteroid Rendezvous (NEAR) Mission (USA)	17 Feb 1996	First spacecraft to orbit and land on an asteroid. 27 Jun 1997: fly-by of asteroid (253) Mathilde. 14 Feb 2000: started orbiting near Earth asteroid, (433) Eros. 12 Feb 2001: spacecraft landed on Eros.
Mars Global Surveyor (USA)	7 Nov 1996	Highly successful Mars remote sensing mission. 12 Sep 1997: reached Mars. Mar 1999: began mapping planet.
Mars Pathfinder (USA)	4 Dec 1996	4 Jul 1997: landed on Mars. 6 Jul 1997: deployed the Sojourner rover.
Cassini–Huygens (USA + Europe)	15 Oct 1997	Mission to Saturn and Titan. 30 Dec 2000: Jupiter fly-by. 1 Jul 2004: Saturn orbit insertion. 14 Jan 2005: Huygens probe lands on Titan.
Deep Space 1 (USA)	24 Oct 1998	22 Sep 2001: close fly-by of comet Borrelly's nucleus. Images returned. 29 Jul 1999: fly-by of (9969) Braille.
Stardust (USA)	7 Feb 1999	Fly-by and cometary dust sample return mission to comet Wild 2. 2 Nov 2002: fly-by of asteroid (5535) Annefrank. 2 Jan 2004: fly-by of comet Wild 2. 15 Jan 2006: capsule carrying cometary dust lands on Earth for analysis.
2001 Mars Odyssey (USA)	7 Apr 2001	11 Jan 2002: entered Mars orbit. Acts as relay for 2003 rover missions.
Genesis (USA)	8 Aug 2001	Solar wind particle sample return mission. 3 Dec 2001: capture experiment deployed. Sep 2004: samples returned to Earth.
Rosetta (ESA)	2003	Comet orbiter and lander. Nominal mission plan: 10 Jul 2006: fly-by of asteroid (4979) Otawara. 24 Jul 2008: fly-by of asteroid (140) Siwa. 29 Nov 2011: orbit entry around comet Wirtanen. Sep 2012: lander deployed. (Note: exact mission plan may change.)
Mars Express (ESA) + Beagle 2	2 Jun 2003	Mars orbiter and lander. 25 Dec 2003: Mars Express enters Mars orbit, and the Beagle 2 spacecraft lands on the surface to look for isotope ratios indicative of life.

APPENDIX B SELECTED PHYSICAL CONSTANTS AND UNIT CONVERSIONS

Table B1 SI fundamental and derived units.

Quantity	Unit	Abbreviation	Equivalent units
mass	kilogram	kg	
length	metre	m	
time	second	s	
temperature	kelvin	K	
angle	radian	rad	
area	square metre	m^2	
volume	cubic metre	m^3	
speed, velocity	metre per second	$m\,s^{-1}$	
acceleration	metre per second squared	$m\,s^{-2}$	
density	kilogram per cubic metre	$kg\,m^{-3}$	
frequency	hertz	Hz	$(cycles)\,s^{-1}$
force	newton	N	$kg\,m\,s^{-2}$
pressure	pascal	Pa	$N\,m^{-2}$, $kg\,m^{-1}\,s^{-2}$
energy	joule	J	$kg\,m^2\,s^{-2}$
power	watt	W	$J\,s^{-1}, kg\,m^2\,s^{-3}$
specific heat capacity	joule per kilogram kelvin	$J\,kg^{-1}\,K^{-1}$	$m^2\,s^{-2}\,K^{-1}$
thermal conductivity	watt per metre kelvin	$W\,m^{-1}\,K^{-1}$	$m\,kg\,s^{-3}\,K^{-1}$

Table B2 Selected physical constants and preferred values.

Quantity	Symbol	Value
speed of light in a vacuum	c	$3.00 \times 10^8\,m\,s^{-1}$
Planck constant	h	$6.63 \times 10^{-34}\,J\,s$
Boltzmann constant	k	$1.38 \times 10^{-23}\,J\,K^{-1}$
gravitational constant	G	$6.67 \times 10^{-11}\,N\,m^2\,kg^{-2}$
Stefan–Boltzmann constant	σ	$5.67 \times 10^{-8}\,W\,m^2\,K^{-4}$
Avogadro constant	N_A	$6.02 \times 10^{23}\,mol^{-1}$
molar gas constant	R	$8.31\,J\,K^{-1}\,mol^{-1}$
charge of electron	e	$1.60 \times 10^{-19}\,C$ (negative charge)
mass of proton	m_p	$1.67 \times 10^{-27}\,kg$
mass of electron	m_e	$9.11 \times 10^{-31}\,kg$
Astronomical quantities:		
mass of the Sun	$M_\odot$	$1.99 \times 10^{30}\,kg$
radius of the Sun	$R_\odot$	$6.96 \times 10^8\,m$
photospheric temperature of the Sun	$T_\odot$	5770 K
luminosity of the Sun	$L_\odot$	$3.84 \times 10^{26}\,W$
astronomical unit	AU	$1.50 \times 10^{11}\,m$

Table B3 Some useful conversions from alternative unit systems to SI units.

Quantity	Unit	SI equivalent
angle	1 degree	$(\pi/180)\,\mathrm{rad}$
pressure	1 bar	$10^5\,\mathrm{Pa}$
temperature	1 °C	1 K
energy	1 erg	$10^{-7}\,\mathrm{J}$
	1 electron volt	$1.60 \times 10^{-19}\,\mathrm{J}$
	1 ton of TNT	$4.18 \times 10^9\,\mathrm{J}$
length	1 foot	0.305 m
	1 mile	$1.61 \times 10^3\,\mathrm{m}$
area	1 square inch	$6.45 \times 10^{-4}\,\mathrm{m}^2$
	1 square mile	$2.59 \times 10^6\,\mathrm{m}^2$
mass	1 pound	0.454 kg
speed, velocity	1 mile per hour	$0.447\,\mathrm{m\,s}^{-1}$

Table B4 The Greek alphabet.

Name	Lower case	Upper case
Alpha	α	A
Beta (bee-ta)	β	B
Gamma	γ	Γ
Delta	δ	Δ
Epsilon	ε	E
Zeta (zee-ta)	ζ	Z
Eta (ee-ta)	η	H
Theta (thee-ta – 'th' as in theatre)	θ	Θ
Iota (eye-owe-ta)	ι	I
Kappa	κ	K
Lambda (lam-da)	λ	Λ
Mu (mew)	μ	M
Nu (new)	ν	N
Xi (cs-eye)	ξ	Ξ
Omicron	o	O
Pi (pie)	π	Π
Rho (roe)	ρ	P
Sigma	σ	Σ
Tau	τ	T
Upsilon	υ	Y
Phi (fie)	ϕ	Φ
Chi (kie)	χ	X
Psi (ps-eye)	ψ	Ψ
Omega (owe-me-ga)	ω	Ω

APPENDIX C THE ELEMENTS

Table C1 The elements and their abundances.

The relative atomic mass, A_r, is the average mass of the atoms of the element as it occurs on Earth. It is thus an average over all the isotopes of the element. The scale is fixed by giving the carbon isotope $^{12}_6C$ a relative atomic mass of 12.0. By convention, the Solar System abundance is normalized to 10^{12} atoms of hydrogen, whereas the CI chondrite abundance is normalized to 10^6 atoms of silicon. To directly compare chondrite abundance to Solar System abundance (by number), you would multiply chondrite abundance by 35.8.

Atomic number, Z	Name	Chemical symbol	Relative atomic mass, A_r	Solar System abundance		CI chondrite abundance by number
				by number	by mass	
1	hydrogen	H	1.01	1.0×10^{12}	1.0×10^{12}	2.79×10^{10}
2	helium	He	4.00	9.8×10^{10}	3.9×10^{11}	2.72×10^9
3	lithium	Li	6.94	2.0×10^3	1.4×10^4	57.1
4	beryllium	Be	9.01	26	2.4×10^2	0.73
5	boron	B	10.81	6.3×10^2	6.8×10^3	21.2
6	carbon	C	12.01	3.6×10^8	4.4×10^9	1.01×10^7
7	nitrogen	N	14.01	1.1×10^8	1.6×10^9	3.13×10^6
8	oxygen	O	16.00	8.5×10^8	1.4×10^{10}	2.38×10^7
9	fluorine	F	19.00	3.0×10^4	5.7×10^5	843
10	neon	Ne	20.18	1.2×10^8	2.5×10^9	3.44×10^6
11	sodium	Na	22.99	2.0×10^6	4.7×10^7	5.74×10^4
12	magnesium	Mg	24.31	3.8×10^7	9.2×10^8	1.074×10^6
13	aluminium	Al	26.98	3.0×10^6	8.1×10^7	8.49×10^4
14	silicon	Si	28.09	3.5×10^7	1.0×10^9	1.00×10^6
15	phosphorus	P	30.97	3.7×10^5	1.2×10^7	1.04×10^4
16	sulfur	S	32.07	1.9×10^7	6.0×10^8	5.15×10^5
17	chlorine	Cl	35.45	1.9×10^5	6.6×10^6	5240
18	argon	Ar	39.95	3.6×10^6	1.5×10^8	1.01×10^5
19	potassium	K	39.10	1.3×10^5	5.2×10^6	3770
20	calcium	Ca	40.08	2.2×10^6	8.8×10^7	6.11×10^4
21	scandium	Sc	44.96	1.2×10^3	5.5×10^4	34.2
22	titanium	Ti	47.88	8.5×10^4	4.1×10^6	2400
23	vanadium	V	50.94	1.0×10^4	5.3×10^5	293
24	chromium	Cr	52.00	4.8×10^5	2.5×10^7	1.35×10^4
25	manganese	Mn	54.94	3.4×10^5	1.9×10^7	9550
26	iron	Fe	55.85	3.2×10^7	1.8×10^9	9.00×10^5
27	cobalt	Co	58.93	8.1×10^4	4.8×10^6	2250
28	nickel	Ni	58.69	1.8×10^6	1.0×10^8	4.93×10^4
29	copper	Cu	63.55	1.9×10^4	1.2×10^6	522
30	zinc	Zn	65.39	4.5×10^4	2.9×10^6	1260
31	gallium	Ga	69.72	1.3×10^3	9.4×10^4	37.8
32	germanium	Ge	72.61	4.3×10^3	3.1×10^5	119

Atomic number, Z	Name	Chemical symbol	Relative atomic mass, A_r	Solar System abundance		CI chondrite abundance by number
				by number	by mass	
33	arsenic	As	74.92	2.3×10^2	1.8×10^4	6.56
34	selenium	Se	78.96	2.2×10^3	1.8×10^5	62.1
35	bromine	Br	79.90	4.3×10^2	3.4×10^4	11.8
36	krypton	Kr	83.80	1.7×10^3	1.4×10^5	45
37	rubidium	Rb	85.47	2.5×10^2	2.1×10^4	7.09
38	strontium	Sr	87.62	8.5×10^2	7.5×10^4	23.5
39	yttrium	Y	88.91	1.7×10^2	1.5×10^4	4.64
40	zirconium	Zr	91.22	4.1×10^2	3.7×10^4	11.4
41	niobium	Nb	92.91	25	2.3×10^3	0.698
42	molybdenum	Mo	95.94	91	8.7×10^3	2.55
43	technetium	Tc[a]	98.91	_[b]	_[b]	_[b]
44	ruthenium	Ru	101.07	66	6.8×10^3	1.86
45	rhodium	Rh	102.91	12	1.3×10^3	0.344
46	palladium	Pd	106.42	50	5.3×10^3	1.39
47	silver	Ag	107.87	17	1.9×10^3	0.486
48	cadmium	Cd	112.41	58	6.5×10^3	1.61
49	indium	In	114.82	6.6	7.6×10^2	0.184
50	tin	Sn	118.71	140	1.6×10^4	3.82
51	antimony	Sb	121.76	11	1.3×10^3	0.309
52	tellurium	Te	127.60	170	2.2×10^4	4.81
53	iodine	I	126.90	32	4.1×10^3	0.90
54	xenon	Xe	131.29	170	2.2×10^4	4.7
55	caesium	Cs	132.91	13	1.8×10^3	0.372
56	barium	Ba	137.33	160	2.2×10^4	4.49
57	lanthanum	La	138.91	16	2.2×10^3	0.4460
58	cerium	Ce	140.12	41	5.7×10^3	1.136
59	praseodymium	Pr	140.91	6.0	8.5×10^2	0.1669
60	neodymium	Nd	144.24	30	4.3×10^3	0.8279
61	promethium	Pm[a]	146.92	_[c]	_[c]	_[c]
62	samarium	Sm	150.36	9.3	1.4×10^3	0.2582
63	europium	Eu	151.96	3.5	5.3×10^2	0.0973
64	gadolinium	Gd	157.25	12	1.8×10^3	0.3300
65	terbium	Tb	158.93	2.1	3.4×10^2	0.0603
66	dysprosium	Dy	162.50	14	2.3×10^3	0.3942
67	holmium	Ho	164.93	3.2	5.2×10^2	0.0889
68	erbium	Er	167.26	8.9	1.5×10^3	0.2508
69	thulium	Tm	168.93	1.3	2.3×10^2	0.0378
70	ytterbium	Yb	170.04	8.9	1.5×10^3	0.2479
71	lutetium	Lu	174.97	1.3	2.3×10^2	0.0367

Atomic number, Z	Name	Chemical symbol	Relative atomic mass, A_r	Solar System abundance		CI chondrite abundance by number
				by number	by mass	
72	hafnium	Hf	178.49	5.3	9.6×10^2	0.154
73	tantalum	Ta	180.95	1.3	2.4×10^2	0.0207
74	tungsten	W	183.85	4.8	8.8×10^2	0.133
75	rhenium	Re	186.21	1.9	3.5×10^2	0.0517
76	osmium	Os	190.2	24	4.6×10^3	0.675
77	iridium	Ir	192.22	23	4.5×10^3	0.661
78	platinum	Pt	195.08	48	9.3×10^3	1.34
79	gold	Au	196.97	6.8	1.3×10^3	0.187
80	mercury	Hg	200.59	12	2.5×10^3	0.34
81	thallium	Tl	204.38	6.6	1.4×10^3	0.184
82	lead	Pb	207.2	110	2.3×10^4	3.15
83	bismuth	Bi	208.98	5.1	1.1×10^3	0.144
84	polonium	Po[a]	209.98	—[c]	—[c]	—[c]
85	astatine	At[a]	209.99	—[c]	—[c]	—[c]
86	radon	Rn[a]	222.02	—[c]	—[c]	—[c]
87	francium	Fr[a]	223.02	—[c]	—[c]	—[c]
88	radium	Ra[a]	226.03	—[c]	—[c]	—[c]
89	actinium	Ac[a]	227.03	—[c]	—[c]	—[c]
90	thorium	Th[a]	232.04	1.2	2.8×10^2	0.0335
91	protoactinium	Pa[a]	231.04	—[c]	—[c]	—[c]
92	uranium	U[a]	238.03	0.32	7.7×10^1	0.0090
93	neptunium	Np[a]	237.05	—[c]	—[c]	—[c]
94	plutonium	Pu[a]	239.05	—[c]	—[c]	—[c]
95	americium	Am[a]	241.06	—[c]	—[c]	—[c]
96	curium	Cm[a]	244.06	—[c]	—[c]	—[c]
97	berkelium	Bk[a]	249.08	—[c]	—[c]	—[c]
98	californium	Cf[a]	252.08	—[c]	—[c]	—[c]
99	einsteinium	Es[a]	252.08	—[c]	—[c]	—[c]
100	fermium	Fm[a]	257.10	—[c]	—[c]	—[c]
101	mendelevium	Md[a]	258.10	—[c]	—[c]	—[c]
102	nobelium	No[a]	259.10	—[c]	—[c]	—[c]
103	lawrencium	Lr[a]	262.11	—[c]	—[c]	—[c]

[a] No stable isotopes.

[b] Detected in spectra of some rare evolved stars.

[c] Too scarce to have been detected beyond the Earth (i.e. abundance value not well known).

GLOSSARY

ablation During the passage of a meteoroid through the Earth's atmosphere, the surface layers of the object become heated to their melting point. The subsequent removal of this material, by atmospheric friction during descent, is termed ablation.

absorption spectrum A graph that shows the relative amount of each wavelength of electromagnetic radiation absorbed when radiation from an ideal thermal source is passed through a thin gas of atoms and/or molecules.

accretion The growth of bodies during the formation of the Solar System, as a result of collisions that are not sufficiently energetic to fragment and disperse the colliding bodies. The term is usually employed to describe the growth of bodies from planetesimal-size upwards, and is used irrespective of whether the colliding bodies are roughly equal in mass or whether one is much smaller (such as a meteoroid or comet striking a planet).

accretional heating Heating by the loss of kinetic energy when a projectile strikes a planetary body. Accretional heat is added to the outside of a body as it grows and, except in a giant impact, the effect is local.

achondrite A stony meteorite that does not contain any chondrules. Meteorites of this nature have generally been formed by igneous processes, but the term may also be applied to samples that represent mechanical mixtures of fragments of igneous rocks and small amounts of chondritic materials.

adiabatic lapse rate The rate of decrease in temperature of a parcel of gas rising through an atmosphere and exchanging no heat with its surroundings. For convection to occur, the rate of decrease of temperature of a planetary atmosphere must exceed the adiabatic lapse rate.

aerodynamic drag The resistance of an object to the passage of an atmosphere over it. It depends on the shape of the object, its surface roughness and the nature of the atmosphere.

albedo The fraction of the solar radiation incident on a body that is reflected by the body.

angular momentum A quantity dependent on the speed of rotation and distribution of mass with respect to the rotation axis that is conserved for rotating bodies. For a parcel of atmosphere of mass m, velocity with respect to the centre of the planet v and a distance r from the centre, the angular momentum is given by mvr.

angular speed/velocity The speed at which an object (e.g. a wheel) turns or at which a body (e.g. a satellite) orbits another body. Defined as the angle turned through in unit time; normal units are radians per second or degrees per second.

aphelion The point on an orbit that lies furthest from the Sun.

aphelion distance The distance from the (centre of the) Sun to the aphelion of an orbit.

asteroid belt Region between the orbits of Mars and Jupiter, from about 2.0 AU to 3.3 AU, where the majority of asteroids are found. Also referred to as the asteroid main belt.

asteroids Small rocky or metallic bodies orbiting the Sun. Most are members of the asteroid belt, which lies between the orbits of Mars and Jupiter, from about 2.0–3.3 AU. Of the many thousands of known asteroids, most are only a few kilometres (or less) in diameter, but about 30 exceed 200 km in diameter. A small proportion have Earth-crossing orbits.

asthenosphere A weak convecting zone of a planetary body, underlying the lithosphere. In this book we use the term to embrace the whole of the convecting part of the mantle.

astronomical unit The astronomical unit is a convenient measure of distance within the Solar System. It is the mean distance between the (centres of) the Sun and the Earth, and is about 149.6 million kilometres.

atmospheric structure The variation of temperature and pressure with height in an atmosphere.

basalt A dark, fine-grained extrusive igneous rock composed of the minerals plagioclase, feldspar, pyroxene, magnetite, and with or without olivine, and containing not more than 53 weight (wt) % SiO_2.

basalt lava Dark-coloured lava containing 45–50% SiO_2, formed by partial melting of chondritic silicates (for example, the peridotite forming the Earth's mantle). Basalts are by far the commonest expression of volcanism on the terrestrial planets, and cover 70% of the Earth's surface.

belt A dark band of Jupiter's atmosphere in which the atmosphere is rising. The term also applies to similar features in the atmospheres of other giant planets.

black-body source A source of radiation with the property that it will absorb completely electromagnetic radiation of any wavelength. The spectrum is a black-body radiation curve. Stars and planets are approximately black-body sources.

bow shock An abrupt boundary between regions of a gas or liquid which are travelling at relative rates greater than the local speed of sound, such as that formed by the solar wind when it encounters a planet, comet, or other body.

breccia Rock composed of sharp-angled fragments embedded in a fine-grained matrix.

calcium–aluminium-rich inclusions (CAIs) Irregularly-shaped entities found in carbonaceous chondrites, generally light in colour and typically of millimetre to centimetre size, composed of minerals of a highly refractory nature. The mineralogy of the inclusions matches very well with what is predicted to form at high temperatures in the condensation sequence.

calderas Large volcanic craters, more than 1 km in diameter, formed by subsidence following the eruption of large volumes of lava or pyroclastic rocks. Major basaltic volcanoes such as Mauna Loa on the Earth and Olympus Mons on Mars are commonly crowned by calderas.

carbonaceous chondrite Meteorite of chondritic composition generally containing a high concentration of carbon (typically between 1% and 5% by mass), in the form of organic compounds. CI carbonaceous chondrites are chemically very primitive, having compositions that closely match that of the Sun (excluding the most volatile material).

carbonatites An unusual rock rich in carbonate minerals thought to be formed by very small degrees of partial melting of the mantle, and which may be erupted as a lava. Eruption temperatures are amongst the lowest of terrestrial lavas, 773 K (500 °C), but the resulting flows have a very low viscosity and are particularly fluid.

Centaurs A class of minor bodies having orbits which cross one or more of the giant planets. Centaurs are believed to be objects from the Kuiper Belt that have undergone orbital evolution.

Chapman scheme A sequence of reactions proposed to account for the formation and destruction of ozone in the Earth's stratosphere.

chemical equilibrium The state of a mixture of substances in a volume where no material or radiation is entering or leaving, where no change in the composition of the mixture (i.e. no chemical change) can occur even if the mixture is left for an infinitely long time, unless the conditions (temperature or pressure) change.

chondrite A meteorite composed of chondrules embedded in a fine-grained silicate matrix.

chondritic composition The composition of material in the Solar System is referred to as chondritic if the relative abundances of the elements (except for the most volatile) are similar to those in chondritic meteorites (strictly CI carbonaceous chondrites).

chondrule A globule of silicate minerals typically about one millimetre across, embedded in a silicate-rich matrix and characteristic of the type of meteorite known as chondrites.

CI carbonaceous chondrites Meteorites of chondritic composition generally containing a high concentration of carbon (typically 1–5% by mass), in the form of organic compounds. CI carbonaceous chondrites are chemically very primitive, having compositions that closely match that of the Sun (excluding the most volatile material).

circumstellar disc (See protoplanetary disc.)

coagulation The process in which dust grains that had condensed in the solar nebula stuck together, when they came into contact as a result of accidental collisions.

column mass (1) The mass per square metre of the solar nebula, measured perpendicular to its plane. (2) The mass of a planet's atmosphere per square metre of surface.

coma The huge gaseous and dusty envelope which surrounds the solid part of a comet (its nucleus). It is derived from evaporation of ices as the comet nears the Sun. It is the coma of a comet that is often seen from the Earth.

comet A minor body composed mainly of water-ice and rocky (silicate) material. Comets originate from the Kuiper Belt or the Oort cloud, and usually have elongated orbits, and a wide range of orbital periods.

cometary nucleus The central body associated with a comet, i.e. a minor body composed mainly of water ice and rocky (silicate) material.

compatible elements Elements with ionic radii that favour their incorporation into the crystal lattices of common silicate minerals. On partial melting, these elements tend to remain in the solid residuum, so that the resulting magma is depleted in these elements.

condensation The joining together of atoms from the solar nebula to form compounds as the temperature dropped, or the formation of liquid droplets or solid particles by cooling of gas.

condensation flow Flow of an atmospheric component from one hemisphere to another, alternating seasonally due to condensation near the 'winter' pole and sublimation near the 'summer' pole.

condensation sequence The order in which solid compounds would form out of an initially hot, gaseous solar nebula during cooling.

conduction The transfer of heat by means of internal kinetic energy on the atomic scale.

continental flood basalt province (CFBP) A region of extensive *basalt lava* sheets erupted in prodigious volumes (of the order of $10^6\,km^3$) in certain parts of the Earth's continents, usually associated with early stages of continental rifting and ocean formation. Individual flows tens to hundreds of kilometres in length combine to form immense flow fields, each representing a single eruption event resulting from the rapid effusion of several hundred cubic kilometres of basalt lava. Individual flows are 10–50 m thick, and stacks of lava sheets may accumulate to 1–2 km thickness.

continuous spectrum A spectrum that is broad and smooth, i.e. the spectral flux density exhibits no sharp changes with wavelength.

convection A process of heat transfer where a fluid in a gravitational field is heated from below to the point where the hotter, less dense fluid rises upwards, displacing the cooler, denser fluid downwards.

convection currents A current of fluid in motion because of convection.

convective ascent region The part of a volcanic eruption column in which hot gas and dust rise convectively through the atmosphere. On Earth, convection carries pyroclastic rocks to heights of 20–40 km.

Coriolis effect A general term used to describe the displacement of an object as a result of its motion in a rotating system. In the context of a rotating planet it describes the E–W movement that an object acquires as it moves N–S.

cosmic sediment A term sometimes used to describe a primitive chondritic meteorite. It conveys the sense that samples of this nature were formed from the settling of dust, mineral fragments, chondrules, CAIs, ices etc. onto the surface of a pre-existing asteroidal body.

cosmic spherule A micrometeoroid of millimetre-size dimensions becomes completely molten as it falls through the Earth's atmosphere, resulting in the formation of a spherical melt droplet. When subsequently collected it is referred to as a cosmic spherule; oceanic sediments contain relatively large concentrations of these materials.

crust A compositionally (and therefore seismically) distinct outer layer that can be recognized on top of the mantle on evolved planetary bodies such as the Earth. It forms the outermost part of the lithosphere.

cryovolcanism Volcanic processes taking place at low temperatures, involving icy magmas with melting points less than ~ 273 K, rather than silicates, which melt at temperatures exceeding ~ 1000 K. Several icy satellites of the outer planets display evidence of cryovolcanism.

dark halo craters Unusual lunar craters a few kilometres in diameter surrounded by aprons of dark deposits thought to be accumulations of fall-out from pyroclastic volcanism similar to Strombolian or Hawaiian eruptions on Earth (i.e. non-impact in origin). They are usually located along cracks or fissures, and may be elongated along them.

decompression melting Partial melting resulting from a decrease in pressure, rather than an increase in temperature. It occurs when a mass of rock (e.g. mantle material) is carried, without losing heat, from a region of high pressure to one of sufficiently lower pressure to initiate melting.

dense cloud The coldest and (apart from circumstellar shells) the densest type of region in the interstellar medium, rich in molecules. Dense clouds give birth to stars, mainly in the form of open clusters.

density The mass per unit volume of a substance under specified conditions of pressure and temperature.

diatomic molecules A molecule consisting of two atoms bound together, e.g. N_2 or CO.

differentiation The process whereby a planetary body (large planetesimal, planetary embryo, planet or satellite) evolves into compositionally distinct layers. Dense material sinks to form the core, and less dense material rises to form the mantle. Sometimes, a chemically distinct crust forms on top of the mantle.

distal ejecta The apron of ejecta surrounding an impact crater beyond the zone of continuous ejecta, which is thin enough for patches of underlying terrain to show through.

eccentricity A measure of the extent of the departure of an ellipse (such as a planetary orbit) from a circle. It is one-half the distance between the two foci, divided by the length of semimajor axis. It is a dimensionless number, which can have any value between 0 and 1.

ecliptic plane The plane of the Earth's orbit around the Sun. Except for Pluto and Mercury, the orbits of the planets lie within a few degrees of the ecliptic plane.

effective temperature (T_e) The effective temperature T_e of a spherical body is defined by:

$$T_e = \left(\frac{L}{4\pi R^2 \sigma} \right)^{1/4} \text{ or } T_e^4 = \frac{L}{4\pi R^2 \sigma}$$

where L is the total radiant energy emitted by a body, R is its radius and σ is Stefan's constant (5.67×10^{-8} W m^{-2} K^{-4}).

effusive volcanism A volcanic eruption in which the dominant products are lava flows rather than pyroclastic rocks.

ejecta Fragmented material excavated and thrown out from a crater across surrounding terrain during the process of impact cratering. See also distal ejecta. The term may also be used to describe pyroclastic rock fragments thrown out by an explosive eruption.

electric dipole A property of an object, e.g. a molecule, that arises from an uneven distribution of electric charge such that the object has oppositely charged ends.

electromagnetic radiation A flow of energy consisting of radiation from all or part of the electromagnetic spectrum.

electromagnetic spectrum A collective term used to describe the various wavelength ranges of electromagnetic radiation. In order of increasing wavelength, these ranges are gamma (γ) rays, X-rays, ultraviolet radiation, visible light, infrared radiation, microwaves and radio waves (see also spectrum).

element partitioning The effect during partial melting where certain elements become either preferentially mobilized and escape into the melt, so becoming more concentrated in the melted material compared with the parent rock; or, alternatively, are preferentially retained in the parent rock mineralogy and so are relatively depleted in the resulting melt.

ellipse The geometric shape that planetary orbits follow (see Kepler's first law). It is characterized by the major and minor axes (or semimajor and semiminor axes), and eccentricity.

emission spectrum A graph that shows the intensity of electromagnetic radiation as a function of wavelength given out (emitted) by atoms or molecules in excited states when they return to the ground state.

energy-level diagram A diagram in which energy increases upwards, and in which short horizontal lines denote the energy of an atom or molecule by virtue of the arrangement of its internal constituents. The gaps between the levels are energies that cannot be attained. The levels display the energies of allowed states of the atom or molecule.

equilibrium constant The constant K that defines the proportions of the various compounds concerned in a chemical reaction that will be present at chemical equilibrium at constant temperature. (Note that for a reaction $n\text{P} + m\text{Q} = a\text{R} + b\text{T}$

$$K = \frac{[\text{R}]^a [\text{T}]^b}{[\text{P}]^n [\text{Q}]^m}$$

where the square brackets [ ] denote concentration.)

eruption columns Column of hot gas, ash and dust which results from an explosive volcanic eruption. Convecting eruption columns may carry pyroclastic material tens of kilometres upwards in an atmosphere, permitting widespread distribution by wind.

escape velocity The minimum speed, in an upward direction, that enables a small body (including molecules and atoms) to just escape from the gravitational field of a far more massive body. (The precise direction of the velocity does not matter, as long as a collision between the bodies is avoided.)

explosive volcanism A volcanic eruption in which the dominant products are usually fragmented solid materials (pyroclastic rocks) rather than flows of molten lava.

fall A meteorite observed during its passage through the atmosphere and subsequently collected.

find A meteorite that has been found by chance or by deliberate search that has *not* been observed during its passage through the atmosphere.

fire fountains Sustained sprays of incandescent basalt lava propelled by escaping gas, which are characteristic of Hawaiian eruptions. Fire fountains may reach heights of over 1 km on Earth, and are thought to have been important during the Moon's volcanic history.

fireball A meteor whose brightness exceeds that of the planet Venus. They are relatively rare though more concentrated periods of fireballs occur during annual meteor showers such as the Perseids or Geminids. Fireballs are also associated with the atmospheric entry of large incoming bodies, which may result in meteorite falls.

fluidized ejecta Impact ejecta that behave as a fluid due to the presence of a volatile such as liquid water. The presence of liquid water in impact ejecta greatly enhances the mobility of the ejected debris, converting the dry fragmental ejecta flows characteristic of lunar craters to fluid debris flows similar to terrestrial mud flows. On Mars, this results in an unusual form of ejecta blanket associated with craters 5–15 km in diameter, with petal-like lobes that end in a low concentric ridge or outward facing escarpment. See also rampart craters.

flux density (F) The rate at which energy in the form of radiation is received from a source, per unit area facing the source.

free-fall velocity The maximum velocity achieved by an object falling freely under gravity through an atmosphere. It is dictated by the object's mass and aerodynamic drag, the atmospheric density, and acceleration due to gravity.

fusion crust The glassy rind that surrounds a meteorite sample, formed by melting at the surface of the sample during atmospheric passage.

Galilean satellites Any one of the four large satellites of Jupiter (Io, Europa, Ganymede, and Callisto), so named because they were discovered by Galileo Galilei (1564–1642).

gamma-rays Electromagnetic waves with the highest frequencies, above the highest frequencies of X-rays. The photon energies are consequently also the highest.

gas chromatography An analytical technique in which volatile substances are separated as they flow in a gas stream through a tube that contains a solid or liquid in which they tend to be absorbed.

gas giants See giant planets.

gas thrust region The lowermost part of a volcanic eruption column, in which hot pyroclastic material is propelled upwards by explosive decompression, like shot from a gun.

giant planets Jupiter, Saturn, Uranus and Neptune are the giant planets. They have radii greater than 24 000 km, densities close to $1 \times 10^3 \, \mathrm{kg \, m^{-3}}$, are predominantly or wholly fluid, and are located in the outer part of the Solar System.

glass Structureless, non-crystalline melted rock with the atomic structure of a liquid, which forms in and around impact craters. Large ponds of melt may pool within craters; small glassy droplets may be sprayed over large areas. Glass is also formed when volcanic rock cools very rapidly, particularly when erupted explosively as small fragments.

granite A light-coloured, coarse-grained igneous rock with abundant feldspar and quartz, and smaller proportions of ferromagnesian minerals. It typically occurs as large bodies which are emplaced at depth within the continental crust.

gravitational focusing The effect of the gravitational field of a planet or sufficiently large planetary body, so that trajectories of smaller bodies are deflected towards the planetary body, thus increasing the likelihood of collision with it.

Great Red Spot A prominent feature in the atmosphere of Jupiter. It is a giant storm rotating anticlockwise.

Great White Spot A large feature observed periodically in the atmosphere of Saturn. It is not long-lasting like the Great Red Spot.

greenhouse effect The rise in atmospheric temperature that occurs when an atmosphere transmits solar radiation at optical wavelengths, with resultant heating of a planetary surface, and absorbs the infrared radiation emitted by the surface.

Hadley cell Convection cell in a planetary atmosphere caused by warm air rising above the equator, travelling polewards, and then sinking to return towards the equator.

heat excess The observed effective temperatures of Jupiter, Saturn and Uranus are higher than would be expected if their sole source of heating were energy from the Sun. The additional energy needed to produce this increase in effective temperature is the heat excess.

heteronuclear A term that describes a molecule that contains atoms of different elements.

homonuclear A term that describes a molecule that contains atoms of only one element.

hot spots Areas with elevated levels of volcanic activity. Hot spots can occur on constructive plate margins, or within plates, e.g. the Hawaiian Islands. The hot spot is thought to be stationary, or nearly so, and to produce volcanoes intermittently as the plate moves over it. It has been suggested that mantle plumes lie beneath hot spots.

hydration-induced melting A process that occurs when volatiles such as hydrous (water-rich) or other gas-rich fluids are added to the mantle resulting in alteration reactions within the constituent minerals. When hydration or similar volatile-induced alteration of the mantle mineral assemblage occurs, it reduces the melting point of the assemblage and thus lead to partial melting. Note: the term 'hydration' is used here in the loosest sense to mean the addition of 'volatile-rich fluids rich in hydrous phases', and not simply just the addition of water.

hydrocode Mathematical computer codes used to model impact cratering processes. The codes use equations derived from the science of forces acting on or exerted by fluids known as hydrodynamics.

hydrothermal alteration Changes that occur within rock when it reacts with (usually warm) water circulating through it.

hypervelocity Relative velocities between planetary bodies of the order of kilometres or tens of kilometres per second.

ices Sometimes used simply to refer to frozen water (the most common type of ice on Earth), but can also be used to describe other volatile elements, compounds or mixtures that exist in a frozen state. Other ices can include methane, ammonia, carbon monoxide, carbon dioxide and nitrogen (either individually or mixed together).

ideal thermal source See black-body source.

igneous A term that refers to rocks formed by the cooling and crystallization of magma (*ignis* is Latin for 'fire').

impact cratering The process of formation of craters on the surfaces of solid planetary bodies through hypervelocity impacts. Impact structures may range in size from millimetres to thousands of kilometres (multi-ringed basins).

impact craters See impact cratering.

impact ejecta Fragmented material excavated and thrown out from a crater across surrounding terrain during the process of impact cratering. See also distal ejecta.

impact flux The rate at which solid objects collide with a larger planetary body to form impact craters. Variations in the size–frequency distribution of the impacting bodies result in variations of the size–frequency distribution of impact craters. The flux may vary with time and location in the Solar System.

incompatible elements Elements such as rubidium and strontium which have ionic radii that are so large that they do not fit readily into the crystal lattices of common silicate minerals such as olivine. On partial melting, they readily segregate into the liquid phase, forming melts that are different in composition from the source mineral.

infrared radiation Electromagnetic waves with frequencies or wavelengths between those of visible light and microwaves.

interplanetary dust Dust-sized meteoroids (about cm-size and smaller) present in the interplanetary medium of the Solar System. The dust, which originates mainly from comets and asteroids, is responsible for faint illuminations in the night sky, such as the zodiacal light. Interplanetary dust grains that enter the Earth's atmosphere may be seen as meteors, or perhaps even collected as micrometeorites.

interplanetary dust cloud The name given to the whole complex of interplanetary dust in the inner Solar System. Its presence can be detected by the zodiacal light.

interplanetary magnetic field (IMF) A magnetic field produced by the Sun which extends into the rest of the Solar System. Its strength varies with solar activity and it entrains charged particles from the solar wind.

ionopause The surface where a planetary ionosphere meets the IMF and deflects the entrained solar wind.

ionosphere The outer layers of a planetary atmosphere where a substantial fraction of gas is present as plasma (separated ions and electrons).

iron meteorite A meteorite composed predominantly of iron–nickel metal. There are two different sorts: (1) those without silicate inclusions, which are generally from the cores of asteroids; and (2) those with silicate inclusions, representing impact melts.

isotope exchange The process in which isotopes are removed from one component to another, with no accompanying change in the relative proportions of the components in question (as for example, where oxygen isotopes exchange between H_2O and CO_2). During this process the isotope compositions of coexisting species change in a predictable way, dependent on temperature and the relative proportions of the components, until isotopic equilibrium is achieved. At this point, although exchange continues, no further changes in isotope compositions take place. Note that although the isotope compositions of individual compounds may become altered during exchange, the isotope composition of the system as a whole remains the same (i.e. isotopes are not created or destroyed, but merely transferred amongst the coexisting species).

isotopic signatures Characteristic compositions of the isotopes of geochemically important elements such as Rb, Sr, C or O that are diagnostic of particular processes or sources.

Jeans mass For a dense cloud to collapse gravitationally its mass must be greater than a critical value known as the Jeans mass.

Kepler's first law First of three laws formulated by German astronomer Johannes Kepler (1571–1630); published in the period 1609–1619. His first law states that planets move in elliptical orbits, with the Sun at one focus of the ellipse.

Kepler's second law This states that a line connecting the Sun to a planet would sweep out equal areas of space in equal intervals of time.

Kepler's third law This states that the square of a planet's orbital period (P), is proportional to the cube of its semimajor axis (a), such that $P^2 = ka^3$.

Keplerian orbit Any orbit that obeys Kepler's laws.

kernel In models of planetary growth, a hypothetical large planetary embryo (bigger than the Earth) formed as a result of runaway growth in the outer Solar System, which sweeps up most of the remaining planetesimals, gas, and dust by gravitational focusing, thereby forming a giant planet.

Kirkwood Gaps A feature of the orbit of the main belt asteroids, whereby particular values of semimajor axis have few asteroids associated with them. These 'gaps' in the semimajor axis distribution result from orbital resonances with Jupiter.

komatiites Primitive basalt lava erupted at temperatures of up to 1600 °C, and which is rich in magnesium ($> 18\%$ MgO by definition, sometimes up to about 35%). Komatiites are formed by high degrees of partial melting of silicates that have chondritic composition. Many lunar basalts resemble komatiites, which are also found in ancient rocks on Earth.

Kuiper Belt The region of the Solar System, beyond the orbit of Neptune (30 AU) containing many icy planetesimals and cometary nuclei. The region is most usually called the Kuiper Belt, following work published in 1951 by G.P. Kuiper. However work by astronomers K.E. Edgeworth (1943) and F.C. Leonard (1930) also remarked upon the likely existence of such a belt (indeed the belt is sometimes called the Edgeworth–Kuiper Belt). The belt is thought to contain between 10^7 and 10^9 bodies.

lava Rock (or, in the icy satellites, a melt produced from ices) that is erupted onto the surface of a planetary body in a molten state. Before it reaches the surface, this melt is more properly known as magma. After it has been erupted as a lava flow and cooled down, the solidified rock may still be described as lava.

lava flow A mass of molten rock (or, in the icy satellites, a melt produced from ices) that is erupted onto the surface of a planetary body. Before it reaches the surface, this melt is more properly known as magma. After it has been erupted it is known as a lava flow. Once it has cooled down, the mass of solidified rock may still be described as lava or lava flow.

lava inflation The process by which the external dimensions of the lava flow expand and thicken as a result of continuing supply of molten material beneath a partially solidified, elastic crust. Supply is maintained through a series of lava tubes, and roofed-over conduits which prevent rapid cooling of the molten material. In this manner, the insulated supply conduits enable single flows to inflate to 10–50 m thickness, and extend for tens or even hundreds of kilometres.

lava tubes Tubes within a lava flow along which lava continues to flow after the walls and roof have become consolidated. The tube can be recognized when lava wholly or completely drains from it.

libration The apparent oscillations of a satellite, with respect to the major body it is orbiting. The motion will result in some tidal heating. Libration of the Moon results in us being able to see (over the period on a month) about 59% of the lunar surface.

lightcurve The variation with time of the brightness received from an asteroid (or another minor body), due to reflected sunlight from its surface. The variation is due to the rotation of the minor body causing the apparent cross-sectional area to change.

lithosphere The outer rigid shell of a planetary body, overlying the asthenosphere. The Earth's lithosphere is about 100 km thick, and comprises the crust and the uppermost mantle.

long-period comets Comets that have orbital periods greater than about 200 years. Long-period comets are thought to be the innermost representatives of the Oort cloud, which is thought to extend to great distances from the Sun. Their orbits are randomly distributed, unlike short-period comets which mostly have orbits concentrated near the ecliptic plane.

low velocity zone 50–100 km thick layer in the Earth's mantle immediately below the lithosphere in which the speed of seismic waves is reduced by a few tenths of 1 km s^{-1}. This is attributed to the presence of a few percent of melt between grain boundaries. Sometimes called the low-speed layer.

magma A hot melt of silicate composition containing dissolved volatiles, which forms the raw material for volcanism before eruption. Magma is not quite synonymous with lava, since lavas are often partially crystallized and have lost large amounts of gas. (Cryovolcanism could be said to involve watery magmas.)

magnetic dipole A source of magnetism that produces a simple magnetic field of the sort that would be formed by a bar magnet.

magnetopause The surface in the sunward direction at which a planetary magnetic dipole field meets the IMF and causes deflection of the solar wind.

magnetosphere A region around a planet where the influence of planetary magnetism (either due to the planet's magnetic dipole or due to its ionosphere) has overcome that of the interplanetary magnetic field.

mantle The compositionally (and therefore seismically) distinct layer of a differentiated planetary body that overlies the core. It may be, as in the case of the Earth, overlain by a chemically and seismically distinct crust.

mantle plumes Localized, hot, buoyant material in the Earth's mantle which is hypothesized to originate near the mantle–core boundary. They are believed to have the form of a cylinder with a radius of about 150 km.

maria (singular mare) Latin word meaning 'seas'. In the early days of lunar observations, it was thought that the dark, circular features visible to the naked eye were oceans similar to the Earth's. The maria are now known to be large impact basins filled with basalt lavas erupted over 3 billion years ago.

mass extinctions An episode of large-scale extinction affecting many different groups of organisms within a short interval of geologic time.

mass spectrometry A means of measuring the abundances of atoms or molecules of different relative molecular mass by magnetic deflection of a beam of ions in a mass spectrometer.

massive solar nebula model An extreme model for the solar nebula, in which the mass of the nebula surrounding the protoSun was of the order of one solar mass ($1M_\odot$).

matrix In a meteorite, fine-grained material within which larger components, such as chondrules and calcium–aluminium-rich inclusions are embedded. The matrix is mostly of silicates, but also contains sulphides and

carbonaceous materials, and small grains of pre-solar material.

mesosphere A layer of the Earth's atmosphere lying between the stratosphere and thermosphere (50–87 km in altitude) in which temperature decreases with altitude.

metallic bonding A form of chemical bond in which a large number of atoms (e.g. an entire crystal) are bound together by electrons which are free to move throughout the assembly. These free electrons give rise to typical metallic properties such as high electrical conductivity.

metallic hydrogen A form of the element hydrogen in a condensed phase (solid or liquid) that displays metallic properties such as high electrical conductivity. It is predicted to occur at very high pressures.

metamorphic A term that refers to rocks whose texture or mineralogy has been changed through the action of heat and/or pressure.

meteor A 'shooting star', or meteor, is the transient, incandescent trail left by a meteoroid (a cometary or meteoroid fragment) burning up in the Earth's atmosphere (at an altitude of between 70 km and 120 km). Very bright meteors are called fireballs.

Meteor Crater Impact crater 1.2 km in diameter in Arizona, USA, formed 50 000 years ago through impact of an iron meteorite ~40 m across. Also known as the Barringer Crater, it has played an important role in understanding crater structures.

meteor shower A high incidence in the number of meteors, as a consequence of the Earth's orbit intersecting that of a meteoroid stream (the dust trail left by a comet).

meteorite Small extraterrestrial body, of silicate or metallic composition, found on the Earth's surface. Most meteorites probably originate in the asteroid belt. Most are only a few centimetres in size.

meteorite fall A meteorite observed during its passage through the atmosphere and subsequently collected.

meteoroid stream A meteoroid stream occurs around the orbit of a comet, and consists of the ejected meteoroids from the comet. The stream thus forms a sort of 'tube' around the orbit, being quite thin at perihelion and more dispersed near aphelion. When the Earth passes through a stream, a meteor shower might be seen.

meteoroids The term referring (mainly) to dust particles ejected from a comet or fragments from an asteroid. If meteoroids burn up in the atmosphere, they produce a meteor, and if they reach the ground, they are referred to as a meteorite.

micrometeorites A dust-sized meteorite.

micrometeoroids A dust-sized meteoroid.

microwaves Electromagnetic waves with frequencies or wavelengths between those of infrared radiation and radio waves.

mid-ocean ridge Term used to refer to the young, hot and therefore shallow part of the ocean floor occurring at a constructive plate boundary. If one side of the ocean has a destructive plate boundary then the ridge is unlikely to actually run along the middle of the ocean.

mid-ocean ridge basalt (MORB) The most common variety of basalt lava on Earth, formed in huge quantities along mid-ocean ridge constructive plate margins, and covering some 70% of the Earth's surface area. It contains about 50% SiO_2.

mineral A naturally occurring compound found as a crystal or fragmentary grain in rock. Most common minerals in the terrestrial planets and meteorites are silicates.

minimum solar nebula model An extreme model for the solar nebula, in which the mass of the nebula surrounding the protoSun was of the order of 0.01 solar mass $(0.01M_\odot)$.

minor bodies Any of the many small rocky, icy or metallic objects in the Solar System, which are not classified as planets or their satellites.

moons See satellites.

Near Earth Asteroids A group of asteroids that have orbits that can pass reasonably close to the orbit of the Earth (generally defined to be within about 0.4 AU of the Earth).

Near Earth Objects (NEOs) Asteroids or inactive short-period comets that have orbits which pass reasonably close to the orbit of the Earth, i.e. they have perihelion distances less than 1.3 AU. A subset of NEOs have orbits that could potentially cross Earth's orbit (i.e. perihelion <1 AU and aphelion >1 AU). Objects of this type may have close approaches with Earth at some time in the future, and are usually referred to as Potentially Hazardous Asteroids. Only three NEOs are larger than 10 km in diameter: (1036) Ganymede, (433) Eros, and (4954) Eric. However, there are likely to be at least several hundred

NEOs larger than 1 km, and at least 100 000 NEOs larger than 100 m in diameter.

nebula hypothesis See nebula theory.

nebula theory The general theory for the formation of the planets that states that they grew from a cloud of gas and dust (the solar nebula) around the primitive Sun.

noble gases Unreactive (inert) gases consisting of single free atoms: helium, He; neon, Ne; argon, Ar; krypton, Kr; xenon, Xe; and radon, Rn.

non-thermal source A source that emits electromagnetic radiation for reasons other than those relating to its temperature.

nuée ardente A French phrase meaning 'glowing cloud' consisting of a very hot, rapidly moving mass of ash and debris from an eruption column, and/or flanks of a volcano. The moving debris entrains air and, when hot magmatic materials are incorporated, travels across the landscape in the form of a glowing cloud.

Oort cloud A spherical cloud of possibly 10^{12} comets (with a total mass about 25 times that of the Earth) that is hypothesized to extend out to tens of thousands of AU from the Sun. Gravitational perturbations caused by passing stars may send comets towards the Sun. These can then become the periodic comets that we normally observe.

orbit The regularly repeated elliptical path of a celestial object or artificial satellite about a star or planet.

orbital evolution The process where the orbital parameters of an orbit change with time. This occurs due to the gravitational influence of other bodies (particularly the giant planets).

orbital inclination The inclination of an orbit is the angle made between the plane of the orbit and the ecliptic plane.

orbital period The time taken, measured in seconds, days, or years, for a celestial object to complete one revolution around its orbit.

orbital resonance The situation in which the orbital periods of two bodies in orbit around a third are in a simple ratio (e.g. 1 : 2).

ordinary chondrite General term meaning the most common kind of chondritic meteorite (see chondrite), and excluding carbonaceous chondrites.

oxidation A chemical change in which a substance is oxidized.

oxidized atmospheres Atmospheres in which compounds tend to be chemically oxidized. Such compounds can be produced, for example, by losing hydrogen ($NH_3 \rightarrow N_2$) or by gaining oxygen ($CO \rightarrow CO_2$).

ozone A gas with molecules that each consist of three atoms of oxygen, O_3.

partial melting Rocks do not have a fixed melting temperature, like pure ice, but melt over a range of temperatures. Partial melting in the low part of the range liberates some components preferentially, leading to liquids (magmas) different in composition from the initial rock. This is an important process in planetary differentiation.

partial pressure The pressure due to a single component of a gaseous mixture. The partial pressure of a component is equal to the total pressure of the gaseous mixture multiplied by that fraction of the molecules in the mixture that are of the component.

peridotite The rock type of which the uppermost part of the Earth's mantle is composed. The deeper mantle probably has a similar chemical composition, but different minerals will exist there because of the extremely high pressure.

perihelion The point on an orbit, which lies closest to the Sun.

perihelion distance The distance from the (centre of the) Sun to the perihelion of an orbit.

phase A substance with a defined chemical composition, and in a particular state (e.g. gas, liquid or solid).

planar deformation features A term used to describe the distinctive shock-produced microscopic structures found in minerals such as quartz as a result of the passage of shock waves through the mineral.

planetary bodies A general term, encompassing planets, their satellites and minor bodies (asteroids and comets). The term would also refer to planetesimals and planetary embryos that existed early in the evolution of the Solar System.

planetary embryo A hypothetical body of something like one-hundredth to one-tenth the mass of a planet, produced as the end-product of runaway growth of planetesimals.

planetesimal A body roughly 100 m to 10 km across, formed by coagulation of dust grains in the solar nebula, or a somewhat larger body produced by accretion of smaller planetesimals. Accretion of smaller planetesimals onto larger ones may have occurred in a runaway fashion, leading to the production of planetary embryos.

planets Celestial bodies that orbit the Sun or another star. Planets can consist of rock and metal (as do the inner planets of the Solar System) or predominantly of liquid and gas (as do the giant outer planets).

plate recycling Outward transfer of heat in a planetary body through the creation of new, hot, lithosphere at the edges of tectonic plates, and the removal (subduction) of old, cold, lithosphere down into the asthenosphere at destructive plate margins. Thanks to plate tectonics, this is the main means of lithospheric heat transfer in the Earth.

plate tectonics A description of the behaviour of the Earth's lithosphere, which is broken into seven major plates and a few minor ones. These plates can be thought of as rigid, and they can move about because of the weakness of the underlying asthenosphere. Plates are added to at constructive plate margins (spreading axes) and destroyed at a globally averaged equivalent rate at destructive plate margins (subduction zones).

plutonic A term used to describe rock which has crystallized at great depth. Magmas can be said to have a plutonic origin if it can be demonstrated they result from partial melting of very deep material (e.g. within the mantle).

polyatomic A molecule that has three or more atoms as constituents.

Potentially Hazardous Asteroids A group of asteroids that have orbits that can cross (or might cross in the foreseeable future) the orbit of the Earth, and thus pose a potential impact threat.

Poynting–Robertson effect A mechanism caused by the slight braking effect due to radiation pressure, which results in a body slowly reducing its semimajor axis, thus slowly spiralling into the Sun. The effect is most pronounced for dust particles in the $10\,\mu m$ size regime (it will take about $10\,000$ years for a $10\,\mu m$ sized dust particle to evolve from the asteroid belt to the orbit of Earth).

primitive A term used to describe a sample of material that appears to have inherited many characteristics unchanged since its formation, such as carbonaceous chondrites, which appear to be representative samples of solar nebula material.

prograde A term used to describe the orbital direction of an object orbiting the Sun if the motion is anticlockwise as seen from the north side of the ecliptic plane. It also describes the direction of rotation (spin) of a planetary body under the same conditions. The opposite term is retrograde. Most bodies in the Solar System have prograde orbits.

protoplanetary disc A disc of gas and dust around a star from which planets may form as a result of condensation, coagulation, and accretion (see also solar nebula).

protosatellite disc A hypothetical disc of gas and dust around a planet (especially a giant planet) from which satellites may have formed as a result of condensation, coagulation, and accretion.

protostar The earliest stage in a star's life during which it is gravitationally contracting and before nuclear fusion has been initiated.

protoSun A term used to describe the very young Sun during the evolution of the solar nebula, and especially before thermonuclear reactions got under way in the solar interior.

pseudotachylite A rare, glassy rock produced by frictional melting during shock metamorphism and other forms of extreme dynamic metamorphism.

pumice Lumps of solidified bubble-rich, glassy froth produced during explosive eruptions.

P-waves Seismic waves of the compressional variety, which can travel through solids and liquids. (A good way to remember the name is to think of it as an abbreviation for 'push–pull-waves'.) See also S-waves.

pyroclastic flow Moving flow of solid, fragmentary volcanic rocks, usually pumice, which spread rapidly from a volcanic vent, commonly as the result of collapse from a dense *eruption column*. They may travel distances of 100 km and have volumes in excess of 1000 km^3.

pyroclastic materials Fragmentary ('fire-broken') volcanic rocks erupted as solid particles rather than molten lava. Fragmentation usually takes place when explosively decompressing gas disrupts magma in a near-surface volcanic vent.

radiation pressure The pressure exerted when visible light (or any other type of radiation) falls on a body.

radio waves Electromagnetic waves with the lowest frequencies/longest wavelengths, extending from the lowest frequency/longest wavelength microwaves.

radiometric dating A way of obtaining the date of formation of a mineral grain or lump of rock, based on comparing the present-day abundance of isotopes produced by radioactive decay with their (presumed) original abundance.

rampart craters Some impact craters on Mars, 5–15 km in diameter, exhibit striking aprons of continuous ejecta, extending about 1 crater diameter, resembling flower petals, which terminate in elevated ridges or 'ramparts'. They are thought to indicate that the target material contained permafrost ice.

rare earth elements A group of 15 related elements spanning the Periodic Table from lanthanum to lutetium (atomic numbers 57 to 71 respectively). The abundances of the REE in meteorites, or any other planetary materials, can give diagnostic information on the formation mechanisms of the materials in question.

rarefaction wave Shock waves produced by hypervelocity impacts set up enormously high transient pressures. Once the shock front has passed, pressures decrease in a rarefaction wave, similar to the behaviour of sound waves.

reducing atmospheres Atmospheres in which compounds tend to be chemically reduced. Such compounds can be produced, for example, by losing oxygen ($CO_2 \rightarrow CO$) or by gaining hydrogen ($O_2 \rightarrow H_2O$).

reduction A chemical change in which a substance is reduced. For our purposes a reduced substance is one in which an element is combined with a maximum quantity of hydrogen, as in CH_4, or in which, in the case of metals, it has a low or zero positive charge or a negative charge.

reflectance spectrum The brightness of a body, due to reflected sunlight from its surface, as a function of wavelength. Often called the 'relative reflectance spectrum' because the spectrum can be plotted as a fraction of some arbitrarily chosen reference value (usually the spectrum is normalized to the value 1, at the wavelength of visible light).

refractory A relative term, describing a substance or element that condenses (and vaporizes) at a high temperature. The term refractories is sometimes used as a noun, to refer to refractory substances or elements in general. The converse is volatile/volatiles.

regolith A layer of loose and broken rock and dust on a planetary surface. The Moon has a regolith some 5–15 m deep that was produced by meteoritic bombardment whereas on Mars the regolith also includes deposits produced by volcanic eruptions and the effects of erosional processes.

relative molecular mass The mass of one molecule of a substance divided by the mass of one atom of the isotope ^{12}C.

retrograde A term used to describe the orbital direction of an object orbiting the Sun if the motion is clockwise as seen from the north side of the ecliptic plane. It also describes the direction of rotation (spin) of a planetary body under the same conditions. The opposite term is prograde.

rhyolite A fine-grained, silica-rich magmatically generated rock.

rifting (or continental rifting) The splitting of a continental mass as a result of plate tectonic processes, and representing the initial stages of continental break-up. Rifting is often accompanied by widespread basaltic volcanism and, in some instances, the generation of continental flood basalt provinces (CFBPs).

rotation period The time that a celestial body takes to make one complete rotation upon its spin axis.

rubble pile A term usually associated with minor bodies, indicating that the body is not made of a solid lump of material, but consists of many small constituent fragments, perhaps relatively loosely bound together.

runaway growth A situation in which the rate of growth of a large planetesimal, which was initially a little larger than all the others in nearby orbits, outpaces that of its competitors because of gravitational focusing.

satellites Celestial bodies that orbit a planet, also referred to as moons.

saturated In cratering terms, a surface is said to be saturated when a new impact crater can only be formed by overprinting and thus obliterating older ones. Lunar surfaces more than about 3.9 billion years old are saturated.

saturation vapour pressure diagram A plot of the saturation vapour pressure (the partial pressure of a component of gas which is in equilibrium with liquid of that component) against temperature.

seasons An effect of the angle between a planetary body's spin axis and its orbital plane. Unless the angle is 90°, northern and southern hemispheres experience solar illumination (and heating) that varies according to where the body is in its orbit.

secondary craters Clumps of ejecta from major impacts are often large enough to form secondary craters when they fall back. Secondary craters up to a few kilometres in diameter are so common that they complicate planetary cratering statistics.

sedimentary A term applied to rocks that have formed from sediment deposited by water or wind.

seismic waves Vibrations transmitted through a planetary body. See P-waves, S-waves.

self-compression An effect of pressure within a planetary body, as a result of its own gravity, causing its internal density to be greater than it would be in the absence of any internal pressure.

semimajor axis The longest dimension of an ellipse is the major axis. The semimajor axis is one-half this length.

semiminor axis The shortest dimension of an ellipse is the minor axis. The semiminor axis is one-half this length.

shatter cones Conical fragments of rock that are formed when the shock waves produced by meteorite impact pass through certain types of rock. The shock waves produce weakened zones radiating out from the apex of the cone, causing shatter cones, which are later revealed when the shocked layers are exposed by erosion.

shock metamorphism Changes induced in rocks and minerals when they are subjected to pressures from shock waves travelling at $10-13\ km\ s^{-1}$ resulting from meteorite, cometary, or asteroidal impact.

shock pressure Hypervelocity impacts between planetary objects set up enormously high-pressure pulses that travel outwards from the impact point. Pressures of up to five million times the Earth's atmospheric pressure may be achieved transiently.

short-period comet Comets that have orbital periods less than about 200 years. Short-period comets do not travel much further from the Sun than the orbits of the outermost planets. Most short-period comets have prograde orbits with quite low inclinations (i.e. mainly in the ecliptic plane), and may originate from the Kuiper Belt.

silicate minerals Minerals containing silicon (Si) and oxygen (O), and usually metallic elements, in its formula. The term silicates may be used to refer to material (such as most rocks) formed principally of silicate minerals.

size–frequency distribution A statistical term that describes the number of bodies of a particular population (for example, the number of asteroids, the number of impact craters etc) which fall within specified size ranges. Often this is described as the total number of bodies that are *greater than* a particular size (i.e. this is a *cumulative* size–frequency distribution).

solar nebula The hypothetical cloud of gas and dust within which the Sun and the other constituents of the Solar System formed, according to the nebula theory. The solar nebula gave rise to a protoplanetary disc as the nebula flattened into a disc.

Solar System (The) The Sun together with the nine planets and all other planetary and minor bodies gravitational bound to the system.

solar system A system of planets or other bodies orbiting a star other than our own Sun.

solar wind A stream of high-speed particles (mainly protons and electrons) that spreads out from the Sun, carrying traces of the Sun's magnetic field with it.

solid-state convection Convection occurring in a solid, without the necessity of melting, at high pressures and temperatures.

spatter Fluid basaltic pyroclasts which accumulate by fallout from a Strombolian volcanic eruption column to form a rampart around the vent. The individual pyroclasts are so fluid when they land that they often mould together forming flattened pancakes.

spectroscopy The production of spectra, and their study.

spectrum A display, versus wavelength or frequency, of the strength of the radiation emitted by, or received from a source.

spidergram A useful method of plotting chemical data to show the variation between two rock types for a wide range of elements. All data are 'normalized' against a chosen standard, typically carbonaceous chondrite, mantle material (i.e. peridotite), or ocean ridge basalt. This is achieved by dividing the abundance of each element of each sample by that of the chosen standard, and then plotting the quotient (usually) in the order of atomic number. Elements present in greater abundance than the standard yield numbers greater than one, those less abundant yield numbers less than one. By connecting the plotted points, a 'spidery' diagram is created, which illustrates immediately any systematic differences between samples.

sporadic meteors Meteors that are not associated with a particular meteor shower, but form the typical meteor 'background'. Typically an observer might see up to about 5 sporadic meteors per hour on a clear night.

stable isotope An atomic nucleus that is not radioactive, but remains stable with time. Most elements have one or more stable isotopes. For an element, the number of protons in the nucleus remains constant, but the number

of neutrons varies, e.g. for carbon the stable isotopes are $^{12}_{6}C$ and $^{13}_{6}C$.

stable isotope compositions The relative proportions of different stable isotopes of a particular element that constitute an individual compound. Some elements have only one stable isotope, e.g. aluminium, $^{27}_{13}Al$, in which case the term has no proper meaning. For most other elements, the stable isotope composition can be described by the ratio of the individual isotopes, e.g. $^{12}C/^{13}C$, or by the use of the δ notation.

star A self-luminous celestial body consisting of a mass of gas held together by its own gravity that, at some stage of its life, produces energy by nuclear fusion.

stony meteorite Meteorite composed predominantly of silicate minerals. Stony meteorites can be subdivided into chondrites and achondrites.

stony–iron meteorites Meteorite composed of roughly equal proportions of iron–nickel metal and silicate minerals.

stratigraphic record A succession of rocks laid down during a specified interval of geologic time. The term stratigraphic record often refers to the entire sequence of strata deposited throughout the Earth's history.

stratosphere A layer of the Earth's atmosphere at 20–50 km altitude, in which the temperature increases with height.

subduction zones See plate tectonics.

suevite A fragmented rock in a matrix of glass found within impact craters and near the site of impact. The shock waves associated with the impact produce extremely high pressures and temperatures in the rocks for a few microseconds. Near the point of impact, these can brecciate and melt the rock.

super-eruptions Extremely large volcanic eruptions that can have effects on a planetary scale.

super-rotation The phenomena in which a planet's atmosphere rotates independently of the surface. It is observed on Venus, where the upper atmosphere has a retrograde rotation with a period of around 4 days, compared with the surface's retrograde rotation of 243 days.

surface heat flow The rate at which heat escapes from the surface of a solid planetary body such as the Earth, usually measured in $mW\,m^{-2}$. Virtually all the Earth's average surface heat flow of $80\,mW\,m^{-2}$ arises partly from primordial heat sources and partly from decay of radioactive isotopes.

S-waves Seismic waves of the shearing variety, which can travel only through solids. (A good way to remember the name is to think of it as an abbreviation for 'shear-waves' or 'shake-waves'.) See also P-waves.

synchronous rotation The circumstance in which a satellite spins on its axis in the same time as it takes to orbit a planet, thereby keeping the same face turned towards the planet at all times. Most major satellites in the Solar System, including our own Moon, are in synchronous rotation as a consequence of tidal action.

T Tauri An early stage of the evolution of stars with masses similar to that of the Sun. T Tauri stars are typically less than 10 Ma old but have diameters several times that of the Sun and are still contracting.

taxonomic classes A term associated with asteroids, where different taxonomic classes are assigned to asteroids based on their reflectance spectra and their albedo. Taxonomic classes are broadly related to the composition of the asteroid.

tectonic/tectonism The process of deformation within the Earth's crust, and its associated structural effects.

tektites Droplets of glassy impact melt up to a few centimetres in size sprayed for thousands of kilometres from terrestrial impact sites, falling out over many thousands of square kilometres. Some tektites can be related to known impact craters, for example the Ries crater in Germany.

terminal fall velocity The maximum velocity achieved by an object falling freely under gravity through an atmosphere. It is dictated by the object's mass and aerodynamic drag coefficient, the atmospheric density, and acceleration due to gravity.

terrestrial fractionation line (**TFL**) A line on a plot of $\delta^{17}O$ against $\delta^{18}O$ along which all samples from the Earth are found to fall, as a result of processes of isotope fractionation within the closed system of the Earth, subsequent to its formation. Samples of other Solar System material have stable isotope compositions that do not fall on this line, except by chance.

terrestrial-like A term applied to planetary bodies to denote that they have many of the characteristics of the terrestrial planets.

terrestrial planets Mercury, Venus, Earth and Mars are the terrestrial planets. They are dominantly rocky objects, with iron-rich cores and silicate mantles, and densities of $3.9–5.5 \times 10^3\,kg\,m^{-3}$. Some large satellites, such as the Moon and Io, have similar properties.

thermal source A source that emits electromagnetic radiation because of its temperature (the higher the temperature, the greater the amount of radiation).

thermal tide Day to night-side atmospheric flow that is caused by a large diurnal temperature difference.

thermosphere The outermost region of an atmosphere, characterized by increasing temperature with altitude.

tidal heating Heating of a planetary body caused when varying tidal forces continually distort it.

transient cavity When a hypervelocity impact takes place, the material at the target site is highly compressed and depressed in a transient cavity. The final shape of the crater depends on the ejection of material from this cavity, and subsequent slumping and settling of material from its walls.

troposphere The lower region of a planetary atmosphere in which mixing occurs by convection.

tuff The compacted equivalent of a volcanic ash deposit which has been generated by pyroclastic processes and deposited either directly by air-fall on land, or else water lain after falling onto lake or sea areas.

turbulence Chaotic stirring motions. In the solar nebula, turbulence probably grew out of inhomogeneities in the initial contraction.

ultraviolet (UV) radiation Electromagnetic waves with frequencies or wavelengths between those of X-rays and visible light.

umbrella region That part of a volcanic eruption column which, having risen convectively through an atmosphere, has reached a level of neutral buoyancy and spread out laterally, forming an overhanging 'umbrella' of ash-laden cloud.

vibrational spectrum Spectrum obtained from the study of the absorption or emission of electromagnetic radiation by molecules when they change their vibrational energy. Changes in vibrational energy are associated with changes in the stretching of chemical bonds and in the bond angles.

vibrational transition A transition in a molecule wherein the molecule goes from a state corresponding to a certain amount of molecular vibrational energy, to a state corresponding to a different amount of molecular vibrational energy.

viscous drag Frictional drag between adjacent components of a rotating disc of gas and/or dust (such as the solar nebula or an accretion disc). This reduces the velocity contrasts dictated by Kepler's third law, which applies only to particles orbiting freely in a non-viscous medium.

viscous relaxation The slow, plastic response of the lithosphere and underlying asthenosphere of a planetary body to some disturbance, such as a major impact. Large local variations in surface height become subdued as the high regions sink under their own weight.

visible light Electromagnetic waves with frequencies or wavelengths between those of ultraviolet radiation and infrared radiation. Our eyes are sensitive to visible light.

volatile In the context of the solar nebula, a relative term describing a substance or element that condenses (and vaporizes) at a low temperature. The term volatiles is sometimes used as a noun, to refer to volatile substances or elements in general. The converse is refractory/ refractories. Within a planetary body that has already formed, the term volatiles refers to gaseous and liquid species, notably those escaping as gases (outgassing) from the interior during volcanic eruptions.

volatile inventory An estimate of the total, accessible inventory of elements in a planetary body that are capable of producing volatile substances. It includes atmospheres, oceans, icy materials and potentially volatile substances that are rendered involatile by chemical combination, as in sedimentary carbonates. It excludes any deeply sited, inaccessible material.

volcanic arcs Arcuate belts of volcanoes produced by magmas generated by partial melting resulting from hydration-induced melting at subduction zones. The horizontal distance between the subduction zone and the volcanic arc is 100–500 km.

volcanic bomb A relatively large (>64 mm) lump of rock thrown out during an explosive eruption.

volcanism The processes associated with the transfer of magma and volatiles from the interior of a planet or planetary body to its surface.

volcanoes Naturally occurring vents or fissures in the Earth's surface through which molten, gaseous and solid materials are erupted.

Widmanstätten pattern The pattern observed when an iron meteorite is cut and polished, and then etched with a mild acid. It results from the coexistence, within the sample, of two different forms of iron–nickel alloys that have separated as a direct consequence of the extremely slow cooling experienced by the iron meteorite parent bodies.

wind speed The speed of motion of part of an atmosphere, relative to the planet's (or satellite's) surface or to the speed of rotation of the central mass of the planet where there is no discernible surface.

wind velocity See wind speed.

xenolith An inclusion or enclave of a pre-existing rock preserved within a later igneous (e.g. volcanic) rock. Xenoliths are often unmelted fragments of the rock from which the magma was derived, or bits of the wall rocks comprising the channels and fissures along which it moved during its surfaceward movement.

X-rays Electromagnetic waves with frequencies or wavelengths between those of gamma (γ) rays and ultraviolet radiation.

yield strength The ability of a material to withstand an applied stress before deforming or breaking. For any given material, yield strength depends on the nature of the applied stress, temperature, pressure, and duration. Rocks are much weaker in tension than compression.

zodiacal light The diffuse light reflected from the interplanetary dust cloud, seen before dawn and after sunset, centre around the ecliptic plane.

zone A light band parallel to the equator in the atmosphere of Jupiter and other giant planets. In zones the atmospheric gases are sinking. Compare belt.

FURTHER READING

de Pater, I. and Lissauer, J.J. (2001) *Planetary Sciences*, Cambridge University Press, Cambridge.

Encrenaz, T. (2003) *The Solar System*, 3rd edn, Springer Verlag, Berlin.

Gilmour, I. and Sephton, M.A. (2004) *An Introduction to Astrobiology*, Cambridge University Press, Cambridge.

Greeley, R. and Baston, R. (2001) *The Compact NASA Atlas of the Solar System*, Cambridge University Press, Cambridge.

Green, S.F. and Jones, M.H. (2004) *An Introduction to the Sun and Stars*, Cambridge University Press, Cambridge.

Jones, M.H. and Lambourne, R.J.A. (2004) *An Introduction to Galaxies and Cosmology* Cambridge University Press, Cambridge.

Shirley, J.H. and Fairbridge, R.W. (1997) *Encyclopedia of Planetary Sciences*, Chapman and Hall, London.

Taylor, S.R. (2001) *Solar System Evolution: a new perspective*, Cambridge University Press, Cambridge.

Weissman, P.R., McFadden, L. and Johnson T.V. (1999) *Encyclopedia of the Solar System*, Academic Press, London.

ACKNOWLEDGEMENTS

The production of this book involved a number of Open University staff, to whom we owe a considerable debt of thanks for their commitment and the high professional standards of their contributions. Jennie Neve Bellamy managed the production of the associated Open University course, ensured authors met deadlines, tracked down copyrights, and generally kept the project on track. Valerie Cliff styled the text for handover to Pamela Wardell and Peter Twomey who copy-edited it and steered the project through the production process. The graphic artwork was prepared by Sara Hack with considerable skill and the design and layout was undertaken in an exemplary fashion by Debbie Crouch. The index was prepared by Jane Henley. We are also grateful to Giles Clark (Open University) and Susan Francis (Cambridge University Press) for their support and help with co-publication.

In addition, we wish to thank the following people who commented on earlier versions of the text: David Hughes (University of Sheffield) and Fred Taylor (University of Oxford), together with anonymous referees appointed by Cambridge University Press. Many other individuals and organizations furnished and/or granted permission for us to use their diagrams or photographs and to them we also express our gratitude.

Grateful acknowledgement is made to the following sources for permission to reproduce material in this book:

Cover photos: NASA.

Figures 1.2, *1.3*, *1.4*, *1.5*, *1.6*, *1.7*, *1.8*, *1.9*, *1.10*, *1.12*, *1.13*, *1.14*, *1.15*, *1.17*, *1.18*, *1.19*, *1.20*, *1.21*, *1.22*, *1.23*, *1.24*, *1.27*, *1.28*, *1.30*, *1.33*, *1.34*, *1.35*, *1.38*, *1.39*, *1.40*, *1.41*, *1.42*, *1.44*, *1.45 and 1.48* NASA; *Figures 1.11*, *1.29 and 1.43* US Geological Survey, Flagstaff, Arizona; *Figures 1.25*, *1.26*, *1.36 and 1.37* Dr Bradford A. Smith, National Space Science Data Center, World Data Center A for Rockets and Satellites, NASA Goddard Space Flight Centre, Greenbelt, Maryland; *Figures 1.31 and 1.32* Courtesy of Calvin J. Hamilton; *Figure 1.46* Copyright © 2002 by Michael Jager (Austria); *Figure 1.47* European Space Agency; *Figure 1.49* Dana Berry, Space Telescope Science Institute, Baltimore.

Figures 2.18, *2.21*, *2.22*, *2.24 and 2.25* NASA; *Figure 2.19* Mike Widdowson/Open University; *Figure 2.23* Courtesy of Calvin J. Hamilton.

Figures 3.1, *3.5*, *3.6*, *3.11a*, *3.13*, *3.14*, *3.16b and 3.26* Steve Self/Open University; *Figure 3.2* Malin *et al.*/NASA; *Figures 3.3*, *3.4*, *3.25*, *3.30*, *3.31*, *3.33*, *3.34*, *3.35*, *3.36*, *3.37*, *3.38*, *3.40 and 3.42* NASA; *Figure 3.7* USGS: Photo by Austin Post; *Figure 3.11b* USGS: Photo by J. D. Griggs; *Figure 3.12* USGS; *Figure 3.15b* Mike Widdowson/Open University; *Figure 3.20* Steve Blake/Open University; *Figure 3.22* US Library of Congress; *Figure 3.27* Mr Frederick J. Doyle, National Space Science Data Center; *Figure 3.28* NASA/JPL/Northwestern University; *Figure 3.29* Gordon H. Pettengill, The Magellan Project and the National Space Science Data Center; *Figures 3.32 and 3.37* US Geological Survey, Flagstaff, Arizona; *Figure 3.41* NASA, JPL, Voyager 2.

Figures 4.1, *4.2*, *4.6*, *4.7*, *4.11*, *4.13*, *4.14*, *4.15*, *4.16*, *4.17*, *4.18*, *4.21*, *4.22*, *4.23*, *4.24*, *4.28*, *4.29*, *4.30*, *4.31 and 4.32* NASA; *Figure 4.3* Courtesy of Buck Sharpton, Lunar and Planetary Institute; *Figures 4.4 and 4.5* Whittaker, E. A., Lunar and Planetary Laboratory, University of Arizona; *Figure 4.9* Institute of Geological

Sciences; *Figure 4.10* © Lunar and Planetary Institute; *Figure 4.12* Giles Graham/ Open University; *Figure 4.19* Robert Hough/Open University; *Figure 4.20* Iain Gilmour/Open University.

Figures 5.1, 5.2, 5.5, 5.14, 5.28, 5.29 and 5.50 NASA; *Figure 5.6* from *Discovering the Universe* by W. J. Kaufmann, published by W. H. Freeman & Co., 1987. Reprinted with permission; *Figures 5.26, 5.29 and 5.42* NASA/JPL/Malin Space Science Systems; *Figure 5.48* University of Iowa.

Figures 6.1, 6.11, 6.15, 6.18, 6.20, 6.21, 6.22, 6.25, 6.27, 6.28, 6.29, 6.33, 6.34, 6.36 and 6.37 NASA; *Figure 6.13* Reprinted 'Voyager 2 Radio Observations of Uranus' with permission from Warwick, J.W., Science, Volume 233. Copyright 1986 American Association of the Advancement of Science; *Figure 6.19* Painting by Duragel, Courtesy of the Observatoire de Paris; *Figures 6.23* and *6.24* European Southern Observatory, Germany; *Figure 6.29* J.T. Trauger (Jet Propulsion Laboratory) and NASA; *Figure 6.32* Reprinted from Science, Vol. 153. Hammel, H.B. *et al.*, 'New measurements of the winds of Uranus', pp.229–235, 2001, with permission from Elsevier Science.

Figure 7.13, 7.14, 7.15, 7.16, 7.17, 7.18, 7.20, 7.22, 7.23, 7.31a, 7.32 and 7.38 NASA; *Figure 7.19* Reprinted 'Toutatis – Asteroid, 4179', with permission from Calvin J. Hamilton. Copyright 1995 American Association of the Advancement of Science; *Figure 7.25* Credit: D. Jewitt and J. Luu, (University of Hawaii); *Figure 7.29* Copyright Johns Hopkins University; *Figure 7.30a and b* David Malin Images; *Figure 7.31b* Copyright © 2000 European Space Agency. All rights reserved; *Figure 7.33* Copyright Don Davis; *Figure 7.37a and b*: NASA/Armagh Planetarium; *Figure 7.40* Juraj Toth (Comenius U. Bratislava), Modra Observatory; *Figure 7.43* J.C. Casado.

Figures 8.2, 8.3, 8.4, 8.15, 8.16, 8.17 and 8.18 Space Telescope Science Institute; *Figure 8.5* Copyright © 2001 Axel Mellinger; *Figure 8.6* Akira Fujii, Tokyo; *Figures 8.7, 8.19, 8.20, 8.21, 8.22 and 8.25* NASA; *Figures 8.8 and 8.10* Anglo Australian Telescope Board. Photograph by David Malin; *Figure 8.11* Ronald Snell; *Figure 8.23* US Geological Survey; *Figure 8.24* Dr Michael H. Carr, National Space Science Data Center.

Figure 9.1 Copyright S. Eichmiller, Altoona, Pennsylvania; *Figure 9.2* Photograph by www.nyrockman.com © 2002; *Figure 9.4* Hoba Meteorite, Namibia. Copyright www.suedafrica.net Online Travel Guide; *Figures 9.5, 9.10 and 9.11* Copyright © Natural History Museum, London; *Figure 9.7a* University of Leiden, The Netherlands; *Figure 9.7b* Dr I.A. Franchi/The Open University; *Figure 9.8* Wasson, J.T. (1985) *Meteorites*. Copyright © 1985 by W.H. Freeman and Company. Used with permission; *Figure 9.9* D. Brownlee, University of Washington; *Figure 9.16* Franchi, I.A., *et al.* (2001) *Philosophical Transactions: Mathematical, Physical and Engineering Sciences*. Reprinted with permission by The Royal Society; *Figure 9.18* Reprinted with permission from, Shu, S.H. *et al.* (1997), 'X-rays and fluctuating X-winds from protostars'. Science, Vol. 227. Copyright © 1997 American Association for the Advancement of Science.

Every effort has been made to trace all the copyright owners, but if any has been inadvertently overlooked, the publishers will be please to make the necessary arrangements at the first opportunity.

FIGURE REFERENCES

Franchi, I.A., Baker, L., Bridges, J., Wright, I.P. and Pillinger, C.T. (2001) Oxygen Isotopes and the early Solar System, *Philosophical Transactions of the Royal Society*, **359**, pp.2019–34.

Hammel, H.B., Rages, K., Lockwood, G.W., Karkoschka, E. and de Pater, I. (2001) New measurements of the winds of Uranus, *Icarus*, **153**, pp.229–35.

Jewitt, D. and Luu, J. (1992) 1992 QB1, International Astronomical Union Circular No. 5611.

Lirer, L., Pescatore, T., Booth, B. and Walker, G.P.L. (1973) Two Plinian pumice-fall deposits from Somma-Vesuvius, Italy, *Geological Society of America Bulletin*, **84**, pp.759–72.

Ledley, T.S., Sundquist, E.T., Schartz, S.E., Hall, D.K., Fellows, J.D. and Killeen, T.L. (1999) Climate Change and Greenhouse Gases, *Proceedings National Academy of Sciences*, **95**, pp.12753–58.

Melosh, H.J. (1989) *Impact Cratering, a Geological Process*, Oxford University Press, Oxford.

Nittler, L.R. (1996) Oxygen and Aluminum Isotopic Ratios in presolar Al_2O_3 from the Tieschitz Meteorite, *Lunar and Planetary Science*, **27**, p.965.

Shu, F.H., Shang, H., Glassgold, A.E., and Lee, T. (1997) X-rays and fluctuating X-winds from Protostars, *Science*, **277**, pp.1475–79.

Wannier, P.G. (1980) Nuclear Abundances and Evolution of the Interstellar Medium, in *Annual review of astronomy and astrophysics*, **18**, *Palo Alto, Calif., Annual Reviews, Inc.*, 1980, pp.399–437.

Warwick, J.W., Evans, D.R., Romig, J.H. *et al*. (1986) Voyager-2 radio observations of Uranus, *Science*, **233**, pp.102–06.

Wasson, J.T. (1985) *Meteorites*, W.H. Freeman and Co., New York.

Weatherill, G.W. (1989) The Formation of the Solar System: consensus, alternatives, and missing factors, in Weaver, H.A. and Danly, L. *The Formation and Evolution of Planetary Systems*, Cambridge University Press, Cambridge.

INDEX

Entries and page numbers in **bold type** refer to key words which are printed in **bold** in the text.